Energy Management Strategies for Multi-Vectored Energy Hubs to Achieve Low Carbon Societies

IEEE Press
445 Hoes Lane
Piscataway, NJ 08854

Energy Management Strategies for Multi-Vectored Energy Hubs to Achieve Low Carbon Societies

Edited by

Shubham Tiwari
Agriculture, Forestry, and Ecosystem Services Research Group, Biodiversity and Natural Resources Program, International institute of Applied Systems Analysis (IIASA), Laxenburg, Austria

Jai Govind Singh
Department of Energy, Environment and Climate Change,
School of Environmental Resource and Development,
Asian Institute of Technology, Khlong Nueng, Thailand

Sivaraman Palanisamy
Department of EEE, CEG Campus, Anna University, Chennai, Tamilnadu, India

Sharmeela Chenniappan
Department of EEE, CEG Campus, Anna University, Chennai, Tamilnadu, India

Rupendra Kumar Pachauri
Electrical Cluster, School of Engineering, University of Petroleum and Energy Studies, Dehradun, Uttarakhand, India

Sanjeevikumar Padmanaban
Department of Electrical Engineering, IT and Cybernetics at the University of South-Eastern Norway Porsgrunn, Norway

IEEE Press

WILEY

Published by John Wiley & Sons, Inc., Hoboken, New Jersey.

Library of Congress Cataloging-in-Publication Data applied for:

Hardback ISBN: 9781394267361

Cover Design: Wiley
Cover Image: © Gerd/stock.adobe.com

Set in 9.5/12.5pt STIXTwoText by Straive, Chennai, India

Contents

About the Editors

Dr. Shubham Tiwari received his Ph.D. (2023) in Sustainable Energy Transition, Asian Institute of Technology (A.I.T), Thailand; M.E (2019), Electrical Power System Management, Asian Institute of Technology (A.I.T), Thailand; and B. Tech (2015) in Electrical Engineering from UCE, RTU, Kota, India. He is currently working as a Research Scholar at International Institute of Applied Systems Analysis (IIASA), Austria, from 2023 and currently leading a spatially explicit system optimization model, BeWhere. He was awarded best student award in both his bachelor's and master's programs. He is a working group member of IEEE standard P2418.5 (Standard for Blockchain in Energy). His research interest includes renewable energy integration, spatial energy system modeling and optimization, energy management, multi-microgrid modeling, and power systems operations and management. He has published more than 17 research articles including journal publication, book chapters, and international conferences. Further, he worked at Madhya Pradesh Madhya Kshetra Vidyut Vitran Company Limited (MPMKVVCL), Bhopal, as an Assistant Engineer and also as a research assistant at AIT.

Dr. Jai Govind Singh received his Ph.D. (2008) in Power and Control, EED, Indian Institute of Technology, Kanpur, India; M. Tech (2003), Power System, EED, Indian Institute of Technology, Roorkee, India; and Bachelor of Engineering (2001) in Electrical Engineering, MNIT, Allahabad, India. In addition, he completed postdoctoral research at KTH Stockholm and University of Queensland, Brisbane, in 2008–2009. Currently, he is working as a Professor in the Department of Energy, and Climate Change, in the School of Environment, Resources and Development (SERD) at Asian Institute of Technology, Bangkok, Thailand. Moreover, Dr. Singh has completed 26 sponsored research projects from ERASMUS+, USAID, DST India, Australian DFAT, World Bank, BP Thailand, IRENA, ADB, EBARA Japan, NSTDA, IEEE PES, ADEME, PEA, Abt Associates, Tetra Tech, etc. He has published more than 169 research papers in reputed journals and conferences. He is having wide research areas such as e-vehicles technologies, smart grid and microgrid design and operation, RE generation and integration, power system operation and control, electricity market restructuring and power trading, DG design and planning, and energy storage technologies.

Mr. Sivaraman Palanisamy (Member'20, Senior Member'21 IEEE) was born in Vellalur, Madurai district, Tamilnadu, India. He completed schooling in Govt. Higher Secondary School, Vellalur, and B.E. in Electrical and Electronics Engineering and M.E. in Power Systems Engineering from Anna University, Chennai, India, in 2012 and 2014, respectively. He has 11 years of industrial experience in the field of renewable energy and its integration studies, solar P.V. systems, wind power plants, energy storage systems, power quality studies and harmonic assessments. He is an expert in power system simulation software like ETAP, PSCAD, DIGSILENT POWER FACTORY, PSSE, PVSYST and MATLAB. Active participant in the IEEE standards association and working group member of IEEE standard P2418.5 (Standard for Blockchain in Energy), P1854 (Guide for Smart Distribution Systems), P2688 (Recommended Practice for Energy Storage Management Systems in Energy Storage Applications), and 3001.9-2023 (Design of Power Systems Supplying Lighting Systems in Commercial and Industrial Facilities). Also a working group member of the IEEE PES task force on Energy storage. He has authored/co-authored/edited ten books in the field of electrical engineering with Elsevier and John Wiley-IEEE Press and published several papers at national and international conferences. Notably, *Fast charging infrastructure for electric and hybrid electric vehicles* by Wiley-IEEE Press, 2023, and *Power systems operation with 100% renewable energy resources* by Elsevier, 2023.

Also, a senior member of the Institute of Electrical and Electronics Engineers (IEEE), a member of the International Council on Large Electric Systems (CIGRE), and a Life Member of the Institution of Engineers (India). I received Professional Engineer (PEng) certification from the Institution of Engineers India.

Google Scholar link: https://scholar.google.co.in/citations?user=XLdd0mgAAAAJ&hl=en&authuser=1

Dr. Sharmeela Chenniappan holds a B.E. in Electrical and Electronics Engineering, M.E. in Power Systems Engineering from Annamalai University, Chidambaram, India, and a Ph.D. in Electrical Engineering from Anna University, Chennai, India, in 1999, 2000, and 2009, respectively. She did P.G Diploma in Electrical Energy Management and Energy Audit at Annamalai University, Chidambaram, in 2010. Currently, she holds the post of Professor in the Department of EEE, CEG campus, Anna University, Chennai, India. She has 20 years of teaching experience and taught various subjects to undergraduate and postgraduate students. She did several research projects and consultancy work in renewable energy, power quality, and design of P.Q. compensators for various industries. She published more than 30 research publications in peer-reviewed journals and 50 papers in international and national conferences in India and abroad. She has discussed power quality's importance for more than 10 years. She guides 14 Ph.D. students (5 completed and 9 ongoing), 58 master's-level students, 10 undergraduate-level students, and 4 AMIE students. She coordinated and organized a short-term course on power quality for Tamilnadu Electricity Board Engineers in different years. She received the best paper award at IEEE POWERCON 2004 conference held in Singapore. She is a senior member of IEEE, a Life Member of CBIP, and a Life Member of the Institution of Engineers (India), ISTE, and SSI.

Google scholar link: https://scholar.google.co.in/citations?user=-YbjPxsAAAAJ&hl=en&authuser=1

Dr. Rupendra Kumar Pachauri received his Ph.D. degree in Renewable Energy from G. B. University, Gautam Buddha Nagar, India, in 2016, and M. Tech degree from EED, Zakir Husain College of Engineering and Technology, Aligarh Muslim University (AMU), Aligarh, India, in 2009. He is currently working as an Senior Associate Professor in the Electrical and Electronics Engineering Department at the University of Petroleum and Energy Studies (UPES), Dehradun, India. He has published more than 100 research papers in internationally reputed Science Citation/Scopus-Indexed journals along with IEEE/Springer conferences. He has published/granted patents also. His fields of research are Solar Energy, Fuel Cell Technology, and Smart Grid Operations.

Dr. Sanjeevikumar Padmanaban (Member'12–Senior Member'15, IEEE) received a Ph.D. degree in Electrical Engineering from the University of Bologna, Bologna, Italy, 2012. He is a Full Professor of Electrical Power Engineering in the Department of Electrical Engineering, Information Technology, and Cybernetics at the University of South-Eastern Norway, Norway. He has authored over 750+ scientific papers and has received the Best Paper cum Most Excellence Research Paper Award from IET-SEISCON'13, IET-CEAT'16, IEEE-EECSI'19, IEEE-CENCON'19, and five best paper awards at ETAEERE'16, sponsored Lecture Notes in Electrical Engineering, Springer book. He is a Fellow of the Institution of Engineers, India, the Institution of Electronics and Telecommunication Engineers, India, and the Institution of Engineering and Technology, United Kingdom. He received a lifetime achievement award from Marquis Who's Who (USA) in 2017 for contributing to power electronics and renewable energy research. He has been listed among the world's top 2 scientists (from 2019) by Stanford University, USA.

He is an Editor/Associate Editor/Editorial Board member for refereed journals, in particular the IEEE Systems Journal, IEEE Transaction on Industry Applications, IEEE Access, *IET Power Electronics*, *IET Electronics Letters*,

Wiley-*International Transactions on Electrical Energy Systems*; Subject Editorial Board Member – *Energy Sources – Energies Journal*, MDPI; and the Subject Editor for the *IET Renewable Power Generation, IET Generation, Transmission and Distribution*, and *FACETS* Journal (Canada).

Google Scholar link: https://scholar.google.co.in/citations?user=KyuMg7IAAAAJ&hl=en&authuser=1

List of Contributors

Ankit Garg
Department of Electrical Engineering
MNIT
Jaipur, Rajasthan, India

Sharmeela Chenniappan
Department of EEE
CEG Campus, Anna University
Chennai, Tamilnadu, India

Prasad Kaparaju
School of Engineering and Built Environment
Griffith University (Nathan Campus)
Nathan, Queensland, Australia

Saeed Khorrami
Department of Electrical Engineering
Sapienza University of Rome
Rome, Italy

Dhandapani Lakshmi
Department of EEE
AMET University
Chennai, Tamilnadu, India

Ravichandran Karthick Manoj
Department of EEE
AMET University
Chennai, Tamilnadu, India

Khaleequr Rehman Niazi
Department of Electrical Engineering
MNIT
Jaipur, Rajasthan, India

Toyese Oyegoke
Chemical Engineering Department
Ahmadu Bello University, Zaria
Nigeria

Rupendra Kumar Pachauri
Electrical Cluster
School of Engineering
University of Petroleum and Energy Studies
Dehradun, Uttarakhand, India

Sivaraman Palanisamy
Department of EEE, CEG Campus
Anna University
Chennai, Tamilnadu, India

Mohammad Parhamfar
Independence Power and Energy Consultant
Isfahan, Iran

Naran M. Pindoriya
Department of Electrical Engineering
Indian Institute of Technology
Gandhinagar, Gujarat, India

Rajasekharan Rajasree
Department of EEE
AMET University
Chennai, Tamilnadu, India

Govindasamy Ramprakash
Department of Electrical and Electronics Engineering
SRM TRP Engineering College
Trichy, Tamilnadu, India

Subhendu Sekhar Sahoo
Department of Electrical Engineering
GIST
Bhubaneswar, Odisha, India

Nilay Kumar Sarker
School of Engineering and Built Environment
Griffith University (Nathan Campus)
Nathan, Queensland, Australia

Divya Sharma
Department of Electrical Engineering
Indian Institute of Technology
Gandhinagar, Gujarat, India

Rengamani Shenbagalakshmi
Department of Electrical & Electronics Engineering
Vel Tech Multi Tech Dr. Rangarajan Dr. Sakunthala Engineering College
Avadi, Chennai, Tamilnadu, India

Jai Govind Singh
Department of Energy
Environment and Climate Change
School of Environmental Resource and Development
Asian Institute of Technology
Khlong Nueng, Thailand

Kesavan Stalin
Department of Marine
AMET University
Chennai, Tamilnadu, India

Jaganathan Subramaniyan
Department of Electrical and Electronics Engineering
SRM TRP Engineering College
Trichy, Tamilnadu, India

Shubham Tiwari
Agriculture, Forestry, and Ecosystem Services Research Group
Biodiversity and Natural Resources Program International institute of Applied Systems Analysis (IIASA)
Laxenburg, Austria

Stephen Oko Gyan Torto
Electrical Cluster
School of Engineering
University of Petroleum and Energy Studies
Dehradun, Uttarakhand, India

Veerappan Vengatesan
Department of Electrical and Electronics Engineering
SRM TRP Engineering College
Trichy, Tamilnadu, India

Preface

The global shift toward low-carbon and resilient energy systems demands innovative approaches to how energy is produced, converted, and utilized. In this context, the concept of Multi-Energy Hubs (MEHs) has emerged as a promising solution. MEHs are integrated systems that connect electricity and natural gas networks with renewable energy sources such as solar, wind, and bioenergy. By incorporating co-generation units and advanced energy storage technologies, MEHs enable the simultaneous and efficient supply of electrical, thermal, and cooling energy – enhancing both system flexibility and sustainability.

This book offers a comprehensive exploration of the architectures and strategies that govern the operation and interaction of MEHs. Special emphasis is placed on Transactive Energy Management and Cooperative Energy Management frameworks, which facilitate coordinated decision-making and energy sharing among hubs. These strategies are essential for scaling up from isolated MEHs to regionally coordinated systems, termed Multi-Energy Networked Energy Hubs (ME-NEHs). Such networks improve overall system reliability, resilience, and economic efficiency.

Beyond foundational principles, the book delves into advanced topics including optimal energy scheduling, the integration of hydrogen and Power-to-X (P2X) technologies, and the role of energy and carbon markets in supporting sustainable energy transitions. It also examines emerging trends such as high renewable energy penetration, associated uncertainties, demand response programs, and electric vehicle integration – highlighting their influence on the design and operation of future energy systems.

Each chapter combines technical modeling with practical insights, offering students and researchers the analytical tools necessary to design, evaluate, and optimize multi-vector energy systems. By bridging theory with application, this book aims to equip the next generation of energy professionals to tackle the complex challenges of a flexible, low-emission energy future.

1

Evaluation of Power/Energy System to the Modern Multi-Vectored Energy Hubs (MV-EHs)

Ankit Garg[1], *Khaleequr Rehman Niazi*[1], *Subhendu Sekhar Sahoo*[2], *and Shubham Tiwari*[3]

[1] *Department of Electrical Engineering, MNIT, Jaipur, Rajasthan, India*
[2] *Department of Electrical Engineering, GIST, Bhubaneswar, Odisha, India*
[3] *Department of AFE, IIASA, Laxenburg, Vienna, Austria*

1.1 Introduction

The requirement for reliable, efficient, and environmentally friendly electricity systems is growing, and this is leading to a major change in the global energy landscape. The primary forces behind this shift are the increasing demand for clean energy, concerns about environmental quality and carbon dioxide emissions, alongside technological advancements that make it possible for more complex and tightly linked energy systems. Historically, energy institutions have operated as isolated units, with separate resources, including mechanisms for heat, gas, including electricity, and other energy vectors. These mechanisms were mostly centralized, including energy generation, distribution, including transportation, followed a straight path from generation to consumers [1]. However, as electrical usage patterns modifications and the critical importance of focusing on warming temperatures increases, there is an urgent need to transition from traditional single-vector electricity infrastructure toward more flexible, distributed, alongside interrelated networks called multi-vectored energy hubs (MV-EHs). The primary energy sources used by conventional power systems were fossil fuels, including natural gas, petroleum, and oil. Concentrated power generation was often used to characterize such structures. These infrastructures operated on a particular essential vector, so this meant that energy functions like power, fuel for transportation, and heating were provided by several networks with minimal or no interaction between one another. For instance, heat was frequently produced using independent systems like boilers driven by gas or

Energy Management Strategies for Multi-Vectored Energy Hubs to Achieve Low Carbon Societies, First Edition.
Edited by Shubham Tiwari, Jai Govind Singh, Sivaraman Palanisamy, Sharmeela Chenniappan, Rupendra Kumar Pachauri, and Sanjeevikumar Padmanaban.

oil, whereas electricity was created at large-scale nuclear power stations and supplied by an electrical grid. Transportation also used gasoline or diesel, which resulted in the creation of another unconnected energy source. Although these systems worked well in the industrialized beginning of the 20th century, they are becoming less and less viable in the face of contemporary demand for energy and environmental issues. Among the main drawbacks of conventional energy systems are significant reductions in energy constitute a common feature of traditional energy systems, especially during the generating and transmission phases. In the process of generating heat, centrally located power stations waste energy; additional expenditures happen throughout long-distance transmission. As they function independently without taking use of the complementarity across energy vectors, distinct transmission networks for heat, gas, and electrical result in inadequacies [2]. Networks of energy reliant on fossil fuels release a large amount of emission of greenhouse gases, which is one of the main causes of climate change in the world. Widespread decline in the environment is a result of the use of nonrenewable energy sources and carbon-intensive extraction, refining, and combustion technologies. With the rising use of renewable energy sources like the sun and the wind, conventional power generation systems are well-equipped to handle the complexity and unpredictability of expanding demand for electricity. Systems that are more adaptable and able to instantly balance supply and demand are needed for these sporadic energy sources. Centralized systems are generally more susceptible to disturbances and harder to adjust to these variations. Integrated technology has an adverse economic impact due to its high purchase costs, maintenance requirements, and operational challenges. The idea of MV-EHs is one possible way to address these issues and the increasing complexity of contemporary energy systems. Heat, gas, and electrical power are the three types of energy that can be generated, stored, and transported using MV-EHs—integrated methods that operate as a unified, flexible infrastructure. Integrated distribution centers aim to optimize the efficient utilization of renewable energy sources by encouraging a decrease in energy consumption, fortifying shock resistance, and cultivating collaboration among various energy channels. The transition to MV-EHs is driven by several key factors: environmental health, safety, and sustainability are the four main components of the Energy Hub System (ESS) and Safety policy. Promote Energy Hub System (EHS) integration and business excellence among compliance workers and stakeholders (clients, suppliers, contractors, and government). A technological or software product is not what an EHS governance structure is. Rather, it is the set of practices, guidelines, and agreements that serve as your company's environmental, health, and security program guidebook [3]. An EHS administration system's goals are to prevent injuries and illnesses associated with work; detect and reduce environmental, chemical, and physical hazards; and enhance collaboration and education that clearly outline the organization's goals for fostering a healthy and secure working atmosphere. Oversee our company's strategy for obtaining sustainable energy, cutting energy expenses, and eliminating harmful environmental effects like spills and leaks. Manage employee health to avoid disease, accidents, and long-term impairment while preserving efficient operations. To recognize safety risks and stop events, accidents, and injuries at work, we must manage our workspace and procedures. The term "MES" refers to the integrating of subsystems for the production, transmission, storage, and application of heat, cooling, gas, and electricity in contemporary energy systems to create the so-called "Energy Internet." The cooperation of various energy systems, powered by computer technology, allows for more intelligent energy administration than is possible when running each type of energy system alone. By utilizing the portability of gas and heat systems as cost-effective virtual battery storage, MES has the capability to significantly increase the adaptability of alternative energy sources because these systems have far more affordable energy storage. Furthermore, when MES is used instead of individual energy sources, it can enable the flexibility of switching between numerous energy vectors in order to improve the overall effectiveness, lower costs, and decrease emissions. Multi-vector energy systems (MESs) are a method of integrating the various energy infrastructures. MESs involve combining several energy sources to work together and enhance one another to create a more effective energy system. One of the most popular current ways to be utilized in modelling MESs is the EH approach, which converts, conditioned, and stores different energy carriers to optimize the system's thermal efficiency and utilization of energy resources. This approach has surely improved MES planning. Nevertheless, there aren't many studies using the EH model that contrast the advantages of coordinating several energy carriers at once with the more conventional approaches of organizing each energy carrier separately. As far as the investigators are aware, no research has compared the advantages of multi energy system (MES)-coordinated planning with the EH approach while taking demand anticipation and renewable energy sources (RES) variables into account.

1.2 Problem Statement

To maximize energy efficiency, environmental responsibility, and dependability, switching from outdated single-vector energy networks to contemporary MV-EHs offers potential and challenges. Even though MV-EHs are becoming more and more popular, the assessment of how they perform is still a difficult and unexplored field. The combining of several energy vectors, such as heat, natural gas, electric power, and others, into a single system is one of the main issues. It is challenging to evaluate the general effectiveness and effectiveness of the system because each vector operates on a separate set of principles, innovations, and developmental periods. A significant research gap exists due to the lack of established frameworks and methodologies to evaluate the interactions and synergies among various energy streams [4, 5]. An important yet unresolved issue is determining how to effectively coordinate these components to achieve optimal efficiency, reduce emissions, and enhance system resilience. Additionally, the unpredictable and intermittent nature of renewable energy sources presents another critical challenge. Although renewable resources such as wind, biomass, and solar energy are integrated into multi-vector energy hubs (MV-EHs), their variability prevents the assurance of a consistent power supply. While advanced management systems and energy storage solutions can help mitigate these fluctuations, their performance within the MV-EH context remains in the early stages of investigation. Additionally, it is yet unclear how best to combine mainstream and environmentally friendly energy sources within these kinds of networks. To fully exploit the future possibilities of MV-EHs, an objective mechanism that takes into consideration the constantly changing dynamics of clean power is necessary. Successful assessment is further hampered by the absence of performance information and real-world examples for MV-EHs. There are many theoretical computer simulations and models available, but there are not many recorded real-world examples of totally integrated MV-EHs. This restricts the amount of scientific information that is available to shed light on the difficulties and achievements of these kinds of systems in terms of functioning. It is challenging to check out conceptual frameworks and assess their scalability when applied to various geographic or macroeconomic contexts in the lack of thorough case studies. Expanding our knowledge of MV-EHs requires creating practical instances and intricate monitoring mechanisms. Furthermore, the regulations and policies are still being developed to allow for the widespread integration of MV-EHs. The intricacies of multi-vector interaction are seldom considered by conventional energy markets and policies, which are frequently created for single-vector systems. There are several obstacles to the implementation of MV-EHs, especially the absence of suitable incentives and legislation. The gap between the theoretical potential of MV-EHs and their practical implementation highlights the challenges of creating a supportive policy environment and the need for regulatory reforms. Key barriers to assessing advanced electrical systems like MV-EHs include managing the variability of renewable energy, integrating multiple energy vectors, the lack of real-world data, and inadequate financial support from regulators. Addressing these shortcomings in research is essential to advancing the development and deployment of MV-EHs as effective and sustainable renewable energy solutions for the future [6].

1.3 Objective

The primary objective of this chapter is to provide a comprehensive analysis of modern energy infrastructure within the context of MV-EHs, focusing on the challenges, strategies, and potential benefits of this innovative paradigm. The specific aims of the chapter include:

Exploring the evolution of traditional energy systems into integrated, multi-vector hubs. The chapter highlights the key differences between MV-EHs and single-vector systems, emphasizing their ability to integrate renewable energy sources and enhance overall efficiency.

Assessing the potential for coordinating heat, gas, and electrical power within MV-EHs. It examines the interactions among these energy vectors and explores how redundancies can be leveraged to optimize energy efficiency, lower costs, and improve system flexibility.

This section will look at how MV-EHs regulate the fluctuation of energy through renewable sources such as wind and solar energy. It will cover how demand handling strategies, energy storage gadgets, and intelligent infrastructure elements can balance supply in addition to demand for electricity. A key objective is to locate and assess the existing methods for MV-EH evaluations, considering aspects like environmental effect, energy conservation, system durability, and commercial viability. Furthermore, fresh perspectives and possible improvements for a thorough and precise review will be provided in the following part. The following portion will mostly focus on evaluating power and energy systems, especially in the setting of modern MV-EHs. A broad range of topics will be addressed, including the integration of several energy vectors (heat, gases, and electricity), the role of energy derived from sustainable sources, and technological advancements that facilitate the operation of MV-EHs. It will look at both theoretical models and real-life examples, giving insights into the practical applications of MV-EHs. The legislative alongside regulatory issues relevant to the ubiquitous implementation alongside usage of MV-EHs will also be covered in this part [7, 8]. The intricacies of assessing and putting into practice MV-EHs, in addition to the possible advantages alongside future possibilities for this changing energy framework, should all become quite clear to consumers (Figure 1.1).

1.4 Theoretical Framework

The growth of contemporary electrical infrastructure has revolved around the continual development of power systems. To generate electricity, conventional energy systems, which are mostly focused on centralized generation after generation,

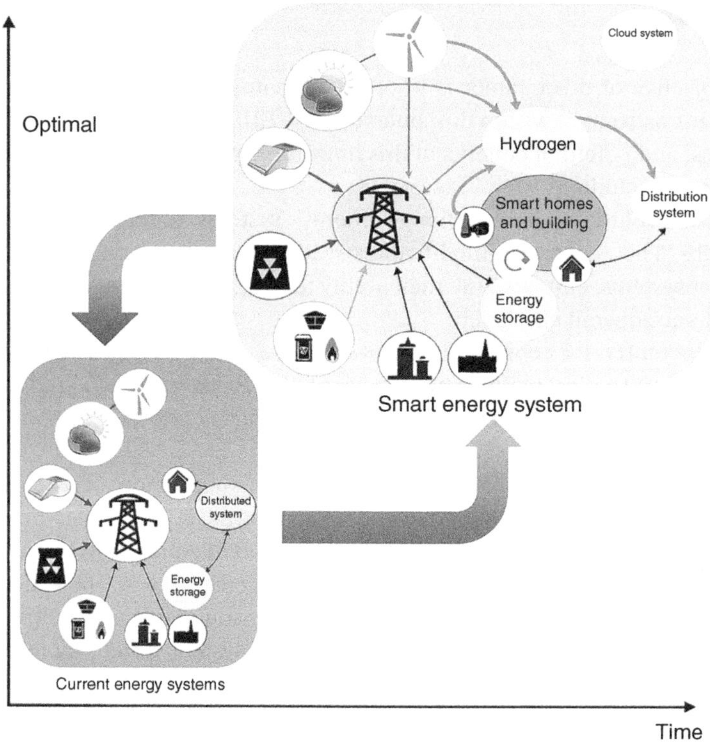

Figure 1.1 Transitions to multi-vector energy hubs. *Source:* [7] / with permission of Elsevier.

have mostly relied on fossil fuels like natural gas, petroleum, and coal [9]. Big hydroelectric dams usually employ these technologies to transmit electricity to end customers over great distances via vast grid networks. Large-scale power delivery is facilitated by such systems, but they have certain drawbacks, such as major transmission inefficiencies and a restricted capacity to integrate clean energy sources. Furthermore, these systems' centralized design makes them more susceptible to disturbances and system breakdowns, which reduces their capacity to withstand the complexity of today's energy demands [10].

Additionally, MV-EHs offer a creative response to the intermittent problem that energy from renewable resources presents. Energy excesses from renewable sources are capable of being stored or transformed among various forms of energy (such as electrical power to gas or heat) or utilized during times of greatest demand through the combination of several energy vectors. As a result, markets can better adapt to fluctuations in demand and supply. By leveraging digital technologies and advanced management systems, MV-EHs can further enhance efficiency through real-time optimization of electricity flow [11]. As the energy sector evolves, MV-EHs are increasingly recognized as vital components of future electrical systems. They support the integration of distributed energy resources, improve overall system flexibility, and pave the way for achieving sustainable energy goals.

These systems offer real-time operational insights, detecting issues such as temperature fluctuations, gas leaks, and equipment malfunctions. Sensors in production environments predict equipment failures, preventing accidents and reducing downtime. This proactive approach boosts operational efficiency and safety [12, 13].

Although the initial cost of implementing IoT and advanced sensors may vary, their long-term benefits—such as minimizing downtime and averting disasters—often outweigh the investment. Smaller organizations can adopt a phased approach, starting with pilot projects in critical areas, to demonstrate value and scalability without requiring significant upfront expenditures.

Technologies that convert electricity into zero-carbon synthetic fuels or chemicals, such as hydrogen, synthetic gas, liquid fuels, or other chemicals, are collectively referred to as power-to-X (P2X). These products can be stored for later use or employed in hard-to-decarbonize industries. For example, hydrogen can be generated through electrolysis and used directly or combined with other substances to produce fuels and chemicals. P2X is a crucial element in the transition to sustainable energy systems.

Additionally, machine learning and artificial intelligence (AI) are increasingly applied to MV-EHs. AI-driven systems can predict energy demand, optimize resource allocation, and maintain grid stability, facilitating renewable energy integration, reducing costs, and improving efficiency [14]. As these technologies advance, they are expected to further enhance the functionality of MV-EHs, solidifying their role in future power systems by providing greater control over energy consumption, promoting decarbonization, and ensuring energy security [15].

Despite the growing body of research on MV-EHs, several challenges hinder a comprehensive understanding of their full potential and practical implementation. A key limitation is the lack of empirical data and real-world applications. Much of the existing research relies on mathematical models or simulations, while large-scale, fully operational MV-EHs are scarce. These real-world implementations could provide critical insights into system performance that are often missed in simulated conditions, such as consumer behavior, market volatility, and infrastructure inefficiencies [16]. This gap between theoretical models and practical applications makes it difficult to evaluate the large-scale effectiveness of MV-EHs.

Another significant limitation is the narrow focus of existing evaluation frameworks. Many studies prioritize specific performance metrics, such as economic feasibility or energy efficiency, while neglecting equally important factors like social acceptability, system resilience, and environmental impacts. Comprehensive assessments that address these multidimensional aspects are necessary to realize the full potential of MV-EHs. Research that focuses on energy utilization optimization, for example, can ignore the trade-offs concerning system availability or the incorporating difficulties brought on by various energy vectors. The fragmented technique leads to insufficient assessments, which complicates the development of comprehensive strategies for the deployment of MV-EH. The decision factors are subject to restraints or limitations. Both asymmetry and restrictions on equality might currently present. There are structural or operational limitations on the OFs. These include the ESS and converter restrictions in addition to the EH operational constraints. Power balancing equations, ramping rate limitations, wind power resource constraint, distribution flow limitations, and other operational restrictions are examples of power system restrictions. The performance limitations of a gas system, such as pipeline limits, compression limitations, well-flow capacity limitations, nodal pressure limits, in addition to nodal balancing equations, are referred to as gas system barriers [17].

1.5 Evaluation Framework

1.5.1 Evaluation Criteria of MV-EHs

It is debatable if several of the electricity structures in use today, which were developed in the latter half of the 20th century, are adequate for the demands of the future. In addition to clogged transmission lines, a lot of facilities are nearing the conclusion of their anticipated lifespan. Furthermore, other concerns like the steadily rising energy demand, reliance on finite fossil fuels, the reorganization of the power sector, and the overarching goal of employing more ecologically friendly and renewable sources of electricity raise an issue about whether incremental improvements to the current systems will be adequate to address all the above problems or if a more profound rethinking of the infrastructure will be required [18] (Figure 1.2).

A strategy that considers many important performance metrics and is multi-dimensional is necessary for the evaluation of MV-EHs. One of the most important criteria is energy efficiency, which evaluates the degree to which the system uses resources across several vectors, such as heat, gas, and electricity while minimizing energy losses. Enhancing the overall performance of the system and decreasing energy waste are possible outcomes of effective coordination across these vectors. Economic viability, which assesses how cost-effective it is to deploy and keep running MV-EHs, is another crucial factor. This covers infrastructure capital expenditures and ongoing costs along with the return on expenditure, especially when considering long-term savings from increased

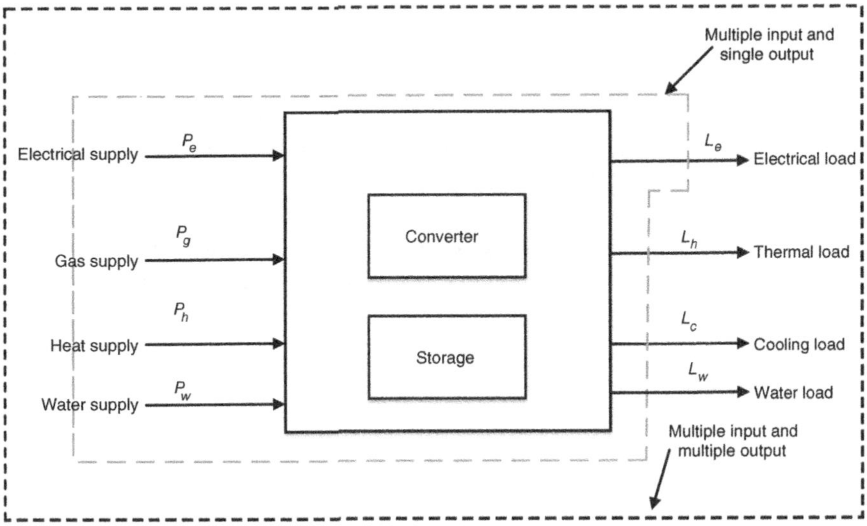

Figure 1.2 Block diagram energy hubs [3].

energy efficiency and the utilization of renewable energy sources. Early in the project, the realization that there are not many well-established tools readily accessible and the comprehensive investigation of numerous energy carrier systems led to the development of a modelling and analytic framework as the main area of concentration. Using the established tools, the second phase, which commenced recently, determines and compares the best system architectures and operating techniques to standard infrastructures. The greenfield strategy is the product of this phase. Finding transition routes and spanning technologies to connect the present-day systems to the chosen ideal structures is the focus of the project's last phase. To meet a people's need for different sources of energy, multi-vector energy hubs are architectural units that mix numerous vectors of electricity that are then transformed into energy storage in a united and synergistic manner. Multi-vector energy hubs completely include multiple resources for electricity by utilizing a variety of powering and heating technologies, such as photovoltaics, wind turbines, the production plants, batteries, and thermal storage systems [19]. These developments also increase the overall range of motion, dependability, and efficiency worldwide of the system that provides energy while lowering operational expenses, grid congestion in the lungs, and the use of fossil fuels with greater permeation of renewable energy sources (RES), as well as carbon footprints. Two energy transmission paths are offered by multi-vector energy networks, giving end users greater freedom to communicate with them as time goes by. By manufacturing and storing energy as needed, or by varying the type of source used to meet requests in reaction to changes in the price of electricity and system circumstances, they may result in modest energy costs. Since MV-EHs are made to facilitate the incorporation of renewable energy sources and reduce the release of greenhouse gases, the environmental effect is another important evaluation criterion. Assessing carbon emissions, enhancing air quality, and reducing reliance on fossil fuels are essential aspects of evaluating MV-EHs. Additionally, resilience and reliability play a critical role in determining the system's ability to deliver a stable energy supply, particularly amidst fluctuations in renewable energy production or extreme weather events. Resilience reflects the system's ability to withstand and recover from disruptions, while reliability measures its capacity to consistently meet energy demands. For an evaluation of MV-EHs to be comprehensive, it must encompass all these factors.

1.5.2 Data Collection

Efficient data gathering is essential to the evaluation of MV-EHs, and the techniques employed differ according on the elements under review. Systems for monitoring energy flows, consumption trends, and system reliability across several vectors are frequently used in real time. These systems automatically gather data on energy supply and consumption using smart meters, sensors, and Internet of Things (IoT) devices. This data may then be evaluated to improve system performance and identify inefficiency. Stakeholder research and interviews are also crucial, especially when evaluating user behavior, social acceptability, and the financial effects on end users. With the aid of these techniques, subjective information that supports quantitative study of energy systems can be collected. To optimize demand responsiveness and maximize the use of the current energy transport infrastructures, the research creates essentially new methods for modelling multi-vector energy carriers. By examining the potential capacity of the present system to alleviate network restrictions and enable demand and capacity development, the best leadership will offer less expensive options. Through performing this, privacy along with supply excellence can be maintained while postponing or, ideally, avoiding the necessary network expenditure. Because of their erratic nature, renewable energy sources are typically limited off during periods of high electrical network congestion and cannot be used to support base loads. This results in a waste of clean energy for society and financial loss for renewable energy-producing firms. Gas networks are an alternate delivery method for renewable energy that the study suggested to offer to end-user clients. The decrease in output constraint during periods of stress on energy systems is an advantage of renewable generation [20]. Apart from the in-the-moment monitoring, past data from current energy systems, like atmospheric trends and grid effectiveness records, are frequently utilized to model and forecast the potential performance of MV-EHs in various scenarios. In addition to being important sources of data, pilot initiatives and case investigations also shed light on how MV-EHs function in actual environments. Investigators can obtain a thorough grasp of the system's effectiveness, resilience, and general impact through the combination of real-time and historical data. This makes it possible to make more accurate assessments along with suggestions for future developments.

1.6 Discussion

Energy Hub is moving away from carbon-intensive, centrally controlled generation and toward intermittent, given-away renewable energy. The relationship between utilities and their consumers is also evolving because of consumer adoption of grid-impacting technology like electric vehicles, solar, and battery backups, and internet-connected thermostats. Utilities can use millions of energy resources distributed through energy hub to meet the grid requirements of a more dependable and clean power system.

The transition towards MV-EHs signifies a substantial change in the functioning of contemporary electrical and electrical systems. More specifically, the emphasis is on MV-EHs' future potential, difficulties, and implications with respect to sustainability, technical integration, and policy frameworks. With the growing pressure on existing energy systems to integrate renewable energy and improve effectiveness while MV-EHs provide a possible way around the drawbacks of classic single-vector schemes. But there are additionally a lot of things to think about, like operational difficulties, technical developments, and the necessity for helpful laws and regulations. The requirement to effectively integrate energy from alternative sources, such as wind, sunlight, and biomass, within the electrical grid represents one of the main forces supporting the emergence of MV-EHs. Because renewable energy is temporary, standard power infrastructure relies primarily on petroleum and gas, find it difficult to integrate it. More flexibility in controlling these shifts is made possible by MV-EHs' capacity to integrate numerous relies directions. To optimize the distribution of energy and minimize waste, excess electricity produced by solar or wind power can be transformed into gasoline or heat and then preserved for consumption later. This feature helps many nations achieve their carbon reduction targets and greatly improves the environment's average energy performance. Furthermore, by lowering dependency on importing of fossil fuels, the addition of renewables inside MV-EHs can further improve energy security. MV-EHs can reduce the vulnerability of municipalities and enterprises to geopolitical threats and price unpredictability in the international market for energy by enabling decentralized energy generation and storage. But to fully enjoy these advantages, techniques for storing energy must continue to progress, especially those that can store huge quantities of electrical energy for extended periods of time. Even while storage systems for batteries have advanced significantly, scaling these advancements up to satisfy the needs of completely interconnected multi-vector energy networks still presents difficulties. The creation and application of cutting-edge technologies that permit smooth communication among various energy pathways is essential to the effective functioning of MV-EHs. One essential element is smart grid technology, which offers the digital framework required for immediate fashion electrical power, maximizing efficiency, control, and monitoring. With the aid of sensors, Internet of Things gadgets, and sophisticated communications networks, MV-EHs can react quickly to shifts in both the supply and the demand of electricity. The network's resilience and effectiveness are increased by this real-time adjustment capability, especially when handling the fluctuating outputs of energy from natural producers. With the goal to make MV-EHs possible, techniques for storing energy, such as heat retention, power sources, especially hydrogen storage, are essential. The system that supplies energy can be able to maintain an equilibrium between the demand and availability for electricity over a range of time periods by keeping energy that takes multiple forms, which includes electrical power in capacitors or excess wind and solar energy translated to hydrogen. These forms of technology are still in the early stages of development; therefore, their widespread adoption is hampered by the substantial initial expenses. The long-term feasibility of MV-EHs will depend heavily on continuous study and progress in energy preservation innovations, especially those aimed at lowering costs and increasing the amount of storage space.

1.6.1 Regulatory and Policy Framework

Another key element in the effective deployment of MV-EHs is the regulatory environment around them. Most energy-related laws and policies in place are based on single-vector systems, which separately control heat, gas, and electricity. These conventional frameworks are put to the test by the multi-vector character of MV-EHs, necessitating the adoption of new laws that facilitate cross-vector integration, decentralized power generation, and the application of revolutionary technologies like P2X and smart grids. Moreover, regulatory structures need to guarantee that MV-EHs are consistent with more general climate objectives, such as decreasing carbon dioxide targets. Clear recommendations on how MV-EHs might support regional or national sustainable energy plans must be provided by organizations. Another of the biggest barriers to the broad implementation of MV-EHs is still regulatory ambiguity, particularly about market regulations, tariff building materials, and cross-sector integration. Addressing the possible social and financial effects on energy users is another aspect of establishing an ecosystem that supports MV-EHs. Legislators need to decide to ensure the switch to MV-EHs does not negatively impact any particular group more than

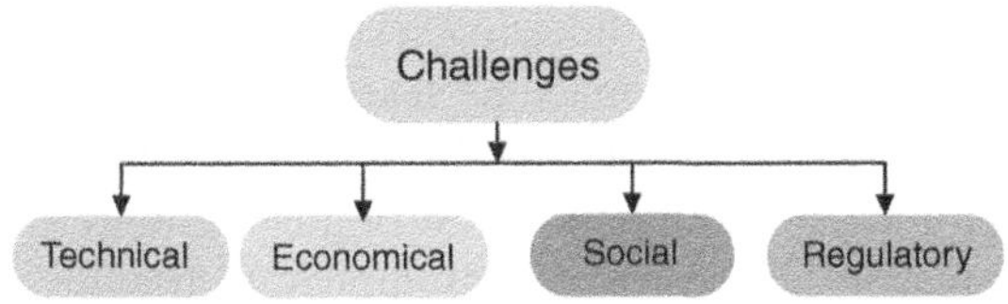

Figure 1.3 Different type of challenges in energy hub. *Source:* Adapted from [16].

others, including low-income homes, by increasing energy bills or creating access problems. Building public trust and the social acceptance of MV-EHs requires clear understanding and stakeholder participation.

1.6.2 Challenges and Future Trends

Difficulties include higher expenses because of shifting loads or coupling component failure. Using these connecting components to enable integrated network functioning has advantages for both the environment and the economy. The problem of preserving energy is central to the shift to a decarbonized and sustainable civilization. Making a guarantee that the electric power generated from these sporadic sources remains accessible for use, when necessary, as is right now the case using resources that come from petroleum and coal, is one of the numerous issues faced by the manufacturing of alternative energy (i.e., wind, sun, and tidal) (Figure 1.3).

That challenge comes in several forms. Generation of renewable energy is erratic due to meteorological and seasonal fluctuations; therefore, supply cannot satisfy end-user consumption. In addition to the power station's supply being placed far from major demand information, national grid limits might also present difficulties when there are higher levels of solar and wind power output in the energy mix. One way to address these issues is through energy preservation, and the market players face both possibilities and challenges because of this relatively recent movement in consumer appetite for increased storage capabilities. As a result, programmers, investors, and companies considering how to incorporate storage technologies throughout their daily activities or portfolios are becoming more interested in the industry. According to our study and years of working with customers, we provide a brief description of the electrical power storage business today in this first of four articles. We also discuss the possibilities and difficulties that come with investing and operating in this sector. In upcoming articles, we will concentrate on energy storage factors, like the state of the economy, the legislative and policy environment, and the present and projected power supply, including storage economic scenarios, including problems that may arise from the energy backup supply chain.

With several developments predicted to influence the way they evolve, the prospects of MV-EHs are bright. By forecasting energy consumption, controlling grid stability, and boosting system adaptability, developments in artificial intelligence (AI) and machine learning are anticipated to be crucial in optimizing the effectiveness of MV-EHs. Furthermore, the economic rationale for MV-EHs will get stronger as the cost of green energy continues to decline, especially in areas with a wealth of environmentally friendly resources. The growing emphasis on employing MV-EHs to decarbonize industries that are difficult to convert to electricity is an additional new trend. For instance, the creation of hydrogen from surplus renewable energy could be crucial in the decarbonization of sectors that presently rely largely on fossil fuels, such as cement and steel manufacturing. In the same vein, by offering more adaptable charging options and bolstering grid stability, the integration of MV-EHs with electric vehicle (EV) technology may help EVs become widely used. With the ability to integrate clean energy sources, boost the reduction of carbon dioxide and increase energy utilization, MV-EHs present a revolutionary approach to contemporary energy management. To reach their full potential, though, they must overcome various technological, financial, and legal obstacles. MV-EHs are expected to be crucial to the worldwide electricity change as long as technology advancements and enabling legislative frameworks keep continuing to evolve. With several developments predicted to influence the way they evolve, the future of MV-EHs is bright. By forecasting energy consumption, controlling grid stability, and boosting system adaptability, developments in artificial intelligence (AI) electricity infrastructure and machine learning are anticipated to be crucial in optimizing the operational efficiency of MV-EHs. The economic case for MV-EHs is becoming increasingly compelling as the average cost of renewable energy continues to decline, particularly in regions rich in renewable resources. Another emerging trend is the focus on leveraging MV-EHs to decarbonize industries that are challenging to electrify. For example, producing hydrogen from surplus renewable energy could play a pivotal role in reducing carbon emissions in sectors heavily dependent on fossil fuels, such as steel manufacturing and building materials production. Similarly, integrating MV-EHs with electric vehicle (EV) infrastructure could support the widespread adoption of EVs by offering more flexible charging solutions and enhancing grid stability.

By accommodating clean energy sources, supporting climate change mitigation, and improving energy efficiency, MV-EHs represent a transformative solution for modern energy management. However, realizing their full potential requires overcoming several technological, financial, and regulatory barriers. With continued advancements in technology and the establishment of supportive policy frameworks, MV-EHs are poised to play a vital role in global energy transitions.

1.7 Conclusion

The ability to synergize various energy sectors—such as the association of the electric power and natural gas networks—to operate under a single unit, or MES—is what will shape global energy systems of the future. This will open up new opportunities that, if taken advantage of, will generate wealth for the electricity and gas industries. This work will aid in the better understanding of the idea of MES planning utilizing the EH approach by scholars and novice researchers. This study examined several research efforts on the planning of energy hubs (EHs) that were conducted to create a useful model for the planning of megawatt-hour plants (MESs). It also examined the energy hub planning model, which contrasted the advantages of coordinating several energy carriers at once with the conventional planning of different energy carriers. The approach used in this investigation considered MES definitions, rationales, and a range of energy vector integration subcategories. To fully grasp the idea of MESs, the advantages and difficulties of MESs were carefully examined. Additionally, using the electrical, gas, and heat network models of each associated to the EH, a thorough review of the MES network that models was conducted. The deterministic alongside unpredictable models, uncertainty modelling techniques, and maximizing the efficiency techniques in the power system, and the practical use of these techniques in EH developing problems, were reviewed methodically. Additionally, MES's DR tactics in a liberalized electrical market were emphasized. Modern energy systems are undergoing a radical change as they transition to MV-EHs, which offer increased sustainability, flexibility, and efficiency. The constraints of conventional single-vector systems are addressed by MV-EHs, which combine many energy vectors—electricity, heat, gasoline, and perhaps even hydrogen—and facilitate the introduction of energy from green sources. This method improves system resilience and energy security in addition to optimizing energy use and cutting waste. Overcoming current obstacles, such as those related to technological breakthroughs, economic viability, and regulatory backing, is essential to achieving the entire potential of MV-EHs. The broad use of MV-EHs will be largely dependent on future developments, including the development of sophisticated energy storage technologies and the integration of electric vehicle infrastructures, especially the use of computational intelligence for real-time optimization. To solve the financial and infrastructure hurdles, additional studies and supportive regulatory frameworks will be necessary. MV-EHs have the potential to be extremely important in accomplishing sustainable development carbon reduction as the world energy landscape keeps on changing in this direction. They will be a key component of future energy systems and make a substantial contribution to a healthier and more resilient electrical future due to their capacity to coordinate a variety of sources of electricity and adapt to changing conditions.

References

1 Bhaskaran Pillai, M. (2020). India's Multivectored Alignment in the East: Aims, Objectives, and Rationale in Perspective (March 16, 2020).

2 Campi, A. (2020). Symbolic diplomacy of place and space in Eurasia: "multi-vector" vs. "third neighbor" policies-impact of nomadic features on foreign policy strategies of Kazakhstan and Mongolia. *Turkic Studies Journal* 2 (3): 54–69.

3 German, T., Kakachia, K., and Jones, S.F. (2023). *Georgia's Foreign Policy in the 21st Century: Challenges for a Small State*. Bloomsbury Publishing.

4 Hu, J., Li, Y., Chen, W. et al. (2022). Performance characteristics of GaInP/InGaAs/Ge triple-junction photovoltaic panels for near-space vehicles via ground experiments. *Applied Energy* 326: 119826.

5 Nyshanbayev, N. and Tarman, B. (2024). The Republic of Kazakhstan's multi-vector foreign policy: re-evaluation under president Tokayev. *Kasetsart Journal of Social Sciences* 45 (3): 915–924.

6 Tiwari, S. and Singh, J.G. (2023). Tri-level stochastic transactive energy management and improved profit distribution scheme for multi-vectored networked microgrids: a multi-objective framework. *Sustainable Cities and Society* 95: 104569.

7 Tiwari, S., Singh, J.G., and Garg, A. (2024). A static robust energy management approach for modelling low emission multi-vectored energy hub including emission markets and power-to-gas units. *Energy* 294: 130827.

8 Palani, V., Vedavalli, S., Veeramani, V.P., and Sridharan, S. (2022). Optimal operation of residential energy hubs include hybrid electric vehicle & heat storage system by considering uncertainties of electricity price and renewable energy. *Energy* 261: 124952.

9 Wang, J. and Zheng, Y. (2024). Data-driven seasonal scenario generation-based static operation of hybrid energy systems. *Energy* 309: 133030.

10 Lu, X., Li, H., Zhou, K., and Yang, S. (2023). Optimal load dispatch of energy hub considering uncertainties of renewable energy and demand response. *Energy* 262: 125564.

11 Karimi, H. and Jadid, S. (2021). Two-stage economic, reliability, and environmental scheduling of multi-microgrid systems and fair cost allocation. *Sustainable Energy, Grids and Networks* 28: 100546.

12 Onishi, V.C., Antunes, C.H., Fraga, E.S., and Cabezas, H. (2019). Stochastic optimization of trigeneration systems for decision-making under long-term uncertainty in energy demands and prices. *Energy* 175: 781–797.

13 Garg, A. and Niazi, K. (2023). A robust energy management strategy for multi carrier energy system. *2023 7th International Conference on Computer Applications in Electrical Engineering-Recent Advances (CERA)*, 1–5. IEEE.

14 Tan, B., Chen, H., Zheng, X., and Huang, J. (2022). Two-stage robust optimization dispatch for multiple microgrids with electric vehicle loads based on a novel data-driven uncertainty set. *International Journal of Electrical Power & Energy Systems* 134: 107359.

15 Rahgozar, S., Seyyedi, A.Z.G., and Siano, P. (2022). A resilience-oriented planning of energy hub by considering demand response program and energy storage systems. *Journal of Energy Storage* 52: 104841.

16 Lekvan, A.A., Habibifar, R., Moradi, M. et al. (2021). Robust optimization of renewable-based multi-energy micro-grid integrated with flexible energy conversion and storage devices. *Sustainable Cities and Society* 64: 102532.

17 Jordehi, A.R., Javadi, M.S., and Catalo, J.P. (2021). Day-ahead scheduling of energy hubs with parking lots for electric vehicles considering uncertainties. *Energy* 229: 120709.

18 George-Williams, H., Wade, N., and Carpenter, R. (2022). A probabilistic framework for the techno-economic assessment of smart energy hubs for electric vehicle charging. *Renewable and Sustainable Energy Reviews* 162: 112386.

19 Sun, Y., Zhang, B., Ge, L. et al. (2020). Day-ahead optimization schedule for gas-electric integrated energy system based on second-order cone programming. *CSEE Journal of Power and Energy Systems* 6 (1): 142–151.

20 Li, Y., Wen, F., Tian, F. et al. (2017). Day-ahead schedule of a multi-energy system with power-to-gas technology. *2017 IEEE Power & Energy Society General Meeting*, 1–5. IEEE.

2

Introduction of Transactive Energy Management in a Multi-Energy Networked System

Ankit Garg[1], Khaleequr Rehman Niazi[1], Subhendu Sekhar Sahoo[2], and Shubham Tiwari[3]

[1]*Department of Electrical Engineering, MNIT, Jaipur, Rajasthan, India*
[2]*Department of Electrical Engineering, GIST, Bhubaneswar, Odisha, India*
[3]*Department of AFE, IIASA, Laxenburg, Vienna, Austria*

2.1 Introduction

The global energy landscape is undergoing a significant transformation driven by the developing interest for supportability, the mix of environmentally friendly power sources, and the requirement for more noteworthy adaptability and effectiveness in energy on the board. Conventional concentrated energy frameworks, where energy age, circulation, and utilization are overseen in a hierarchical way, are progressively tested by the ascent of distributed energy resources (DERs), like sunlight-powered chargers, wind turbines, and energy stockpiling frameworks. These assets, frequently claimed and worked by individual buyers or limited-scope makers, bringing intricacy into the energy framework, requiring new ways to deal with the powerful connections between various energy transporters and partners. One such methodology that has acquired unmistakable quality is transactive energy [1]. Transactive energy alludes to a structure where financial and control instruments are utilized to facilitate the age, circulation, and utilization of energy inside an organized framework. This coordination happens through a progression of exchanges, where energy and related administrations are purchased, sold, or traded between different members—going from enormous service organizations to individual buyers, frequently alluded to as "prosumers" when the two of them produce and consume energy. The vital thought behind transactive energy is to use market standards and constant information to accomplish a harmony among organic markets, in this way improving the proficiency, dependability, and maintainability of energy frameworks [2]. The significance of

Energy Management Strategies for Multi-Vectored Energy Hubs to Achieve Low Carbon Societies, First Edition.
Edited by Shubham Tiwari, Jai Govind Singh, Sivaraman Palanisamy, Sharmeela Chenniappan, Rupendra Kumar Pachauri, and Sanjeevikumar Padmanaban.

transactive energy in current energy frameworks couldn't possibly be more significant. As the infiltration of sustainable power sources like sunlight-based and wind-expanded, the changeability and vulnerability related to these sources present huge difficulties to matrix soundness. Dissimilar to customary power establishes that it can be dispatched on request, sustainable power sources are frequently irregular and erratic. This changeability requires a more unique and adaptable way to deal with energy on the board, where the organic market can be adjusted continuously [3]. Transactive energy gives an answer by empowering the network to answer vacillations in energy creation and utilization through decentralized direction and computerized market exchanges. One of the central standards of transactive energy is decentralization. Rather than the unified control frameworks of conventional energy matrices, where choices are made by a focal power, transactive energy frameworks disperse control among different entertainers inside the organization. Every member in a transactive energy framework has the independence to pursue choices in view of neighborhood conditions, for example, energy accessibility, utilization examples, and market costs [4]. This decentralized methodology considers more granular and responsive control, which is essential in dealing with the intricacies of current energy frameworks that consolidate a different blend of fuel sources and advancements. One more key quality of transactive energy is the utilization of market-based components to work with exchanges. In a transactive energy framework, energy costs are resolved progressively and founded on organic market conditions. Members can exchange energy and related benefits constantly, answering cost flags that mirror the present status of the energy organization. This market-based approach not just guarantees that energy assets are assigned proficiently yet in addition gives monetary motivations to members to enhance their energy creation and utilization. For instance, during times of popularity or low stock, costs might rise, empowering shoppers to lessen their energy use or shift it to times when energy is more bountiful and less expensive [5].

The job of cutting-edge correspondence advancements and information examination is additionally fundamental to the working of transactive energy frameworks. Constant information on energy creation, utilization, and economic situations are fundamental for settling on informed choices and executing exchanges. Innovations like brilliant meters, sensors, and the Internet of Things (IoT) empower the assortment and correspondence of this information, while cutting-edge examination and computerized reasoning (simulated intelligence) are utilized to handle the information and enhance dynamic cycles. These advances on the whole empower the consistent activity of transactive energy markets, guaranteeing that energy streams are overseen proficiently and that the lattice stays stable even notwithstanding changes. Transactive energy is especially significant regarding multi-energy organized frameworks, which include the combination of various energy transporters, like power, gas, and nuclear power. In such frameworks, transactive energy can work with the coordination between these transporters, enhancing the general energy stream and lessening failures. For example, abundance power produced from inexhaustible sources can be changed over into heat or put away for some time in the future, contingent upon economic situations and framework needs. This adaptability upgrades the strength of the energy organization, improving it prepared to deal with disturbances or changes in energy accessibility. Besides, transactive energy upholds the mix of prosumers into the energy organization. As customers become progressively enabled to produce their own energy, whether through roof sunlight-based chargers, home batteries, or different means, they additionally gain the capacity to take part in energy markets. Transactive energy frameworks permit prosumers to sell overabundant energy back to the lattice, buy energy when required, and even take part in sought-after reaction programs where they are made up for diminishing their energy utilization during peak periods. This dynamic investment of customers not just upgrades the proficiency of the energy framework but also democratizes energy creation and utilization, adding to a more reasonable and evenhanded energy future. Transactive energy addresses a change in perspective in how energy frameworks are overseen and worked. By utilizing decentralized control, market-based systems, and high level correspondence advancements, transactive energy empowers the productive and adaptable administration of present-day energy organizations, especially those coordinating various energy transporters and appropriated energy assets. As the energy scene keeps on developing, the standards and practices of transactive energy play had a basic impact in guaranteeing the maintainability, unwavering quality, and versatility of future energy frameworks [6].

2.2 Problem Statement

Overseeing multi-energy organized frameworks, which include the combination of different energy transporters like power, gas, and nuclear power, presents a remarkable arrangement of difficulties. The essential intricacy emerges from the need to organize the progression of these different energy shapes, each with its unmistakable qualities, inside a solitary interconnected network. This coordination should guarantee that energy is productively delivered, appropriated, and consumed while keeping up with the security and unwavering quality of the whole framework. One of the key difficulties is framework intricacy. Multi-energy networks are innately more intricate than single-energy frameworks because of the cooperation between various energy transporters. For instance, power can be changed over into heat or put away as synthetic energy, yet the transformation processes include misfortunes and require refined control systems to guarantee productivity. Moreover, the powerful idea of an organic energy market across these transporters adds layers of intricacy, as variances in a single piece of the framework can affect the others. Dealing with these associations progressively requires progressed demonstrating, recreation, and advancement devices [7, 8]. Another huge test is information for the executives and correspondence. The coordination of numerous energy transporters requires a strong correspondence foundation to gather, process, and send huge measures of information progressively. This information is fundamental for checking framework execution, gauging energy interest and supply, and empowering dynamic estimating in transactive energy markets. In any case, the sheer volume and variety of information from various energy transporters and organized parts can overpower conventional information the executives frameworks [9]. Guaranteeing information precision, practicality, and security is critical, however troublesome, especially in enormous, geologically scattered networks. Market plan and guideline additionally present difficulties. Multi-energy frameworks require new market structures that can oblige the exchanging of different sorts of energy and related administrations. These business sectors must boost the effective utilization of energy assets while guaranteeing fair access and rivalry among members [10]. Notwithstanding, planning such business sectors is perplexing, especially while thinking about the requirement for administrative systems that address the remarkable qualities of every energy transporter and their associations. Finally, framework versatility and dependability are basic worries. The reliance on energy transporters implies that a disappointment in one piece of the organization can have flowing impacts across the whole framework. Creating methodologies to improve the flexibility of multi-energy organizations, for example, consolidating energy stockpiling, requesting reaction, and adaptable age sources, is indispensable yet testing because of the requirement for coordination across various energy spaces. These difficulties highlight the requirement for creative arrangements, like transactive energy the board, to deal with the intricacies of multi-energy organized frameworks and guarantee their solid and effective activity.

2.3 Objective

The essential target of this part is to investigate the coordination of transactive energy to the executives inside multi-energy arranged frameworks. This includes understanding how transactive energy standards can upgrade the coordination and streamlining of different energy transporters in a bound together framework [11]. The part has dug into the center credits and instruments of transactive energy, making sense of how decentralized control and market-based exchanges can be applied to deal with the perplexing connections between power, gas, and nuclear power inside a multi-energy organization. Another focus is on analyzing the advantages of transactive energy the executives, like superior proficiency, adaptability, and flexibility of energy frameworks. The section has featured how these advantages add to better treatment of the inconstancy and vulnerability related to environmentally friendly power sources. The part has likewise tended to the difficulties related to executing transactive energy in multi-energy organizations, including specialized, administrative, and market-related issues. Understanding these difficulties is significant for creating compelling methodologies for the fruitful mix. Ultimately, the part expects to give bits of knowledge into future patterns and examination headings in transactive energy to the executives for multi-energy frameworks [12].

Scope: This section gives a thorough prologue to the idea of transactive energy of the board and its application in multi-energy organized frameworks. The extension incorporates a point-by-point investigation of the essential standards, traits, and components of transactive energy, with an emphasis on how these ideas can be coordinated into frameworks that include various energy transporters, such as power, gas, and nuclear power. The part takes care of the hypothetical underpinnings of transactive energy, including decentralized control, market-based exchanges, and ongoing dynamic cycles, as well as commonsense perspectives connected with the execution and activity of transactive energy frameworks [13] (Figure 2.1).

2.4 Conceptual Framework

Transactive energy is a forward-looking approach to energy management that combines economic and control mechanisms to balance supply and demand within a networked energy system. It uses ongoing business sector exchanges to organize the age, conveyance, and utilization of energy among different members, including conventional utilities, DER proprietors and buyers who may likewise go about as "prosumers" (the two makers and customers of energy). The expression "transactive" alludes to the utilization of exchanges—robotized trades of energy and monetary value—to oversee energy streams inside the framework [14]. At its core, transactive energy is built on several key principles: 1. Decentralization: Not at all like conventional energy frameworks, where choices are made halfway, transactive energy frameworks disseminate control among different partners. Every member has the independence to make choices in view of the neighborhood data, for example, energy costs, accessibility, and individual utilization needs [15]. This decentralization considers more responsive and granular administration of energy assets. 2. Market-Based Control: Transactive

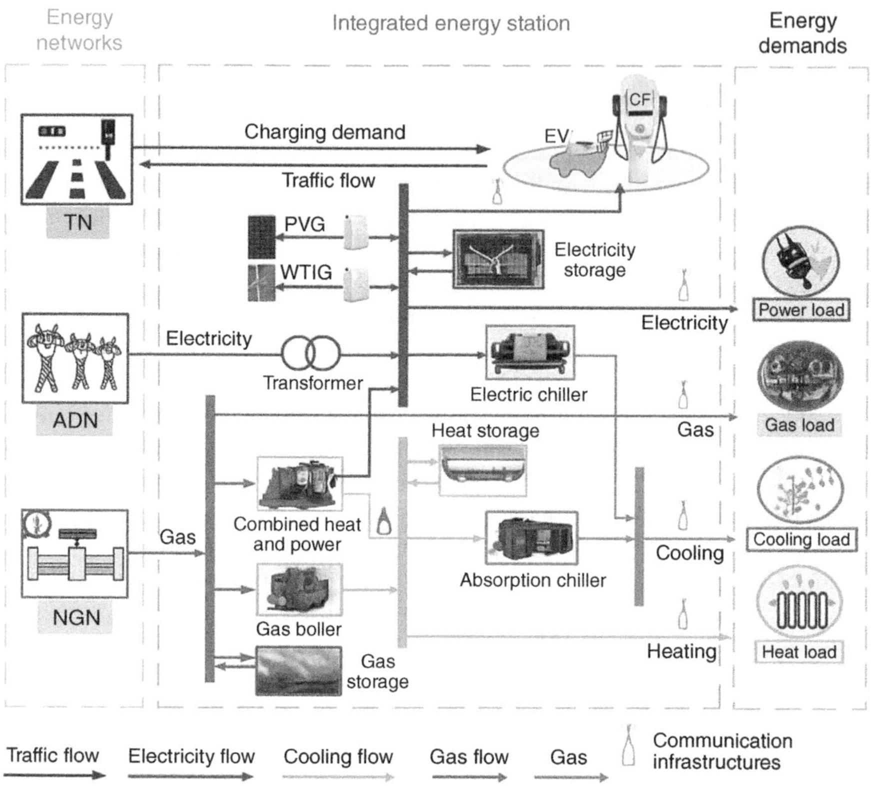

Figure 2.1 Multi-energy network system. *Source:* Xie et al. (2020) / with permission of Elsevier.

energy depends on market components to decide energy costs powerfully, in light of constant states of market interest. Members can purchase, sell, or exchange energy as per these market costs, which helps in productively designating assets across the organization. This market-based approach boosts members to streamline their energy use, adding to the by-and-large framework proficiency. 3. Real-Time Direction: Transactive energy frameworks work continuously, utilizing progressed correspondence advances and information examination to screen and answer changes in the energy network quickly. This continuous ability is essential for coordinating variable environmentally friendly power sources, for example, sunlight-based and wind, which are intrinsically irregular and erratic. The significance of transactive energy lies in its capability to change conventional energy frameworks, making them more versatile, productive and strong. By empowering decentralized navigation and making a commercial center for energy exchanges, transactive energy can streamline the utilization of dispersed energy assets and lessen the requirement for costly framework speculations. In addition, it upholds the joining of environmentally friendly power sources, assisting with lessening fossil fuel byproducts and advancing maintainability [16]. In a multi-energy organized framework, where power, gas, and nuclear power are interconnected, transactive energy assumes a basic part in planning these different energy transporters. It empowers the adaptable administration of energy streams, guaranteeing that the right kind of energy is accessible where and when it is required, while additionally improving expenses and limiting waste [17]. As energy frameworks become progressively intricate and decentralized, the standards of transactive energy will be fundamental for dealing with the developing scene, eventually prompting a more economical and proficient energy future.

2.5 Multi-Energy Networked System

A multi-energy organized framework alludes to an incorporated energy framework that interfaces and directs different sorts of energy transporters, like power, gas, and nuclear power, inside a brought-together foundation. Dissimilar to conventional energy frameworks that ordinarily center around a solitary energy transporter, multi-energy frameworks are intended to improve the cooperation between various energy structures, utilizing their interesting properties to upgrade generally productivity, dependability, and manageability. In a multi-energy organized framework, every energy transporter assumes an unmistakable part. Power is a flexible and generally utilized energy structure, controlling everything from private machines to modern hardware. It is additionally the essential conductor for incorporating environmentally friendly power sources like sunlight-based and wind into the lattice. Nonetheless, power age from renewables is much of the time variable and irregular, requiring the reconciliation of other energy transporters to supply guaranteed steady and solid energy (Figure 2.2).

Gaseous petrol is one more basic part of multi-energy frameworks. It is frequently utilized for warming, modern cycles, and as a reinforcement fuel for power age, particularly during periods when sustainable power sources are inadequate. Gas can likewise be changed over into power through gas turbines, creating it as an adaptable energy transporter that can supplement power in a multi-energy framework [18]. Nuclear power alludes to the energy related to heat. It is ordinarily utilized in private and modern warming, cooling, and boiling water frameworks. Nuclear power can be produced from different sources, including direct utilization of petroleum gas, power (through heat siphons or electric warmers), or waste intensity recuperation from modern cycles. In a multi-energy framework, nuclear power can be put away for some time in the future or changed over once again into power, contingent upon

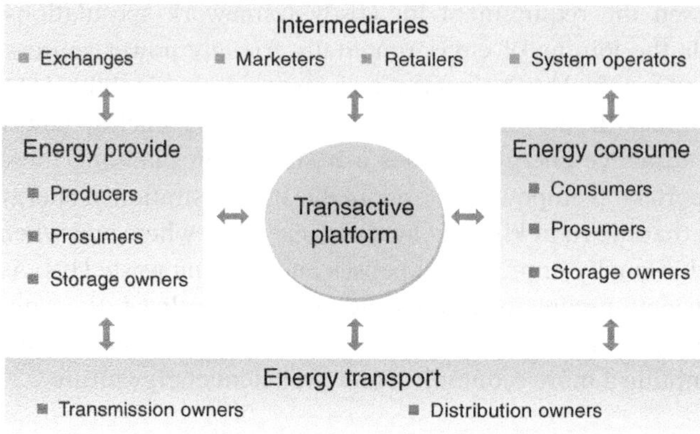

Figure 2.2 Transactive energy framework. *Source:* [18] / with permission of Elsevier.

the requirements of the framework. The communication between these different energy transporters makes multi-energy frameworks especially beneficial. For example, overabundance of power created from sunlight-based chargers during the day can be utilized to deliver intensity or hydrogen (through electrolysis), which can be put away and changed over once again into power or utilized straightforwardly when required. Additionally, squander heat from modern cycles can be caught and appropriated through area warming organizations, lessening the general energy interest. These collaborations empower a more proficient utilization of energy assets, as the framework can progressively distribute energy where it is most required, in light of constant interest and supply conditions. This interconnectedness additionally upgrades the framework versatility, as the accessibility of different energy transporters gives elective pathways to energy conveyance in the event of disturbances. A multi-energy organized framework is a comprehensive way to deal with energy the board that incorporates and improves the utilization of various energy transporters [3]. By organizing the collaborations between power, gas, and nuclear power, these frameworks can further develop by and large energy productivity, diminish ozone-harming substance emanations and increment the adaptability and unwavering quality of the energy supply, making them a pivotal part of the progress to manageable energy frameworks.

2.6 Integration of Transactive Energy Management

Coordinating transactive energy, the executives into a multi-energy organized framework includes implanting market-based control and decentralized dynamic instruments across various energy transporters—like power, gas, and nuclear power to enhance the general framework execution. This coordination is vital to guaranteeing that energy streams are proficiently made due, progressively adjusted and lined up with constant organic market conditions across the organization [19].

This study evaluates the effectiveness of transactive energy (TE) in a smart grid environment by analyzing three microgrids. Each microgrid integrates a combination of energy generation resources and storage systems. Specifically, the energy generation resources consist of power generation units (PGUs) and solar photovoltaic (PV) systems, supported by electrical storage systems within each microgrid. The research focuses on electrical loads, which can be supplied either by the energy resources within their respective microgrids or through external sources, such as the main power grid or neighboring microgrids. Furthermore, the microgrids have the capability to sell surplus energy to the main power grid and the local energy market (other microgrids) or purchase energy from these sources when their internal generation is insufficient to meet demand.

2.6.1 Objective function

$$Obj = \sum_t (EP_{n,t} \cdot EPP_t - ES_{n,t} \cdot EPS_t) + \sum_t (F^{PG}_{n,t} \cdot P^{PG}_t), \forall n \tag{2.1}$$

2.6.2 Constraints

To effectively manage energy transactions, generation, and consumption, certain constraints must be adhered to across all components of the system.

2.6.2.1 Power Balance Constraint

$$EP_{n,t} + E^{PV}_{n,t} + EPG_{n,t} + E^{Bd}_{n,t} + E^{Tin}_{n,t} = E^{L}_{n,t} + E^{S}_{n,t} + \frac{E^{Bc}_{n,t}}{\eta_{Bc}} + E^{T_{out}}_{n,t} \forall n, \forall t \tag{2.2}$$

2.6.2.2 Power Generation Constraint

$$F^{PG}_{n,t} \leq X^{PG}_{n,t}.S^{PG}_{n} \forall n, \forall t \tag{2.3}$$

$$E^{PG}_{n,t} \leq (F^{PG}_{n,t} - b_{PG}.X^{PG}_{n,t})/a_{PG} \forall n, \forall t \tag{2.4}$$

2.6.3 PV Constraints

$$E^{PV}_{n,t} \leq S^{PV}_{n}.SR_t.\eta_{PV} \forall n, \forall t \tag{2.5}$$

2.6.4 Battery Storage Constraints

$$I_{n,t}^{Bc} + I_{n,t}^{Bd} \leq 1 \forall n, \forall t \tag{2.6}$$

$$S_n^{bs} \cdot \gamma_{bmin} \leq E_{n,t}^{b} \leq S_n^{bs} \forall n, \forall t \tag{2.7}$$

$$E_{n,t}^{b} = E_{o}^{b} + (E_{n,t}^{bc} - E_{n,t}^{bd}) \cdot \Delta t \forall n, t = 1 \tag{2.8}$$

$$E_{n,t}^{b} - E_{n,t-1}^{b} = (E_{n,t}^{bc} - E_{n,t}^{bd}) \cdot \Delta t \forall n, t \geq 2 \tag{2.9}$$

$$S_n^{bs} . \gamma_{min}^{bc} \cdot X_{n,t}^{bc} \leq E_{n,t}^{bc} \leq S_n^{bs} \cdot \gamma_{max}^{bc} \cdot X_{n,t}^{bc} \tag{2.10}$$

$$S_n^{bs} \cdot \gamma_{min}^{bd} \cdot X_{n,t}^{bd} \leq E_{n,t}^{bd} \leq S_n^{bd} \cdot \gamma_{max}^{bd} \cdot X_{n,t}^{bd} \tag{2.11}$$

2.6.5 Market Constraints

$$XE_{n,t}^{in} + XE_{n,t}^{out} \leq 1 \forall n, \forall t \tag{2.12}$$

$$ET_{n,t}^{in} \leq P \cdot XE_{n,t}^{in} \forall n, \forall t \tag{2.13}$$

$$ET_{n,t}^{out} \leq P \cdot XE_{n,t}^{out} \forall n, \forall t \tag{2.14}$$

$$\sum_n ET_{n,t}^{in} = \sum_n ET_{n,t}^{Out} \tag{2.15}$$

2.6.6 Working Cases of Microgrids

This study examines two cases of microgrid operation within the framework of transactive energy (TE) management to assess its impact on reducing energy costs. The first case analyzes microgrid operations without any energy exchange between them. In contrast, the second case explores interconnected microgrid operations, where energy can be exchanged among microgrids under this configuration.

2.6.6.1 First Case

In this case, each microgrid's electrical energy is supplied by its internal generation resources and energy obtained from the main power grid.

$$MInObj_I = \sum_n Obj_{I,n} \tag{2.16}$$

Table 2.1 Comparison results of cases.

Microgrid Index	Cost in Case I ($)	Cost in Case II	Cost saving ($)	Cost saving (%)
1	25.051	21.356	3.695	14.75
2	27.458	27.506	−0.048	−0.1748
3	26.565	25.145	1.42	5.345
Total cost	79.074	74.007	5.067	6.407

2.6.6.2 Second Case

Under this case, electricity can be exchanged between the microgrids and the power grid within the framework of the TE structure. Each microgrid has the flexibility to connect with other microgrids to optimize shared benefits (Table 2.1).

$$MinObj_{II} = \sum_{n} Obj_{II,n} \tag{2.17}$$

2.6.7 Benefits of Integration

1) **Enhanced Proficiency:** Transactive energy on the board permits every energy transporter to be utilized all the more productively by adjusting energy creation and utilization through market-driven cost signals. For example, during times of low power interest, surplus energy can be utilized to create nuclear power or put away as gas, which can be subsequently changed back over completely to power during top interest times. This advances asset uses as well as diminishes squander and functional expenses.
2) **Increased Adaptability:** By coordinating transactive energy the board, multi-energy frameworks gain the adaptability to switch between various energy transporters in view of continuous circumstances. For instance, assuming power costs ascend because of popularity, the framework can focus on the utilization of gas or nuclear power, consequently balancing out the general energy organization [20]. This adaptability is especially significant in dealing with the fluctuation of environmentally friendly power sources.
3) **Improved Strength:** The decentralized concept of transactive energy enhances the flexibility of multi-energy systems In case of a disturbance in one piece of the organization, for example, a blackout, the framework can rapidly reconfigure itself by utilizing other accessible energy transporters, guaranteeing persistent energy supply. This versatility is critical for keeping up with matrix security, particularly despite the environment-related related difficulties.

2.6.8 Challenges of Integration

1) **Complex Framework Coordination:** Incorporating transactive energy from the executives into a multi-energy framework requires modern coordination between various energy transporters. This includes complex demonstrating and constant control calculations to guarantee that the collaborations between power, gas, and nuclear power are upgraded without causing irregular characteristics or shortcomings in the framework.
2) **Data Administration and Correspondence:** Compelling coordination request strong information from the executives and correspondence framework. The framework should deal with huge measures of information from different sources, like savvy meters, sensors, and market stages, to pursue informed choices. Guaranteeing information precision, security, and practicality is a huge test, especially in enormous-scope organizations.
3) **Regulatory and Market Difficulties:** The reconciliation of transactive energy executives into multi-energy frameworks frequently faces administrative and market obstructions. Existing guidelines may not completely oblige the decentralized and market-driven nature of transactive energy frameworks, requiring new structures and strategies to work with their reception. Furthermore, making fair and straightforward business sectors for various energy transporters includes complex plan and execution challenges. While the combination of transactive energy the executives into multi-energy frameworks offers huge advantages with regard to productivity, adaptability, and versatility, it likewise presents difficulties that should be painstakingly made to do. Tending to these difficulties requires progressed mechanical arrangements, administrative help, and cooperative endeavors among partners to understand the maximum capacity of transactive energy in multi-energy organizations [21].

2.7 Discussion

The implementation of transactive energy management in multi-energy networked systems has shown promising results in various real-world applications. These case studies illustrate how integrating transactive energy principles can enhance the efficiency, flexibility, and resilience of energy networks.

A) Brooklyn Microgrid (New York, USA)
   The Brooklyn Microgrid project is a prominent illustration of transactive energy the board applied to a restricted, multi-energy framework.

This people-group-driven drive permits members to exchange overabundance sun-oriented energy produced from roof boards inside a nearby organization. By utilizing blockchain innovation for secure and straightforward exchanges, the microgrid empowers ongoing energy trades among prosumers and customers. This arrangement not only upgrades the utilization of privately produced sustainable power, yet in addition coordinates battery capacity to adjust the organic market. The undertaking shows the way that transactive energy can work with shared exchanging and upgrade network versatility in a local area level [22].

B) Local (London, UK)
The nearby task in London includes a multi-energy network that coordinates power, gas, and nuclear power to upgrade the general framework execution. Using progressed control frameworks and market-based exchanges, Nearby deals with the appropriation of energy across these various transporters. For example, an overabundance of power from nearby sustainable sources is utilized to create nuclear power for local warming, while petroleum gas is used as a reinforcement and for top shaving. The task features the advantages of transactive energy in planning different energy transporters to further develop effectiveness and decrease functional expenses.

C) Smart Grid Projects in Germany
A few brilliant network drives in Germany exhibit the reconciliation of transactive energy for the executives in a multi-energy setting. One such task includes the utilization of constant evaluating signs to oversee power, gas, and nuclear power across private and modern areas. Savvy meters and progressed examinations are utilized to screen energy utilization and change use designs in light of economic situations. By organizing the progression of various energy transporters, these ventures expect to lessen top interest, lower energy expenses, and upgrade the combination of environmentally friendly power sources. The German tasks represent the capability of transactive energy to make adaptable and responsive energy frameworks.

D) Energy Hub in the Netherlands
The Energy Center Point project in the Netherlands centers around coordinating power, gas, and nuclear power through a focal control framework that utilizes transactive energy standards. The center point facilitates energy creation and utilization across different areas, including private, business, and modern applications. By utilizing continuous information and market-based exchanges, the framework enhances the utilization of energy assets and works on generally speaking effectiveness. The energy center shows the way that transactive energy the board can work with the

consistent collaboration between various energy transporters, prompting a stronger and more versatile energy organization. These contextual analyses outline the assorted uses of transactive energy on the board in multi-energy frameworks. Every Case features the expected advantages of incorporating market-based controls and decentralized decision production to improve framework proficiency, adaptability, and flexibility. As these advancements keep on developing, they offer important bits of knowledge and illustrations for the more extensive reception of transactive energy the executives in multi-energy arranged frameworks.

2.7.1 Advantages

2.7.1.1 Enhanced Efficiency

Transactive energy is when the executives altogether work on the effectiveness of multi-energy frameworks by streamlining the utilization of different energy transporters. By utilizing constant market-based exchanges, the framework guarantees that energy is utilized where it is generally required and at the most reduced cost. For instance, abundance inexhaustible power can be used to produce nuclear power or put away as gas, along these lines decreasing waste and limiting functional expenses. This unique change helps in adjusting the organic market successfully across various energy transporters.

2.7.1.2 Expanded Adaptability

One of the critical qualities of transactive energy is its capacity to give adaptability in overseeing energy assets. Multi-energy frameworks benefit from the capacity to switch between various energy transporters in view of continuous circumstances. This adaptability is vital for obliging the changeability of environmentally friendly power sources and adjusting to variances in energy interest. By considering changes in the utilization of power, gas, and nuclear power, transactive energy improves the framework's ability to answer changing circumstances and upgrade in general execution.

2.7.1.3 Further Developed Strength

Transactive energy the executives improve the versatility of multi-energy networks by decentralizing control and empowering fast reaction to interruptions. In the event of a disappointment or deficit in one energy transporter, the framework can rapidly adjust by depending on other accessible transporters. For example, in the event that there is a blackout, the framework can utilize put-away gas or nuclear power to keep up with supply progression. This capacity adds to a more steady and dependable energy framework, lessening the gamble of inescapable blackouts and guaranteeing a predictable energy supply [23].

2.7.2 Disadvantages

2.7.2.1 Complex Framework Joining

Incorporating transactive energy from the board into multi-energy frameworks includes complex coordination between various energy transporters. The requirement for cutting-edge demonstrating, control calculations, and continuous information handling can make framework mix testing. Dealing with the communications between power, gas, and nuclear power requires modern innovation and aptitude, which can be troublesome and expensive to execute. This intricacy can likewise prompt expected issues in guaranteeing consistent activity and interoperability among different parts of the framework.

2.7.2.2 Information About Executives and Security

Viable transactive energy on the board depends on the assortment and examination of tremendous measures of information from different sources, like brilliant meters, sensors, and market stages. Guaranteeing the precision, security, and opportune handling of this information is a huge test. Information the board frameworks should deal with different and voluminous data while safeguarding against digital dangers and guaranteeing protection. The intricacy of overseeing and getting information can present dangers and increment the general expense of executing transactive energy arrangements.

2.7.2.3 Administrative and Market Boundaries

In the fruitful execution of transactive energy the board frequently faces administrative and market-related difficulties. Existing guidelines may not completely oblige the decentralized and the market-driven nature of transactive energy frameworks, requiring refreshes or new systems. Furthermore, making fair and proficient business sectors for numerous energy transporters includes complex planning and administrative contemplations. Tending to these obstructions requires cooperative endeavors among partners and policymakers to foster suitable guidelines and market structures. While transactive energy the board offers huge benefits regarding productivity, adaptability, and versatility, it likewise presents difficulties connected with framework incorporation, information for the executives, and administrative issues. Adjusting these advantages and limits is fundamental for the successful execution of transactive energy in multi-energy arranged frameworks.

2.7.3 Challenges

2.7.3.1 Technical Challenges

1) **System Intricacy:** Incorporating the transactive energy of the board into multi-energy arranged frameworks presents huge specialized intricacy.

Organizing different energy transporters—like power, gas, and nuclear power—requires modern control frameworks and ongoing information examination. Creating and executing progressed calculations that can deal with the connections between various energy types while keeping up with framework strength is an imposing test [12]. The requirement for exact demonstrating, reproduction, and streamlining instruments adds to the intricacy of dealing with these interconnected frameworks.

2) **Data Incorporation and Security:** Powerful transactive energy the executives depend on the consistent combination of information from different sources, including brilliant meters, sensors, and market stages. Guaranteeing that this information is precise, ideal, and secure represents a huge specialized challenge. The framework should deal with immense measures of data while safeguarding against potential digital dangers and information breaks. Carrying out vigorous network safety gauges and guaranteeing the dependability of information, the executives frameworks are basic yet testing errands.
3) **Interoperability:** Accomplishing interoperability among various innovations and parts inside a multi-energy framework is a significant specialized obstacle. Energy frameworks frequently utilize assorted innovations and correspondence conventions, which can cause similarity issues. Guaranteeing that different frameworks and gadgets can cooperate agreeably is fundamental for the fruitful organization of transactive energy for the executives. Creating normalization and interoperability systems is important yet can be testing and tedious.

2.7.3.2 Regulatory Challenges

1) **Regulatory Structures:** Existing administrative systems may not completely support the decentralized and market-driven nature of transactive energy on the board. Adjusting or making new guidelines that oblige the intricacies of multi-energy frameworks is a critical test. Administrative bodies should attempt to foster arrangements that work with the coordination of transactive energy while guaranteeing reasonableness, straightforwardness, and purchaser insurance.
2) **Market Guideline:** Laying out and controlling business sectors for different energy transporters is mind-boggling. Administrative specialists need to resolve issues, for example, market access, rivalry, and evaluating instruments to make a fair and proficient commercial center. Adjusting the interests of different partners, including utilities, customers, and prosumers, requires cautious thought and can be testing.
3) **Policy Arrangement:** Organizing approaches across various areas and locales is essential for the fruitful execution of transactive energy by the board. Guaranteeing arrangements between energy, ecological, and financial approaches can

be troublesome, especially in areas with assorted administrative conditions. Successful strategy coordination is important to help the joining of transactive energy and accomplish wanted results. While transactive energy the executives offers significant advantages, its execution faces a few difficulties. Tending to specialized, monetary, and administrative issues is fundamental for understanding the maximum capacity of transactive energy in multi-energy organized frameworks.

2.7.4 Future Directions

The eventual fate of transactive energy and multi-energy frameworks is ready for critical progressions driven by arising advances and developing exploration regions. One key pattern is the rising combination of computerized reasoning (man-made intelligence) and AI (ML) into transactive energy the executives. Simulated intelligence and ML calculations can upgrade prescient capacities, advance constant direction, and work on the general effectiveness of multi-energy frameworks overwhelmingly of information from different sources. Blockchain innovation is one more encouraging region, offering secure and straightforward stages for shared energy exchanging. Blockchain can work with decentralized exchanges, empowering more productive and solid energy trades inside multi-energy frameworks. As this innovation develops, it might assume an essential part in upgrading the versatility and reception of transactive energy cases. Energy capacity progressions will likewise shape the fate of transactive energy. Developments in battery innovation, for example, longer-enduring and more savvy stockpiling arrangements, will empower better administration of discontinuous environmentally friendly power sources like sunlight-based and wind. This will permit multi-energy frameworks to turn out to be stronger and more versatile, further supporting the mix of assorted energy transporters. Examination into strategy and administrative structures will be basic in tending to the difficulties related to transactive energy frameworks. As these frameworks fill in intricacy, creating adaptable, forward-looking guidelines will be fundamental for guaranteeing that they work effectively and decently. The fate of transactive energy and multi-energy frameworks will be molded by mechanical developments, computer based intelligence coordination, progressions in energy capacity, and strong administrative turns of events, preparing for more practical, productive, and tough energy organizations.

2.7.5 Possible Improvements and Innovations in Transactive Energy Management

To enhance the effectiveness of transactive energy management, several key improvements and innovations can be pursued.

a) **Advanced Artificial Intelligence:** Coordinating more complex artificial intelligence and AI calculations can fundamentally work on the exactness of interest in determining constant direction and dynamic valuing of cases. These advances can assist with streamlining energy dispersion across numerous transporters by breaking down complex information examples and making prescient changes, prompting more proficient and adjusted energy frameworks.
b) **Enhanced Correspondence Organizations:** The arrangement of cutting edge correspondence advancements, for example, 5G and edge registering, can empower quicker and more solid information trade across the energy organization. These headways would work with continuous coordination between various energy transporters and gadgets, lessening inertness and working on the responsiveness of the framework.
c) **Improved Interoperability Guidelines:** Creating normalized conventions for correspondence and information trade between different energy frameworks and gadgets is urgent. This would guarantee consistent interoperability, permitting different innovations and parts inside a multi-energy framework to cooperate effectively. Normalization would likewise uphold the adaptability of transactive energy frameworks across various areas and markets.
d) **Blockchain Combination:** Growing the utilization of blockchain innovation can give a safer and more straightforward structure for energy exchanges. By guaranteeing the respectability of distributed energy exchanging and lessening the gamble of extortion, blockchain can increment trust among members and improve the general proficiency of transactive energy markets.
e) **Advanced Energy Stockpiling Arrangements:** Advancements in energy capacity, like more proficient and practical batteries or new types of warm and synthetic capacity, could essentially work on the ability to store and oversee energy. Better capacity choices would consider more noteworthy adaptability in adjusting the organic market, especially in frameworks with high sustainable power entrance [18]. The effectiveness of transactive energy management can be significantly enhanced through advances in AI, communication technologies, interoperability standards, blockchain, and energy storage, paving the way for more resilient and efficient multi-energy systems.

2.8 Conclusion

This part has given a far-reaching investigation of transactive energy the board and its mix into multi-energy organized frameworks. The conversation started with a prologue to transactive energy, an idea that epitomizes the utilization of market-based instruments and decentralized decision-production to oversee energy streams. By adjusting energy creation and utilization to ongoing cost signals, transactive energy empowers more proficient, adaptable, and strong energy organizations. We dug into the construction of multi-energy frameworks, which incorporate different energy transporters like power, gas, and nuclear power. These frameworks are intended to upgrade asset use, lessen functional expenses, and improve energy security. The incorporation of transactive energy the executives inside these organizations takes into account dynamic acclimations to energy streams, working on the general execution of the framework. Key benefits of transactive energy the executives were featured all through the part. One of the essential advantages is improved effectiveness. By working with ongoing energy trades and advancing the utilization of various energy transporters, transactive energy diminishes waste and brings down costs. Also, the adaptability presented by transactive energy frameworks permits them to adjust to vacillations in energy organic market, which is especially important in dealing with the fluctuation of environmentally friendly power sources. Besides, the strength of energy networks is worked on through the decentralized idea of transactive energy, which empowers speedy reactions to disturbances and keeps up with matrix solidness. Notwithstanding, the part additionally tended to the difficulties related to carrying out transactive energy for the executives. Specialized difficulties, like framework intricacy and information reconciliation, require modern control frameworks and hearty network safety measures. Financial difficulties incorporate the high introductory expenses of execution and the vulnerabilities encompassing the business sector plan and speculation chances. Administrative difficulties include adjusting existing structures to oblige the decentralized and market-driven nature of transactive energy, as well as guaranteeing strategy arrangement across various areas and purviews. Certifiable contextual analyses gave substantial instances of how transactive energy the board has been applied in different settings. Projects like the Brooklyn Microgrid, Neighborhood in London, and shrewd matrix drives in Germany and the Netherlands show the capability of transactive energy to upset energy the board in multi-energy frameworks. That's what these cases show, in spite of the difficulties, the advantages of transactive energy are critical and can prompt more maintainable, proficient, and versatile energy organizations. The significance of transactive energy the board with regards to multi-energy frameworks couldn't possibly be more significant. As the energy scene keeps on developing, driven by the requirement for maintainability and the reconciliation of sustainable power sources, transactive energy gives an indispensable instrument to overseeing progressively complex energy organizations. It offers a method for adjusting market interest progressively, lessening expenses, and improving framework dependability, making it a basic part of future energy frameworks. For the users, the primary action items from this section are the significant advantages that transactive energy the executives can bring to multi-energy frameworks, the provokes that should be tended to for fruitful execution, and the future capability of this field. Understanding these perspectives is critical for partners associated with energy arranging, strategy making, and mechanical turns of events. As transactive energy the executives keep on developing, it will assume a focal part in molding the eventual fate of maintainable and versatile energy frameworks.

References

1 Alizadeh, A., Esfahani, M., Cao, B. et al. (2024). A distributed-decentralized tri-layer game-based transactive energy framework including adaptive proximal ADMM and scale-independent DRO. *IEEE Access* 12: 82068–82082.

2 Alizadeh, A., Esfahani, M., Dinar, F. et al. (2024). A cooperative transactive multi-carrier energy control mechanism with P2P energy+ reserve trading using Nash bargaining game theory under renewables uncertainty. *Applied Energy* 353: 122162.

3 Bhaskaran Pillai, M. (2020). India's Multivectored Alignment in the East: Aims, Objectives, and Rationale in Perspective (March 16, 2020).

4 Alshammari, O., Basem, A., Ghodratallah, P. et al. (2024). Transactive energy framework in fuel cell based multi-carrier energy hubs based on conditional value-at-risk. *International Journal of Hydrogen Energy* 81: 292–304.

5 Azimi, M., Salami, A., Javadi, M.S., and Catalo, J.P. (2024). Optimal and distributed energy management in interconnected energy hubs. *Applied Energy* 365: 123282.

6 Barati, A., Karimi, H., and Jadid, S. (2024). Multi-objective operation of interconnected multi-energy systems considering power to gas and gas to power systems. *International Journal of Electrical Power & Energy Systems* 158: 109986.

7 Chen, P., Ye, Y., Wang, H. et al. (2024). Holistic coordination of transactive energy and carbon emission right trading for heterogenous networked multi-energy microgrids: a fully distributed adaptive consensus ADMM approach. *Sustainable Energy Technologies and Assessments* 64: 103729.

8 Cordieri, S.A. (2024). Exact and heuristic algorithms for the integrated planning of multi-energy systems.

9 Machele, I.L., Onumanyi, A.J., Abu-Mahfouz, A.M., and Kurien, A.M. (2024). Interconnected smart transactive microgrids—a survey on trading, energy management systems, and optimisation approaches. *Journal of Sensor and Actuator Networks* 13 (2): 20.

10 Mohsenzadeh-Yazdi, H., Kebriaei, H., and Aminifar, F. (2024). Multi-agent reinforcement learning in a new transactive energy mechanism. *IET Generation, Transmission & Distribution* 18 (18): 2943–2955.

11 Pu, X., Xiao, H., Pei, W. et al. (2024). Optimal energy management of networked multi-energy microgrids based on improved multi-agent federated reinforcement learning. *CSEE Journal of Power and Energy Systems* 1–12. https://ieeexplore.ieee.org/document/10609314.

12 Rodriguez-Gil, J.A., Mojica-Nava, E., Vargas-Medina, D. et al. (2024). Energy management system in networked microgrids: an overview. *Energy Systems* 1–32. https://link.springer.com/article/10.1007/s12667-024-00676-6.

13 Seyednouri, S., Safari, A., Quteishat, A. et al. (2024). Stochastic energy management of a multi-microgrid system with battery/supercapacitor energy storages considering demand response and transactive energy. *Renewable Energy Focus* 48: 100531.

14 Tiwari, S. and Singh, J.G. (2024). A cooperation based transactive energy management for networked energy hubs considering improved payoff allocation mechanism. *Sustainable Energy Technologies and Assessments* 65: 103777.

15 Tooki, O.O. and Popoola, O.M. (2024). A comprehensive review on recent advances in transactive energy system: concepts, models, metrics, technologies, challenges, policies and future. *Renewable Energy Focus* 50: 100596.

16 Xia, Y., Xu, Q., Ding, Y. et al. (2024). Generalized Nash equilibrium analysis for peer-to-peer transactive energy market considering coupling distribution network constraints. *IEEE Transactions on Industrial Informatics* 20 (6): 8125–8137.

17 Xie, X., Qian, T., Li, W. et al. (2024). An individualized adaptive distributed approach for fast energy-carbon coordination in transactive multi-community integrated energy systems considering power transformer loading capacity. *Applied Energy* 375: 124189.

18 Xiong, H., Luo, F., Yan, M. et al. (2024). Distributionally robust and transactive energy management scheme for integrated wind-concentrated solar virtual power plants. *Applied Energy* 368: 123148.

19 Onishi, V.C., Antunes, C.H., Fraga, E.S., and Cabezas, H. (2019). Stochastic optimization of trigeneration systems for decision-making under long-term uncertainty in energy demands and prices. *Energy* 175: 781–797.

20 Garg, A. and Niazi, K. (2023). A robust energy management strategy for multi carrier energy system. *2023 7th International Conference on Computer Applications in Electrical Engineering-Recent Advances (CERA)*, 1–5. IEEE.

21 Tan, B., Chen, H., Zheng, X., and Huang, J. (2022). Two-stage robust optimization dispatch for multiple microgrids with electric vehicle loads based on a novel data-driven uncertainty set. *International Journal of Electrical Power & Energy Systems* 134: 107359.

22 George-Williams, H., Wade, N., and Carpenter, R. (2022). A probabilistic framework for the techno-economic assessment of smart energy hubs for electric vehicle charging. *Renewable and Sustainable Energy Reviews* 162: 112386.

23 Sun, Y., Zhang, B., Ge, L. et al. (2020). Day-ahead optimization schedule for gas-electric integrated energy system based on second-order cone programming. *CSEE Journal of Power and Energy Systems* 6 (1): 142–151.

3

Energy Management Strategies for Optimal Scheduling of Multi-Energy Network Hubs

Divya Sharma and Naran M. Pindoriya

Department of Electrical Engineering, Indian Institute of Technology Gandhinagar, Gujarat, India

Nomenclature

ζ, t	Set and index of time (hours)
M, k	Set and index of EHs
i, j, Ω	Index and set of buses
$V_{i,t}$	ith bus voltage (kV)
$V_{j,t}$	jth bus voltage (kV)
Z	Impedance of power distribution network (Ω)
$\delta_{ij,t}$	Power angle difference between bus ith and jth buses (degree)
$P_{i,t}^{EH}$	Total amount of power sold to EHs (MW)
λ_t	Price of power purchased from the DSO by EHs (Rs. /MWh)
$P_{i,t}^{GRID}$	Power purchased by the DSO from the electricity market (MW)
δ_t	Power purchasing price of the DSO (Rs. /MWh)
$P_{i,t}^{D}$	Electrical demand at the ith bus (MW)
$S_{ij,t}$	Apparent power flow between the ith and jth buses (MW)
$P_{i,t}^{EH}$	Electrical demand of EH (MW)
OF^{EH}	Objective function of EH
$P_{k,t}^{G}$	Amount of power purchased from the DSO by the kth EH (MW)
$G_{k,t}$	Natural gas consumption in the kth EH (MW)
$P_{k,t}^{trade}$	Electrical power shared between EHs (MW)
$H_{k,t}^{trade}$	Heating power shared between EHs (MW)
$C_{k,t}^{trade}$	Cooling power shared between EHs (MW)

Energy Management Strategies for Multi-Vectored Energy Hubs to Achieve Low Carbon Societies, First Edition.
Edited by Shubham Tiwari, Jai Govind Singh, Sivaraman Palanisamy, Sharmeela Chenniappan, Rupendra Kumar Pachauri, and Sanjeevikumar Padmanaban.

γ_t	Gas price (Rs. /MWh)
β_t	Electrical power sharing price between EHs (Rs. /MWh)
α_t	Heating power sharing price between EHs (Rs. /MWh)
φ_t	Cooling power sharing price between EHs (Rs. /MWh)
$P_{k,t}^{WT}$	Power output of WT in the *k*th EH (MW)
$P_{k,t}^{PV}$	Power output of PV in the *k*th EH (MW)
$I1_{k,t}^{ch}, I1_{k,t}^{dch}$	Binary variables for HES of the *k*th EH
$I_{k,t}^{ch}, I_{k,t}^{dch}$	Binary variables for BESS of the *k*th EH
OF^{DSO}	Objective function of the DSO
$P_{ij,t}$	Active power flow between the *i*th and *j*th buses (MW)
$Q_{ij,t}$	Reactive power flow between the *i*th and *j*th buses (MW)
$H_{k,t}^{CHP}$	Heat power output of CHP unit in the *k*th EH (MW)
$G_{k,t}^{CHP}$	Gas consumed by CHP unit in the *k*th EH (MW)
$P_{k,t}^{CHP}$	Electrical power output of CHP unit in the *k*th EH (MW)
$\eta_{gh}^{CHP}, \eta_{ge}^{CHP}$	Gas-to-heat and gas-to-electricity conversion efficiencies of the CHP unit, respectively
$C_{k,t}^{AC}$	Cooling power output of AC in the *k*th EH (MW)
$H_{k,t}^{AC}$	Heat power consumption by AC in the *k*th EH (MW)
η^{AC}	Efficiency of AC
$C_{k,t}^{EC}$	Cooling power output of EC in the *k*th EH (MW)
$P_{k,t}^{EC}$	Electrical power consumption by EC in the *k*th EH (MW)
η^{EC}, η^{EB}	Efficiency of EC and EB, respectively
$P_{k,t}^{EB}$	Electrical power consumption by EB in the *k*th EH (MW)
$H_{k,t}^{EB}$	Heat power generated by EB in the *k*th EH (MW)
$H_{k,t}^{HF}$	Heat power output of HF in the *k*th EH (MW)
$G_{k,t}^{HF}$	Gas consumed by HF in the *k*th EH (MW)
η^{HF}	Efficiency of HF
CoP	Coefficient of performance (CoP) of EHP
$C_{k,t}^{EHP}$	Cooling power output of EHP in the *k*th EH (MW)
$H_{k,t}^{EHP}$	Heat power output of EHP in the *k*th EH (MW)
$P_{k,t}^{EHP}$	Electrical power consumption by EHP in the *k*th EH (MW)
$P_{k,t}^{ch}, P_{k,t}^{dch}$	Electrical power of BESS in charging and discharging mode in the *k*th EH (MW)
$H_{k,t}^{ch}, H_{k,t}^{dch}$	Heat power of HES in charging and discharging mode in the *k*th EH (MW)
$E_{k,t}^{BESS}$	Electrical energy stored in BESS in the *k*th EH (MWh)
$Q_{k,t}^{HES}$	Heat energy stored in HES in the *k*th EH (MWh)
$\eta_{B,ch/B,\,dch}$	Charge/discharge efficiency of BESS
$\eta_{ch/dch}$	Charge/discharge efficiency of HES

w_1, w_2	Weights of the MO problem
T^{ab}	Ambient temperature
L	Length of the pipe
c_p	Specific heat capacity of water
m	Mass flow rate of water
k	Thermal conductivity of the pipe
T^{in}, T^{out}	Temperatures at the inlet and outlet of the pipe, respectively

3.1 Introduction

3.1.1 Background

The rise in environmental concerns, depletion of fossil fuel resources, escalating fossil fuel prices, and unregulated growth in energy demand have necessitated the restructuring and upgrading of existing energy systems. Ensuring the decarbonization of energy systems is crucial to fulfilling climate protection duties. The widespread adoption of renewable energy sources (RES) as a replacement for fossil fuels is the primary strategy for achieving global carbon neutrality. The intermittent and stochastic behavior of RES leads to the development of sustainable multi-energy systems (MES) that incorporate a high level of flexibility and reliability. The energy systems of the future are going to adopt the sustainable MES [1].

To optimize the performance of MES, effective planning, monitoring, and scheduling are essential and necessitate a comprehensive energy management system. The energy hub (EH) possesses significant potential for implementing such energy system models and transitioning toward sustainable MES. An EH is a novel and promising method for effectively managing MES by converting, storing, and transforming different energy vectors. The integration of numerous energy vectors enhances the flexibility of the local energy system and expands the capacity to accommodate RES. MES maximizes the interaction between multiple energy vectors, such as heating, cooling, and electricity, at different levels (e.g., within a district, city, or region), in contrast to traditional energy systems that treat sectors as independent entities. This offers a significant opportunity to improve performance in terms of technology, economy, and environment during the planning and operational stages [2].

The conventional monopolistic energy supply system, where a single utility company provides energy to fulfill the energy demands of all consumers, is steadily being substituted by hierarchical MES due to ongoing market reforms and the swift advancement of information and communication technologies (ICT). Within a hierarchical MES, multiple suppliers, users, and energy generation facilities come together within a single system at various tiers, resulting in a dynamic and competitive energy market [3]. MES with varying source-load characteristics exhibit advantageous synergies and reciprocal energy support, enabling the increased integration of RES over time and across different spatial dimensions [4]. However, the coordination and control of the MES face significant hurdles because of the complexity and the variety of energy devices involved. One of the biggest problems facing the global energy sector is scheduling MES optimally, which may effectively actualize the complementarity and interconnection between various energy vectors, encourage local energy consumption, and raise the percentage of uncertain RES [5]. The optimal scheduling model of MES is difficult to solve using the conventional approaches [6] due to the several issues in MES, including interaction among EHs, robust coupling of energy vectors, and high intermittency over access to renewable energy. Hence, it is essential to develop the most cost-effective scheduling strategy for MES.

3.1.2 Related Work

The authors in Ref. [7] examined how a multi-microgrid system operates using a non-cooperative game-theoretic approach. Microgrids (MGs) integrate RES without taking into account their demand response (DR) program and uncertainties. In Ref. [8], an optimization model based on competition is developed between the distribution network (DN) and MGs. Several technical limitations and energy storage (ES) are taken into account, while the concepts of DR and MES have been neglected. In Ref. [9], bi-level, multi-follower-based optimal operation of multi-microgrids is proposed. This study does not take into account ES and multi-energy concept. A bi-level stochastic multi-carrier optimization framework is proposed in Ref. [10]. However, internal energy trade among EHs is not considered. In Ref. [11], a robust energy management framework for EH considering emission markets is developed. The authors in Ref. [12] proposed a cooperative theory-based energy management framework for multi-energy MGs. In Ref. [13], a cooperative strategy for the optimal management of single-carrier MGs is suggested. The study of demand response in networked MGs for community energy management is done in Ref. [14]. However, RES's stochastic modeling is not taken into account. The authors in Ref. [15] proposed stochastic optimization for optimum scheduling of multi-EHs. However, DR, equitable cost distribution, EVs, and energy trade between EHs have all been neglected. The authors in Ref. [16] presented a two-stage stochastic optimization model for MGs that takes into account DR and EV modeling. But neither the multi-energy concept nor the energy exchange between the MGs is taken into consideration. In Ref. [17], an EH concept-based optimization strategy is developed to manage a group of buildings cooperatively. The authors in Ref. [18] proposed a hierarchical optimization model for small- and medium-sized regional multiple energy systems, with time-dependent tariffs and adaptable operating modes. In Ref. [19], an optimal operating framework was built for an EH-based MES, with the goal of minimizing the overall operation cost of electricity, natural gas, and heat. An EH manager's market transaction problem was demonstrated in Ref. [20], when the manager had to decide on the best level of participation in upstream energy markets and the best pricing to charge customers for heat and electricity. A day-ahead dynamic optimal operating model based on the EH concept has been proposed in Ref. [21] for an MG, taking into account the real-time pricing. The research in Refs. [17–21] focused on the optimal scheduling, operation, and transaction problems from a single EH operator aspect, while ignoring the interaction with other EHs. In recent times, it has become common practice to investigate the interactions of various EHs using game-theoretic methodologies. In Refs. [22, 23], the problem is presented as an ordinal potential game, in which each EH selfishly and locally chooses its course of action in order to minimize its operation cost. The research in Refs. [24, 25] developed a non-cooperative game to handle the interaction among EHs, and a Nash equilibrium is identified to ensure the optimal operation of EHs. The authors in Ref. [26] extended the work in Refs. [20–23] by incorporating uncertainty regarding load demand and energy pricing, as well as focusing on competition among various EHs using game theory. The interaction of various parks with different energy flows is also described as a non-cooperative game model in Ref. [27]. A natural paradigm to investigate the disputes and cooperation of various EHs in a distributed manner is provided by game theory, which takes into account their self-interest and information privacy [28]. But the non-cooperative game theory-based methods mentioned above in Refs. [22–27] tend to provide non-Pareto optimum solutions because they overemphasize the independence of EHs while ignoring their potential for collaboration. A joint optimal scheduling framework for several community energy exchange systems was developed in Ref. [29], and the distribution of energy sharing advantages is rationalized based on enhanced Shapley values. A prediction approach based on the producer–consumer model was presented in Ref. [30] for the real-time control of P2P energy trade. The majority of the aforementioned literature have developed optimization models using non-cooperative games with the goal of minimizing individual costs, yet it is challenging to determine the optimal global solution while maintaining local objectives.

The widespread use of smart gadgets has made it possible for several EHs to cooperate by exchanging some information. Moreover, because different EHs have varied configurations and demands, it is possible that these EHs can trade energy with one another [31]. This not only lowers the wastage of excess energy but also increases the operational flexibility of MES. In addition, it may help these EHs more than merely interacting in a non-cooperative manner with the utility providers. It is equally vital to take into account the scenario in which several small autonomous EHs operate cooperatively. Since this thought is to be a key component of the energy business model of the future [32, 33]. In cooperative game theory, as contrasted with non-cooperative games, players create coalitions and seek to maximize total profit. The cooperative game optimization approach assures an optimal global solution while maintaining local objectives. Since gas, electrical, and thermal systems are interdependent in MES, their coordinated operation ensures reliability. Thus, cooperative game theory is best suited for managing multi-energy networked EHs. This chapter proposes an optimal scheduling model for networked multi-energy EHs in both cooperative and non-cooperative modes.

3.2 System Architecture and Problem Formulation

3.2.1 System Architecture

The system architectures of networked EHs operating in centralized and decentralized modes are illustrated in Figures 3.1 and 3.2. Each of the three EHs possesses distinct components within its structure. Each EH contains various types of energy conversion devices, such as a combined heat and power (CHP) unit, electric boiler (EB), electric chiller (EC), absorption chiller (AC), electric heat pump (EHP), heat furnace (HF), battery energy storage system (BESS), heat energy storage (HES), and three kinds of energy consumption (electricity, heating, and cooling). Each EH is connected to the electrical grid, which is managed by the distribution system operator (DSO) and gas network.

In the centralized system illustrated in Figure 3.1, each of the three EHs is coordinated and managed by an EH coordinator who has access to a wide range of information relating to the EHs. In contrast to a centralized system, a decentralized system (as shown in Figure 3.2) involves the sharing of limited information among the EHs through a virtual EH coordinator. Since the privacy of EHs is upheld in a decentralized mode. Each EH is connected not only to the electrical grid but also to neighboring EHs, allowing for the exchange of energy with the DSO and among neighboring hubs in a decentralized system, contrasting with a centralized system. Energy generated within the EHs will initially be utilized to fulfill their electricity, cooling, and heating demands. Any excess or shortfall of energy will be traded with other EHs and DSOs. In a decentralized manner of operation, every EH is working together and aiming to reduce the overall operational expenses. However, in a centralized system, each EH operates independently in a non-cooperative manner, aiming to reduce its operational costs. Figure 3.3 shows the schematic diagram of EH.

The mathematical modeling of various components of EHs is discussed below:

1) **Wind Turbine (WT):** The WT converts wind power into electrical power. The power output of a wind unit can be affected by wind speed and has essential

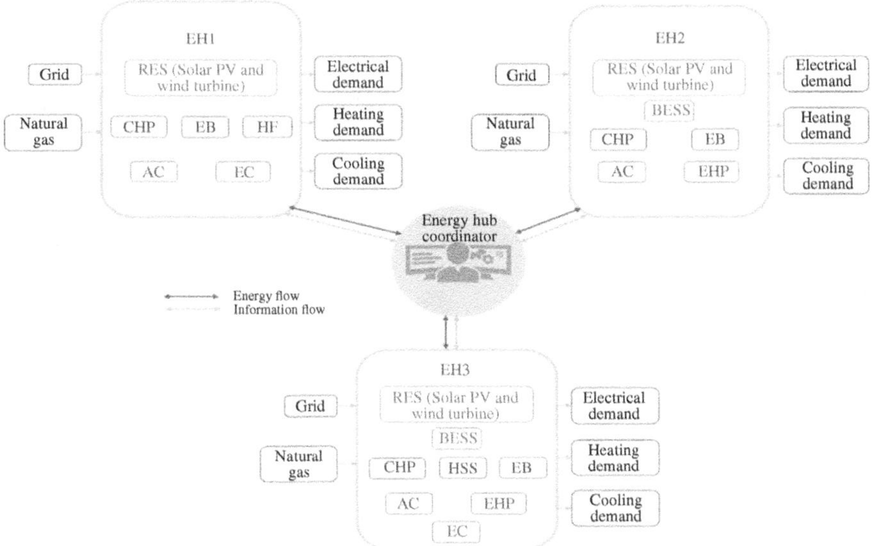

Figure 3.1 Centralized EH system architecture.

parameters such as cut-in (V_{cut-in}), rated ($V_{k,t}$), and cut-off ($V_{cut-off}$) speeds [34], as expressed by Eq. (3.1).

$$P_{k,t}^{WT} = \begin{cases} 0 & V_{k,t} < V_{cut-in} \\ P^{WT,Max}\left(\frac{V_{k,t}-V_{cut-in}}{V_{rated}-V_{cut-in}}\right) & V_{cut-in} \le V_{k,t} < V_{rated} \\ P^{WT,Max} & V_{rated} \le V_{k,t} < V_{cut-off} \\ 0 & V_{k,t} \ge V_{cut-off} \end{cases} \quad \forall\, t\, \varepsilon\, \zeta \qquad (3.1)$$

2) **Solar Photovoltaic (PV):** Solar PV converts solar power into electrical power. The power output of solar PV can be calculated using solar radiation, system efficiency, and surface area, as determined by Eq. (3.2) [35].

$$P_{k,t}^{PV} = \eta^{PV} A^{PV} \mathrm{I}_{k,t}(1 - 0.005(\mathrm{T}_{out,k,t} - 25)) \qquad \forall\, t\, \varepsilon\, \zeta \qquad (3.2)$$

where A^{PV} and η^{PV} are module area and energy conversion efficiency of solar PV, respectively. $T_{out,k,t}$ and $I_{k,t}$ are outside air temperature and solar irradiation at a time instant t, respectively.

3) **CHP Unit:** Natural gas is used as the CHP unit's fuel source. The unit then uses the gas to generate both heat and electricity. The modeling of a back-pressure-type CHP unit is as follows:

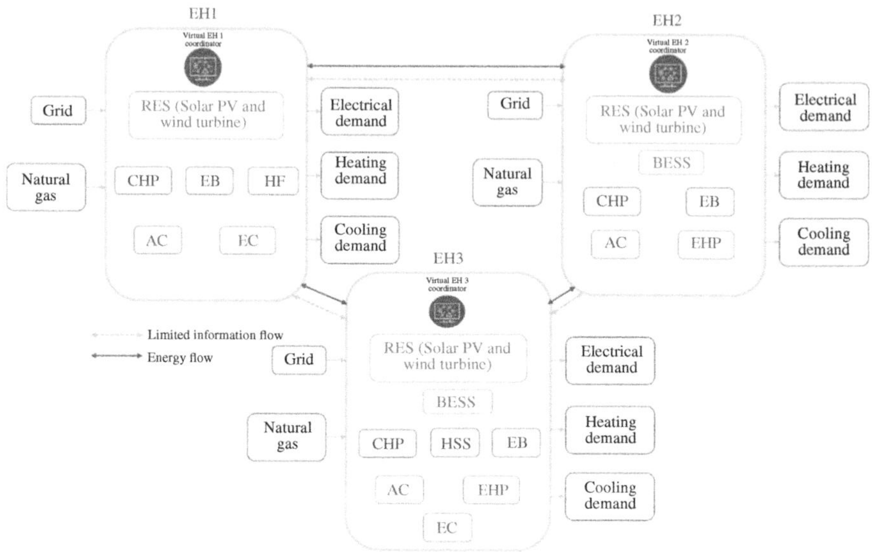

Figure 3.2 Decentralized EH system architecture.

The heat power output of the unit:

$$H_{k,t}^{CHP} = \eta_{gh}^{CHP} G_{k,t}^{CHP} \qquad \forall\, t\, \varepsilon\, \zeta \tag{3.3}$$

$$H_k^{CHP,min} \leq H_{k,t}^{CHP} \leq H_k^{CHP,max} \qquad \forall\, t\, \varepsilon\, \zeta \tag{3.4}$$

The electrical power output of the unit:

$$P_{k,t}^{CHP} = \eta_{ge}^{CHP} G_{k,t}^{CHP} \qquad \forall\, t\, \varepsilon\, \zeta \tag{3.5}$$

$$P_k^{CHP,min} \leq P_{k,t}^{CHP} \leq P_k^{CHP,max} \qquad \forall\, t\, \varepsilon\, \zeta \tag{3.6}$$

The minimum and maximum constraints of the CHP unit's heat and power output are represented by Eqs. (3.4) and (3.6), respectively.

4) **Absorption Chiller:** AC absorbs heat power and transforms it into cooling power (Eq. 3.7). AC output is constrained by Eq. (3.8).

$$C_{k,t}^{AC} = \eta^{AC} H_{k,t}^{AC} \qquad \forall\, t\, \varepsilon\, \zeta \tag{3.7}$$

$$C_k^{AC,min} \leq C_{k,t}^{AC} \leq C_k^{AC,max} \qquad \forall\, t\, \varepsilon\, \zeta \tag{3.8}$$

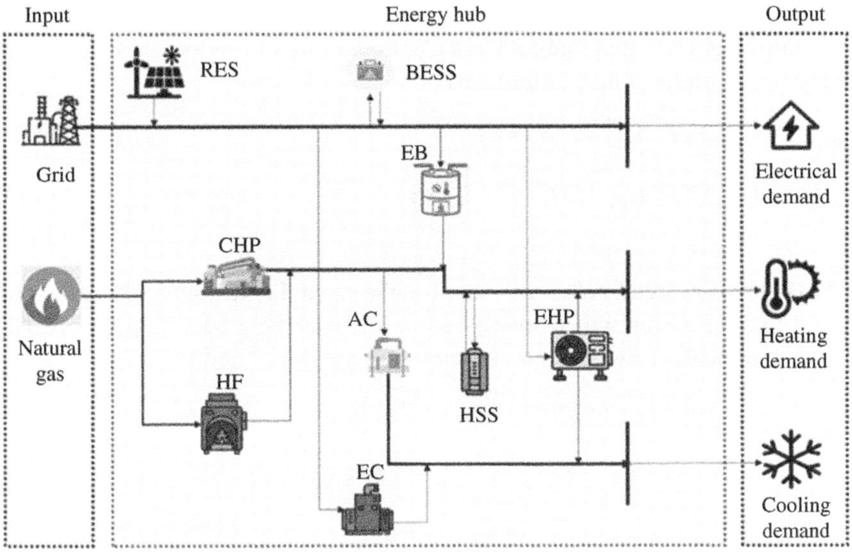

Figure 3.3 The schematic diagram of an individual EH.

5) **Electric Chiller:** EC converts electrical power to cooling power for meeting cooling demand (Eq. 3.9). Equation (3.10) represents the output minimum and maximum limits of EC.

$$C_{k,t}^{EC} = \eta^{EC} P_{k,t}^{EC} \qquad \forall\, t \,\varepsilon\, \zeta \tag{3.9}$$

$$C_{k}^{EC,min} \leq C_{k,t}^{EC} \leq C_{k}^{EC,max} \qquad \forall\, t \,\varepsilon\, \zeta \tag{3.10}$$

6) **Electric Boiler:** EB utilizes the passage of electric current through a heating element in order to generate heat, thereby facilitating the transformation of water into steam at elevated temperatures and pressures. The process of electrical power conversion to heat power is employed by EB in order to fulfill the heating demand (Eq. 3.11). The heat power output of EB is bounded by Eq. (3.12).

$$H_{k,t}^{EB} = \eta^{EB} P_{k,t}^{EB} \qquad \forall\, t \,\varepsilon\, \zeta \tag{3.11}$$

$$H_{k}^{EB,min} \leq H_{k,t}^{EB} \leq H_{k}^{EB,max} \qquad \forall\, t \,\varepsilon\, \zeta \tag{3.12}$$

7) **Heat Furnace:** The process of heat generation in HF involves the utilization of natural gas as a fuel source (Eq. 3.13). Equation (3.14) represents the bottom and upper limits of heat output for HF.

$$H_{k,t}^{HF} = \eta^{HF} G_{k,t}^{HF} \qquad \forall\, t\, \varepsilon\, \zeta \tag{3.13}$$

$$H_{k}^{HF,min} \leq H_{k,t}^{HF} \leq H_{k}^{HF,max} \qquad \forall\, t\, \varepsilon\, \zeta \tag{3.14}$$

8) **EHP:** It is powered by electricity and has the capability to produce both heating and cooling power, which vary based on the specific mode of operation. The mathematical formulation can be expressed as:

$$C_{k,t}^{EHP} + H_{k,t}^{EHP} = \mathrm{CoP} \times P_{k,t}^{EHP} \qquad \forall\, t\, \varepsilon\, \zeta \tag{3.15}$$

$$H_{k}^{EHP,min} \leq H_{k,t}^{EHP} \leq H_{k}^{EHP,max} \qquad \forall\, t\, \varepsilon\, \zeta \tag{3.16}$$

$$C_{k}^{EHP,min} \leq C_{k,t}^{EHP} \leq C_{k}^{EHP,max} \qquad \forall\, t\, \varepsilon\, \zeta \tag{3.17}$$

9) **BESS:** BESS units are needed to accommodate the electricity supply and demand. The storage process is enabled at times of either lower generation costs or off-peak demand and from BEES to be used at times of high generation costs, and peak demand. The mathematical formulation is as follows:

$$E_{k,t}^{BESS} = E_{k,(t-1)}^{BESS} + \left(\mathrm{I}_{k,t}^{\mathrm{ch}} \eta_{\mathrm{B,ch}} \mathrm{P}_{k,t}^{\mathrm{ch}} - \frac{\mathrm{P}_{k,t}^{\mathrm{dch}} \mathrm{I}_{k,t}^{\mathrm{dch}}}{\eta_{\mathrm{B,dch}}} \right) \Delta \mathrm{t} \qquad \forall\, t\, \varepsilon\, \zeta \tag{3.18}$$

$$\mathrm{I}_{k,t}^{\mathrm{ch}} + \mathrm{I}_{k,t}^{\mathrm{dch}} \leq 1 \qquad \forall\, t\, \varepsilon\, \zeta \tag{3.19}$$

$$E_{k}^{BESS,min} \leq E_{k,t}^{BESS} \leq E_{k}^{BESS,max} \qquad \forall\, t\, \varepsilon\, \zeta \tag{3.20}$$

10) **HES:** During off-peak hours, HES accumulates excess heat energy and releases it during peak hours. The mathematical formulation of HES is as follows:

$$Q_{k,t}^{HES} = Q_{k,(t-1)}^{HES} + \left(\eta_{\mathrm{ch}} \mathrm{H}_{k,t}^{\mathrm{ch}}\, \mathrm{I1}_{k,t}^{\mathrm{ch}} - \frac{\mathrm{H}_{k,t}^{\mathrm{dch}} \mathrm{I1}_{k,t}^{\mathrm{dch}}}{\eta_{\mathrm{dch}}} \right) \Delta \mathrm{t} \qquad \forall\, t\, \varepsilon\, \zeta \tag{3.21}$$

$$I1_{k,t}^{ch} + I1_{k,t}^{dch} \leq 1 \qquad \forall\, t\, \varepsilon\, \zeta \tag{3.22}$$

$$Q_{k}^{HES,min} \leq \mathrm{Q}_{k,t} \leq Q_{k}^{HES,max} \qquad \forall\, t\, \varepsilon\, \zeta \tag{3.23}$$

3.3 Problem Formulation

In the following section, the optimal operation of EHs coupled to the distribution and thermal networks is formulated.

3.3.1 DSO Objective Function

The goal of DSO is to maximize daily profit. The formulation of the DSO's objective function is as follows:

$$OF^{DSO} = Max\sum_{i=1}^{33}\sum_{t=1}^{24} P_{i,t}^{EH}\lambda_t - P_{i,t}^{GRID}\delta_t \tag{3.24}$$

The first term represents the profit from selling power to EHs, while the second term represents the cost of obtaining power from the upstream network. If the DSO purchases power from EHs, this term will be negative and regarded as a cost to the DSO.

3.3.2 EH Coordinator Objective Function

There are five terms in the cost function as expressed in Eq. (3.25). The first term denotes how much it costs to buy energy from the grid, the second term stands for how much it costs to buy natural gas from the gas network, and the other three terms denote how much it costs to purchase power (electrical, heating, and cooling) from internal trading with EHs.

$$OF^{EH} = Min\sum_{k=1}^{3}\sum_{t=1}^{24}\left(P_{k,t}^{G}\lambda_t + G_{k,t}\gamma_t + P_{k,t}^{trade}\beta_t + H_{k,t}^{trade}\alpha_t + C_{k,t}^{trade}\varphi_t\right) \tag{3.25}$$

3.3.3 Electrical Network

The active and reactive power flows through the DN are given by Eqs (3.26) and (3.27), respectively.

$$P_{ij,t} = \frac{|V_{i,t}^2|\ \cos\theta}{|Z|} - \frac{|V_{i,t}||V_{j,t}|\ \cos(\theta + \delta_{ij,t})}{|Z|} \qquad \forall\, i,j\,\varepsilon\,\Omega \qquad \forall\, t\,\varepsilon\,\zeta \tag{3.26}$$

$$Q_{ij,t} = \frac{|V_{i,t}^2|\ \sin\theta}{|Z|} - \frac{|V_{i,t}||V_{j,t}|\ \sin(\theta + \delta_{ij,t})}{|Z|} \qquad \forall\, i,j\,\varepsilon\,\Omega \qquad \forall\, t\,\varepsilon\,\zeta \tag{3.27}$$

Active power balance: The amount of active power purchased from upstream network should be equal to the active power demand of DN, active power demand of EHs, and active power flow through the network.

$$P_{i,t}^{GRID} = P_{i,t}^{D} + P_{i,t}^{EH} + P_{ij,t} \qquad \forall\, i,j\,\varepsilon\,\Omega \qquad \forall\, t\,\varepsilon\,\zeta \tag{3.28}$$

Reactive power balance: The amount of reactive power purchased from upstream network should be equal to the reactive power demand of DN, reactive power demand of EHs, and reactive power flow through the network.

$$Q_{i,t}^{GRID} = Q_{i,t}^{D} + Q_{i,t}^{EH} + Q_{ij,t} \qquad \forall\, i, j\, \varepsilon\, \Omega \qquad \forall\, t\, \varepsilon\, \zeta \tag{3.29}$$

The apparent power flow through the DN is determined by Eq. (3.29), and Eq. (3.30) restricts the apparent power flow between the buses in the DN.

$$S_{ij,t} = \sqrt{(P_{ij,t})^2 + (Q_{ij,t})^2} \qquad \forall\, i, j\, \varepsilon\, \Omega \qquad \forall\, t\, \varepsilon\, \zeta \tag{3.30}$$

$$S_{ij}^{min} \leq S_{ij,t} \leq S_{ij}^{max} \qquad \forall\, i, j\, \varepsilon\, \Omega \qquad \forall\, t\, \varepsilon\, \zeta \tag{3.31}$$

The voltage magnitude of DN buses should follow the upper and lower constraints of the voltage range as expressed in Eq. (3.31).

$$V_{i}^{min} \leq V_{i,t} \leq V_{i}^{max} \qquad \forall\, i, j\, \varepsilon\, \Omega \qquad \forall\, t\, \varepsilon\, \zeta \tag{3.32}$$

The DSO's purchase of active and reactive powers from the upstream network is constrained by Eqs. (3.32) and (3.33).

$$P^{GRID,min} \leq P_{i,t}^{GRID} \leq P^{GRID,max} \qquad \forall\, i, j\, \varepsilon\, \Omega \qquad \forall\, t\, \varepsilon\, \zeta \tag{3.33}$$

$$Q^{GRID,min} \leq Q_{i,t}^{GRID} \leq Q^{GRID,max} \qquad \forall\, i, j\, \varepsilon\, \Omega \qquad \forall\, t\, \varepsilon\, \zeta \tag{3.34}$$

3.3.4 Thermal Network

Heat Loss: The temperature drop throughout the pipelines is given by:

$$T^{out,t} = (T^{in,t} - T^{ab})e^{-kL/c_p m} + T^{ab} \qquad \forall\, t\, \varepsilon\, \zeta \tag{3.35}$$

Nodal Flow Balance: The flow balance for any node in the network at any time instant is such that the sum of the inlet flow at the node is equal to the sum of the consumption at the node and the outward flow from the respective node. The following equation describes the node flow balance:

$$\sum_{In} m_t^{pipe} = \sum_{out} m_t^{pipe} + m_t^{node} \qquad \forall\, t\, \varepsilon\, \zeta \tag{3.36}$$

Nodal Energy Balance: The energy balance at each time instant is the sum of energy inflows equals the sum of energy outflows from the node in the pipeline.

$$\left(\sum m_{out}\right) T^{out,t} = \sum m_{in} T^{in,t} \qquad \forall\, t\, \varepsilon\, \zeta \tag{3.37}$$

3.3.5 Supply-demand Balance in EHs

The supply–demand balance of electrical, heating, and cooling power in all three types of EHs is represented below and should be satisfied for 24 hours of the scheduling period.

EH1: It has RES, a CHP unit, EC, EB, HF, and AC. The following three equations should be satisfied at each time instant.

$$P^{CHP}_{EH1,t} + P^{G}_{EH1,t} + P^{PV}_{EH1,t} + P^{WT}_{EH1,t} = P^{EC}_{EH1,t} + P_{d1,t} + P^{EB}_{EH1,t} + P^{trade}_{12,t} + P^{trade}_{13,t}, \quad \forall t \varepsilon \zeta \tag{3.38}$$

$$H^{CHP}_{EH1,t} + H^{EB}_{EH1,t} + H^{HF}_{EH1,t} = H^{AC}_{EH1,t} + H_{d1,t} + H^{trade}_{12,t} + H^{trade}_{13,t}, \quad \forall t \varepsilon \zeta \tag{3.39}$$

$$C^{AC}_{EH1,t} + C^{EC}_{EH1,t} = C_{d1,t} + C^{trade}_{12,t} + C^{trade}_{13,t}, \quad \forall t \varepsilon \zeta \tag{3.40}$$

EH2: It comprises RES, a CHP unit, EB, HF, EHP, BESS, and AC. At each time instant, the three equations listed below should be satisfied.

$$\begin{aligned} P^{CHP}_{EH2,t} + P^{G}_{EH2,t} + P^{PV}_{EH2,t} + P^{WT}_{EH2,t} + P^{dch}_{EH2,t} I^{dch}_{EH2,t} &= P_{d2,t} + P^{EB}_{EH2,t} + P^{EHP}_{EH2,t} \\ &+ P^{ch}_{EH2,t} I^{ch}_{EH2,t} + P^{trade}_{21,t} + P^{trade}_{23,t}, \quad \forall t \varepsilon \zeta \end{aligned} \tag{3.41}$$

$$H^{CHP}_{EH2,t} + H^{EB}_{EH2,t} + H^{EHP}_{EH2,t} = H^{AC}_{EH2,t} + H_{d2,t} + H^{trade}_{21,t} + H^{trade}_{23,t}, \quad \forall t \varepsilon \zeta \tag{3.42}$$

$$C^{AC}_{EH2,t} + C^{EHP}_{EH2,t} = C_{d2,t} + C^{trade}_{21,t} + C^{trade}_{23,t}, \quad \forall t \varepsilon \zeta \tag{3.43}$$

EH3: It has an AC unit, BESS, HES, HF, EB, EC, EHP, and a CHP unit. At every time instant, the following three Eqs. (3.41–3.43) must be met.

$$\begin{aligned} P^{CHP}_{EH3,t} + P^{G}_{EH3,t} + P^{PV}_{EH3,t} + P^{WT}_{EH3,t} + P^{dch}_{EH3,t} I^{dch}_{EH3,t} &= P_{d3,t} + P^{EB}_{EH3,t} + P^{EHP}_{EH3,t} \\ &+ P^{ch}_{EH3,t} I^{ch}_{EH3,t} + P^{EC}_{EH3,t} + P^{trade}_{31,t} + P^{trade}_{32,t}, \quad \forall t \varepsilon \zeta \end{aligned} \tag{3.44}$$

$$\begin{aligned} H^{CHP}_{EH3,t} + H^{EB}_{EH3,t} + H^{EHP}_{EH3,t} + H^{dch}_{EH3,t} I1^{dch}_{EH3,t} &= H_{AC,s,t} + H_{D,s,t} \\ &+ H^{ch}_{s,t} I1^{ch}_{s,t} + H^{trade}_{31,t} + H^{trade}_{32,t}, \quad \forall t \varepsilon \zeta \end{aligned} \tag{3.45}$$

$$C^{AC}_{EH3,t} + C^{EHP}_{EH3,t} + C^{EC}_{EH3,t} = \mathrm{C}_{d3,t} + C^{trade}_{31,t} + C^{trade}_{32,t}, \quad \forall t \varepsilon \zeta \tag{3.46}$$

3.3.6 Multi-Objective Optimization Formulation for DSO and EH Coordinator

Multi-objective (MO) optimization problems always maximize or minimize multiple objective functions simultaneously. The scalarization method is an appropriate solution to solve these problems. In this method, weight is determined before optimization, and the MO function creates a single solution. The objective function's weight determines the fitness function solution and performance priority. A high weight for an objective function indicates a higher importance than those with lower weights. There are three ways to determine weights: equal weight, rank order centroid weights, and rank-sum (RS) weights. A multi-objective problem is formulated for the DSO and EH coordinator, where the DSO tries to maximize profit and the EH coordinator wants to minimize the operational cost of EHs. To solve the proposed MO problem, an equal weight approach is used. Equation (3.47) illustrates the MO to SO formulation for the proposed framework.

$$Min\{Obj\} = Min\{OF^{DSO}, OF^{EH}\} = Min\{w_1 OF^{DSO} + w_2 OF^{EH}\} \tag{3.47}$$

Subjected to Eqs. (3.1)–(3.46),
where

$$w_1 + w_2 = 1 \tag{3.48}$$

In this study, objectives OF^{DSO} and OF^{EH} are first subjected to a normalization technique [36] and then assigned specific weights. Through the process of normalizing, the assigned weights are adjusted to appropriately represent the relative significance of each chosen objective function. This leads to an enhancement in the accuracy and pertinence of optimization. The mathematical expression for normalizing the objective functions is as follows:

$$OF' = \frac{(OF^{max} - OF)}{OF^{max} - OF^{min}}, \tag{3.49}$$

where minimum and maximum values of OF are represented by OF^{min} and OF^{max}, respectively.

3.3.7 Bargaining Game Between EHs

In the community, EHs recognize that cooperation might benefit them even when they belong to different operators. In contrast to independent EH operations, cooperative EHs can coordinate their local energy supplies and demands while simultaneously conducting energy trading with one another. As a result, cooperation among EHs can produce a better economic outcome (or lower operating costs) than isolation based solely on self-interest. However, the crucial point for any collaboration is how to share the lowered costs fairly.

One frequent strategy for coalition optimization is to reduce the overall cost of each member from the standpoint of a global controller. However, the privacy and autonomy of the involved EHs are compromised by this method. Additionally, it can result in an unfair allocation of the coalition's advantages. This chapter employs a negotiating cooperative game theory-based technique to solve this problem because it can maximize community performance overall and ensure that the profit of cooperation is divided equitably between involved EHs [37, 38].

First, the payoff of the kth EH is defined as x_k in this bargaining game. Meanwhile, the kth EH payoff is defined as y_k, in the case that no collaboration is reached between EHs. In order to get the optimal cooperation between EHs through bargaining, the following optimization problem must be solved:

$$Max \sum_{k=1}^{M} ln(x_k - y_k) \tag{3.50}$$

Subjected to $x_k \geq y_k, \forall k \,\varepsilon\, M$.

The feasible set of the bargaining game only contains payoffs that are better than the payoff at the dispute point in order to encourage participants to cooperate with one another. The Nash bargaining solution, which is the optimal solution of Eq. (3.50), can effectively encourage cooperation between EHs by achieving an optimal trade-off between Nash efficiency and fairness [27]. In this case, the beginning disagreement point is the minus non-cooperative operation cost C_k^{non} of the kth EH, or $y_k = -C_k^{non}$, and the reward is the minus operation cost C_k of the kth EH in the time horizon, or $x_k = -C_k$. Subsequently, the cooperative model for bargaining among EHs is as follows:

$$Max \sum_{k=1}^{N} ln\left(C_k^{non} - C_k\right) \tag{3.51}$$

Subjected to $C_k \leq C_k^{non}, \forall k \,\varepsilon\, M$,
where the cost reduction of the kth EH through negotiation is represented by $C_k^{non} - C_k$. Each EH aims to minimize its operation cost in this cooperative bargaining model when compared to the non-cooperative option.

3.3.8 Economic Scheduling Model of Cooperative EHs

In order to reduce operating expenses, these networked EHs form a cooperative society where they can share their excess energy with other EHs that are in need. EH's expenses can be divided into three categories, namely, purchasing gas, purchasing electricity from the grid, and trading energy (or revenue) with other EHs. $P_{k,t}^{trade}$ represents the amount of electricity exchanged, $H_{k,t}^{trade}$ represents the heat transferred between EHs, and $C_{k,t}^{trade}$ represents the cooling energy exchanged between EHs by the kth EH. When the kth EH buys energy at timeslot t from other EHs, then $P_{k,t}^{trade} > 0$, $H_{k,t}^{trade} > 0$, and $C_{k,t}^{trade} > 0$. Otherwise, if the kth EH sells electricity to other EHs, then $P_{k,t}^{trade} < 0$, $H_{k,t}^{trade} < 0$, and $C_{k,t}^{trade} < 0$. Equation (3.49) is a detailed formulation of the kth EH's cost:

$$C_k = \sum_{t=1}^{24} \left(P_{k,t}^{G} \lambda_t + G_{k,t}\gamma_t + P_{k,t}^{trade} \beta_t + H_{k,t}^{trade} \alpha_t + C_{k,t}^{trade} \varphi_t \right) \tag{3.52}$$

In order to ensure that the community cooperation yields benefits, the kth EH will negotiate the trade cost with other EHs. As stated above, the following optimization problem must be solved in order to reach the optimal coordination in the bargaining-based cooperative community:

$$Max \sum_{k=1}^{M} ln \left(C_k^{non} - \sum_{t=1}^{24} \left(P_{k,t}^{G} \lambda_t + G_{k,t}\gamma_t + P_{k,t}^{trade} \beta_t + H_{k,t}^{trade} \alpha_t + C_{k,t}^{trade} \varphi_t \right) \right) \tag{3.53}$$

Subjected to Eqs. (3.1)–(3.23) and (3.38)–(3.46).

It should be noted that the non-cooperative cost C_k^{non} of each EH $k \varepsilon M$ is considered the known input parameter in the objective function. As a requirement for entering the cooperative community, it is anticipated that every EH $k \varepsilon M$ would frankly declare its value of C_k^{non}. The Vickrey-Clarke-Grove mechanism, or other mechanisms, can provide the motivation to ensure this true assertion [39–41].

3.4 Results and Discussion

The case study, as depicted in Figure 3.4 [42], is a thermal-electricity MES system with three EHs. The topology of the proposed system is described using the IEEE 33-bus system and the three-node radiant thermal network. The IEEE test system's specifications are drawn from Ref. [43]. The wholesale market price is taken from the Indian Energy Exchange. In order to solve the MINLP model under study and verify that the suggested framework is optimal, the General Algebraic Modeling System (GAMS) is used, and the scheduling period is considered for 24 hours a day at 1-hour resolution. In Table 3.1, the performance indices of EH components are listed. The numerical findings used to validate the proposed scheme are presented in this section. The EHs have different kinds of energy conversion devices such as a CHP unit, EB, EC, BESS, HF, AC, and HES [44]. The characteristics of these components are studied in Section III. Figures 3.5 and 3.6 show the solar and wind power generation in all three EHs. Figure 3.7 depicts electricity and natural gas pricing. As illustrated, the gas price follows a three-part tariff structure, whereas the electricity price follows a ToU pattern. Figures 3.8–3.10 illustrate the hourly demand profiles of electricity, heating, and cooling of all EHs for a typical day. The voltage at the buses should be between 12.027 and 13.293 kV in the electricity distribution network. The maximum amount of power that EHs can share

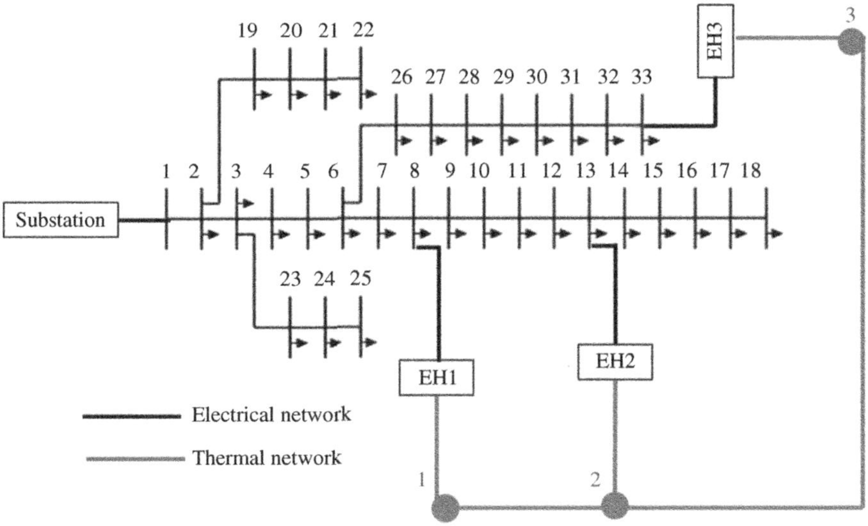

Figure 3.4 HE-MES topology for networked EHs.

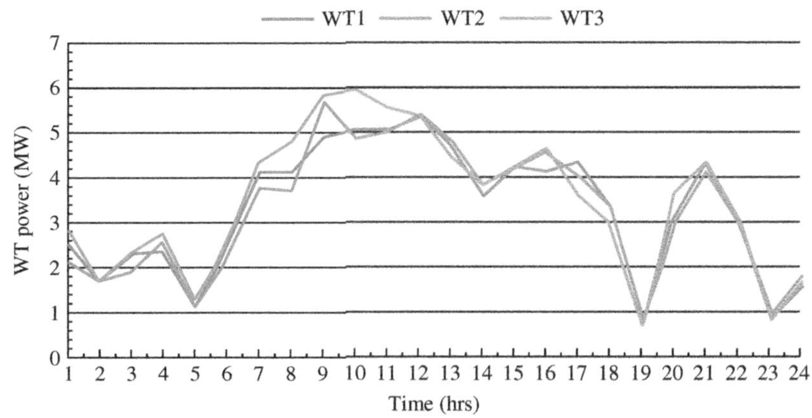

Figure 3.5 WT output profile in EHs.

(electricity, heating, and cooling) is assumed to be 3 MW. Additional information regarding the power and thermal networks, as well as EHs, can be found in Ref. [45]. The study is carried out using the GAMS software package on a PC with a 3 GHz Core i7 and 16 GB RAM. In order to assess the effectiveness of the suggested EH-based integrated multi-energy infrastructure, the following two cases are taken into consideration.

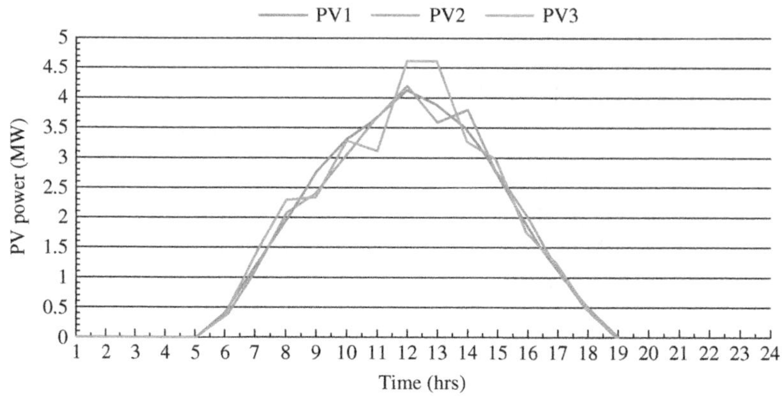

Figure 3.6 Solar PV output profile in EHs.

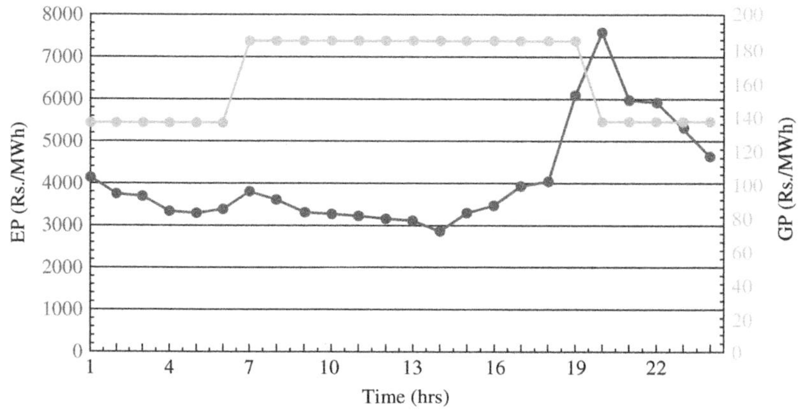

Figure 3.7 Electricity and gas prices.

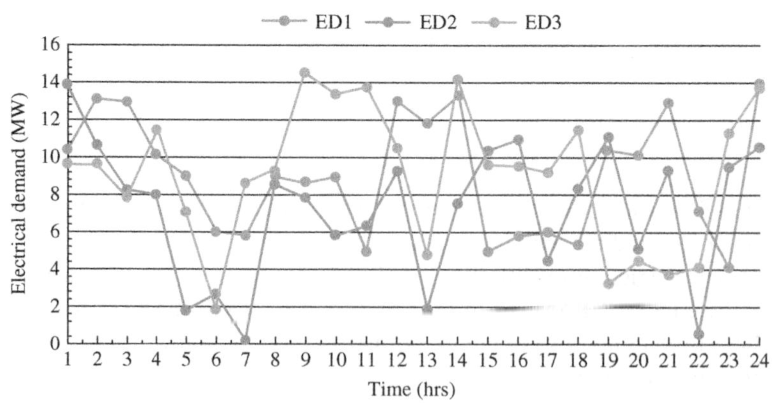

Figure 3.8 Electrical demand profile in EHs.

Table 3.1 Technical characteristics of EH's component.

Parameter	Value	Parameter	Value
$\eta_{B,ch/B,dch}$	0.9	η_{CB}	0.95
$\eta_{ch/dch}$	0.95	CoP	2.5
$E_{EH2}^{BESS,min}$, $E_{EH2}^{BESS,max}$	0.6 MWh, 3 MWh	η_{ec}	0.98
$E_{EH3}^{BESS,min}$, $E_{EH3}^{BESS,max}$	0.6 MWh, 3 MWh	η_{gh}	0.9
$Q_{EH3}^{HES,min}$, $Q_{EH3}^{HES,max}$	0.6 MWh, 3 MWh	η_{eh}	0.98
η_{gh}^{CHP}, η_{ge}^{CHP}	0.35, 0.65	$V_{slack,t}$	12.66 kV
V_i^{min}	12.027 kV	V_i^{max}	13.293 kV
λ_t	9000 Rs./MWh	β_t	5000 Rs./MWh
$P_{EH1}^{CHP,max}$	23 MW	$P_{EH2}^{CHP,max}$	19 MW
$P_{EH3}^{CHP,max}$	20 MW	$C_{EH1}^{EC,max}$	5 MW
$C_{EH1}^{AC,max}$	5 MW	$C_{EH2}^{EHP,max}$	10 MW
$C_{EH2}^{AC,max}$	5 MW	$C_{EH3}^{AC,max}$	10 MW
$C_{EH3}^{EC,max}$	8 MW	$C_{EH3}^{EHP,max}$	9 MW
$H_{EH1}^{HF,max}$	35 MW	$H_{EH1}^{EB,max}$	9 MW
$H_{EH2}^{EB,max}$	20 MW	$H_{EH2}^{EHP,max}$	10 MW
$H_{EH3}^{EHP,max}$	10 MW	$H_{EH3}^{EB,max}$	25 MW
$P_{EH1}^{PV,max}$	6 MW	$P_{EH2}^{PV,max}$	6 MW
$P_{EH3}^{PV,max}$	6 MW	$P_{EH1}^{WT,max}$	6 MW
$P_{EH2}^{WT,max}$	6 MW	$P_{EH3}^{WT,max}$	6 MW
c_p	4.2 J/g°C	T^{ab}	10 °C

Case 1: Non-cooperative mode of operation: EHs do not communicate with one another.

Case 2: Cooperative mode of operation: EHs share information.

3.4.1 Case 1: Non-cooperative Operation of EHs

In a non-cooperative mode of operation, EHs do not interact with each other, as depicted in Figure 3.1, and EHs try to maximize their individual operating cost without sharing any information with one another. The EH coordinator oversees the operation of all three EHs and is in charge of information sharing and scheduling that works best for them.

In the EH1 CHP unit, the electricity grid and RES are responsible for providing electrical power. Figure 3.11 shows the electrical power scheduling of EH1

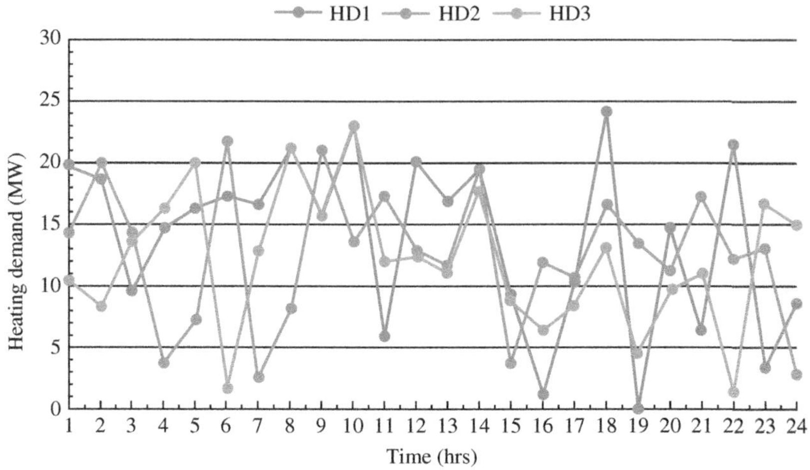

Figure 3.9 Heating demand profile in EHs.

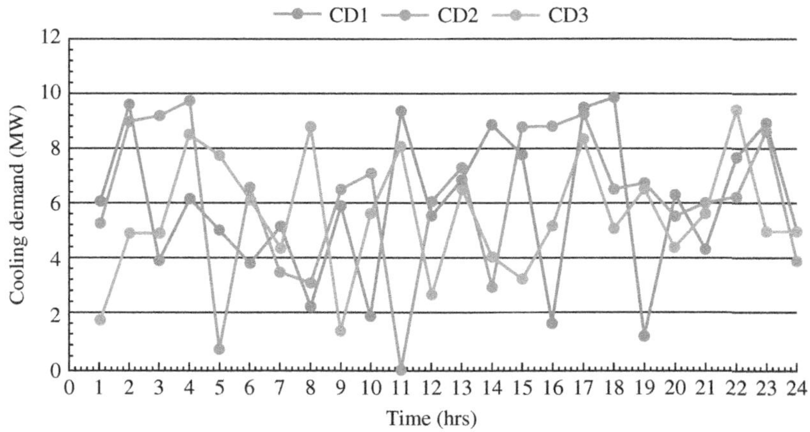

Figure 3.10 Cooling demand profile in EHs.

components. As can be observed, at 2 a.m. and 7 p.m., power is purchased from the grid only. During the rest of the hours, the electrical power demand of EH1 and the electricity consumption by EB and EC are met by the CHP unit and RES. Figure 3.12 illustrates the EH1 components' heat power scheduling. Heat demand in EH1 is met by the CHP unit, EB, and HF. All three units operate optimally to meet the heat demand of EH1 and heat consumption by AC as per their efficiency. As can be seen, the heat demand at 4 and 7 p.m. is relatively low, and it is primarily met by the CHP unit. Figure 3.13 depicts the cooling power scheduling of

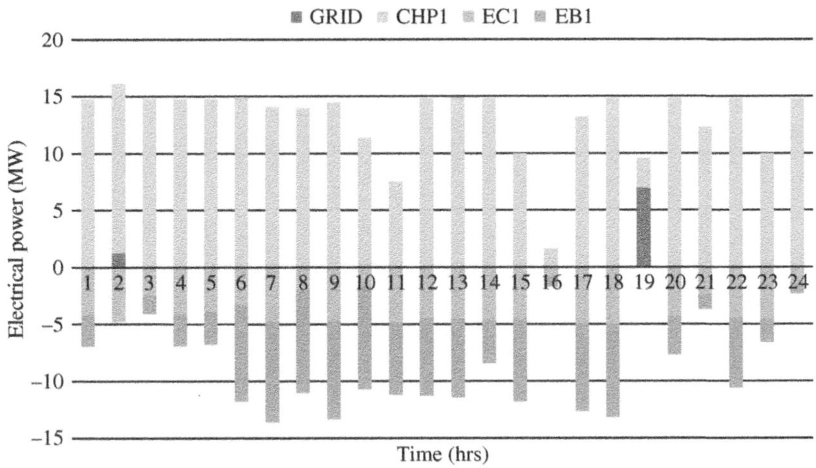

Figure 3.11 Electrical power scheduling in non-cooperative mode of EH1.

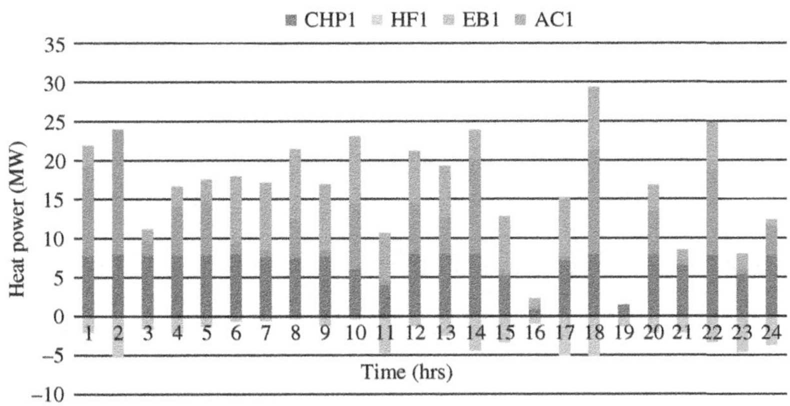

Figure 3.12 Heat power scheduling in non-cooperative mode of EH1.

EH1 components. AC and EC satisfy the cooling power need of EH1. At 7 p.m., the cooling demand is minimal and can only be fulfilled by AC.

The CHP unit, BESS, electricity grid, and RES facilitate electrical power in the EH2. Electrical power scheduling of EH2 components is illustrated in Figure 3.14. The quantity of power purchased from the grid is higher in EH2 than in EH1. During the majority of the hours, the electrical power demand of EH2, electricity consumption by EB, and EHP is supplied by the CHP unit, BESS (Figure 3.21), and RES (Figures 3.5 and 3.6). At 7 a.m. electricity demand in EH2 is very low, so during this hour power generation from RES (solar and wind) is stored in

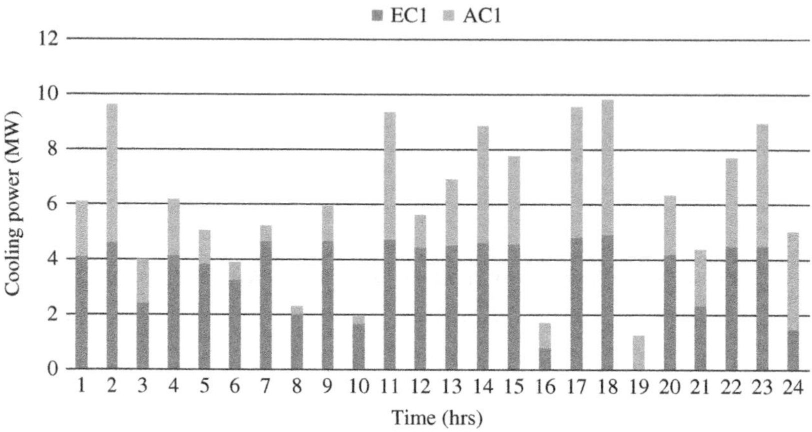

Figure 3.13 Cooling power scheduling in non-cooperative mode of EH1.

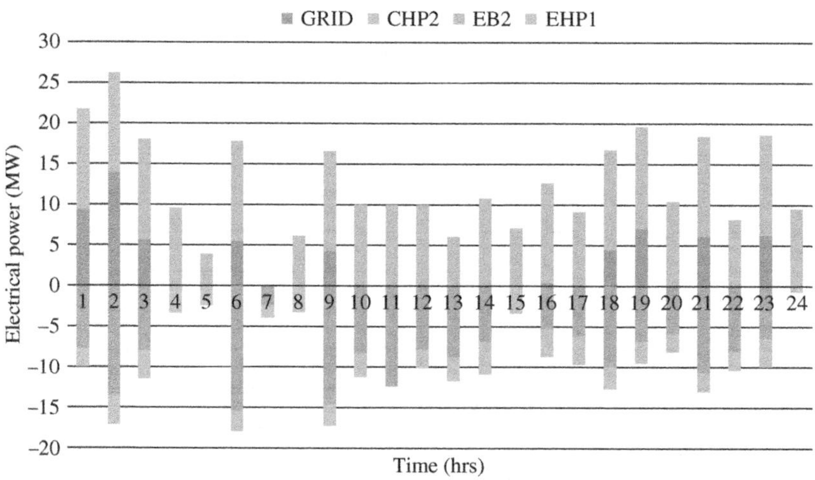

Figure 3.14 Electrical power scheduling in non-cooperative mode of EH2.

BESS. Heat power scheduling of EH2 components is shown in Figure 3.15. Heat demand in EH2 is satisfied by the CHP unit, EB, and EHP. All three units perform ideally to meet the heat demand based on their efficiency. Only the CHP unit is meeting the demand for heating at 4 a.m., 3 p.m., and 12 a.m. During these hours, cooling demand is quite low, resulting in low heat consumption in AC and low heat demand in EH2. Figure 3.16 shows the cooling power scheduling of EH2 components. EHP and AC meet EH2's cooling demand. EHP meets the majority of cooling demand because of its high coefficient of performance. Figure 3.21

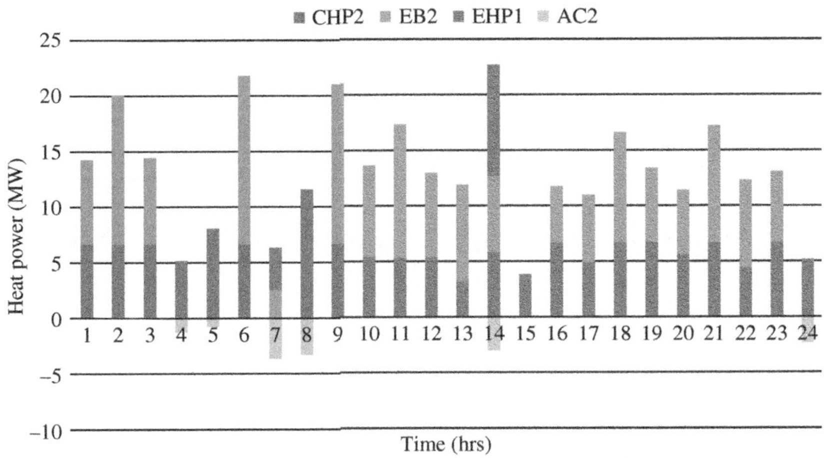

Figure 3.15 Heat power scheduling in non-cooperative mode of EH2.

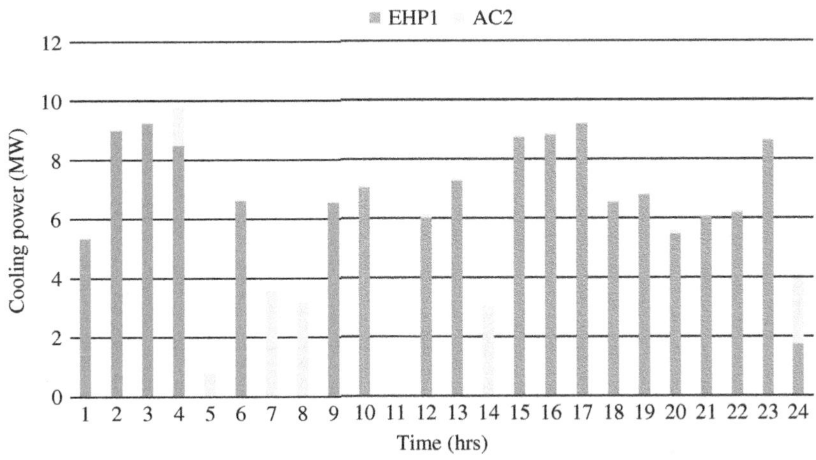

Figure 3.16 Cooling power scheduling in non-cooperative mode of EH2.

shows the charge–discharge status of BESS. BESS in EH2 charges between 4 a.m. and 5 a.m., 7 a.m., and 10 a.m. since electricity demand is minimal; consequently, power generated by RES accumulates in BESS. Due to high power consumption and low solar PV generation, it discharges at 6 a.m., 9 a.m., 4 p.m., and 6–7 p.m. Figure 3.22 depicts the state of charge (SoC) of the BESS in EH2, following the charge–discharge pattern.

Electrical power is facilitated in the EH3 via the CHP unit, BESS, RES, and the electrical grid. Figure 3.17 shows the EH3 components' electrical power

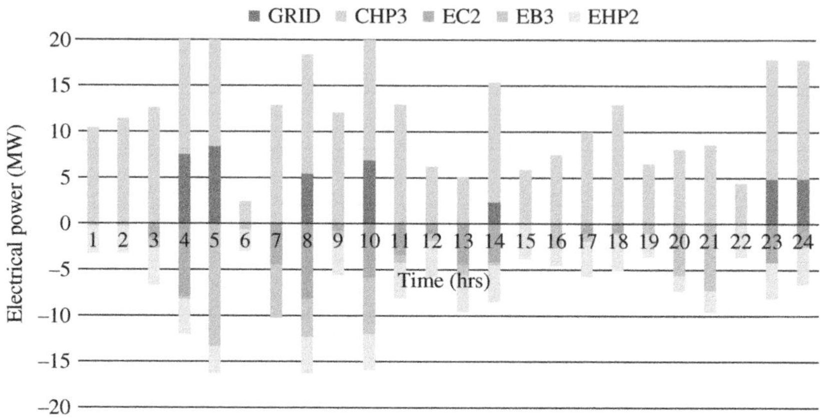

Figure 3.17 Electrical power scheduling in non-cooperative mode of EH3.

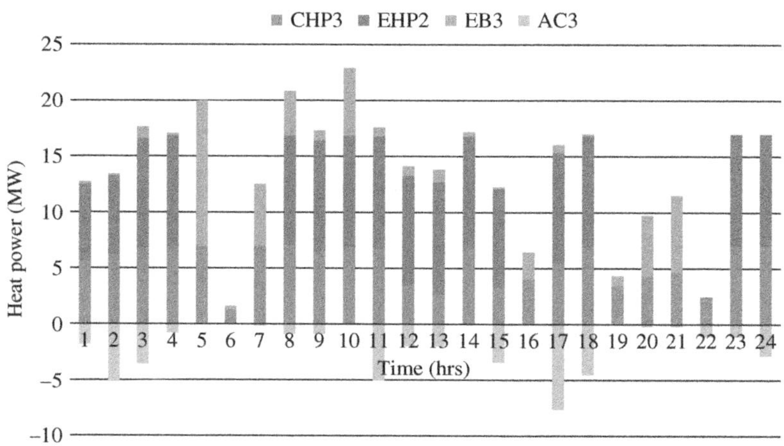

Figure 3.18 Heat power scheduling in non-cooperative mode of EH3.

scheduling. The quantity of power purchased from the grid in EH3 is greater than that in EH1, although it is less in EH2. It can be analyzed from Figure 3.17 that the CHP unit and RES meet the majority of EH3's electrical power needs, as well as EB, EHP, and EC power consumption. The heat power scheduling of the EH3 components is presented in Figure 3.18. Heat demand in EH3 is met by the CHP unit, EB, and EHP. All three units operate optimally to meet the heat demand as per their efficiency. Heating demand is quite low between 6 a.m. and 10 p.m., so only the CHP unit operates during these hours. Figure 3.19 shows the cooling power scheduling of EH3 components. Cooling demand in EH3 is met by EHP,

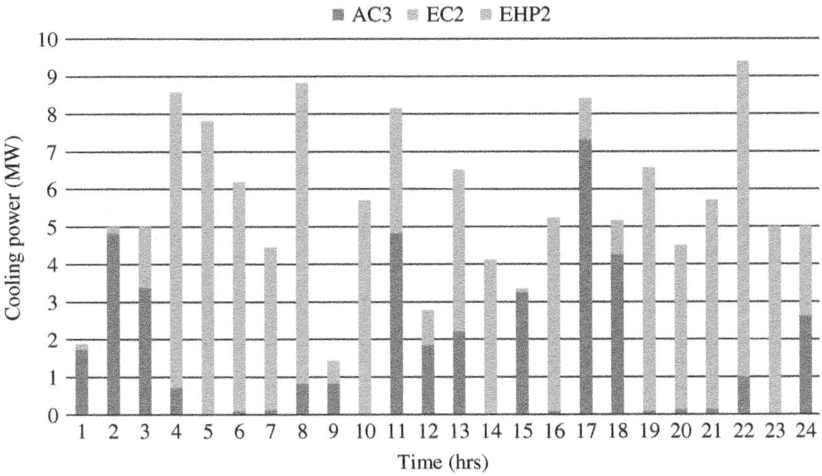

Figure 3.19 Cooling power scheduling in non-cooperative mode of EH3.

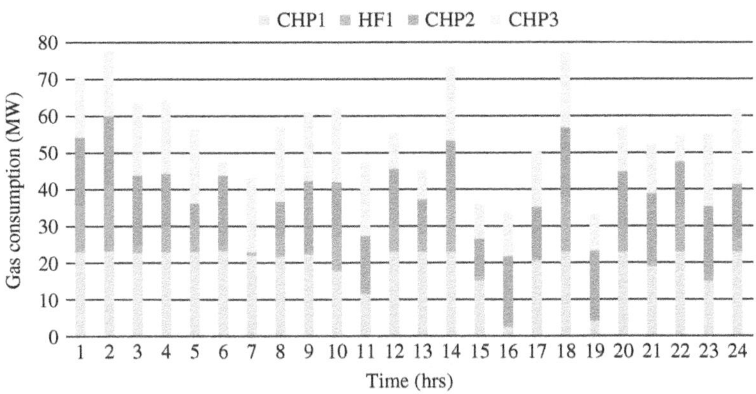

Figure 3.20 Gas consumption of EHs in non-cooperative mode.

EC, and AC. Figure 3.21 shows the charge–discharge pattern of BESS and HES in EH3. Because of the lower power demand during these hours, the BESS in EH3 charges at 6 a.m., 9 a.m., and 7 p.m. It discharges at 5 a.m., 11 a.m., 4 p.m., and 6–7 p.m. At 1 a.m., 3 a.m., 9 a.m., 11 a.m., 1 p.m., and 9 p.m., HES in EH3 charges. It discharges between 7 a.m. and 8 a.m., 11 p.m. and 12 a.m., and at 2 p.m. and 6 p.m. Figure 3.22 represents the SoC of BESS and HES in EH3 subsequent to charge–discharge status. Figure 3.20 demonstrates the quantity of gas consumed by three CHP units and HF in EH1, EH2, and EH3. The total operational cost of all three EHs operating in the non-cooperative mode is depicted in Figure 3.23. It

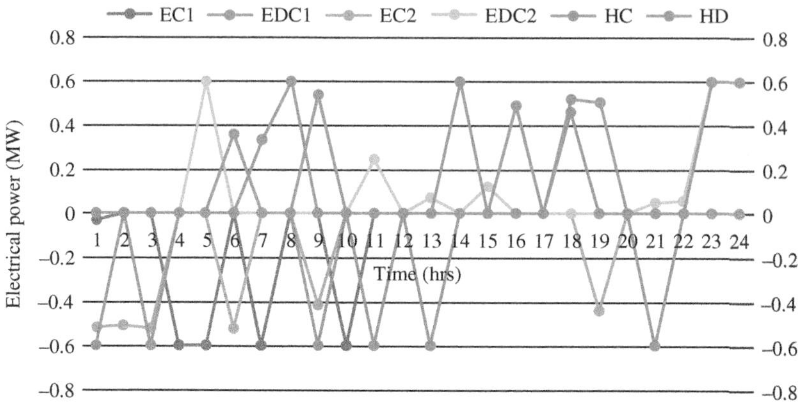

Figure 3.21 BESS and HES charge and discharge profile in non-cooperative mode.

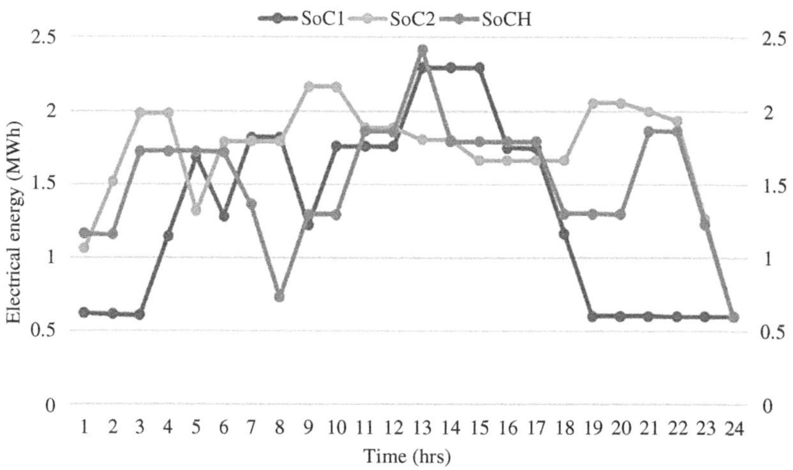

Figure 3.22 SoC profile of BESS and HES in non-cooperative mode.

can be noticed that the operational cost of EHs is relatively lower from 11 a.m. to 5 p.m. because less electricity is extracted from the upstream network.

3.4.2 Case 2: Cooperative Operation of EHs

In a cooperative operation, EHs engage with one another in order to maximize overall operating costs by sharing a limited quantity of information. Each EH is maintained by a virtual EH coordinator, as shown in Figure 3.2.

The amount of power acquired from the upstream network, power generated by the CHP unit, and the amount of electricity used by EB and EC are presented

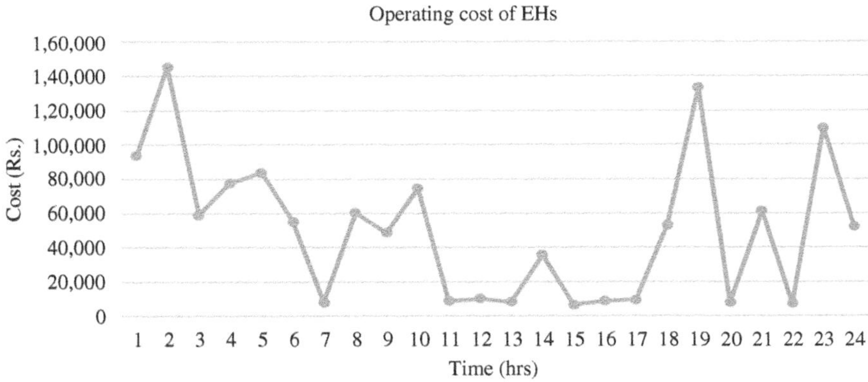

Figure 3.23 Operating cost of EHs in non-cooperative mode.

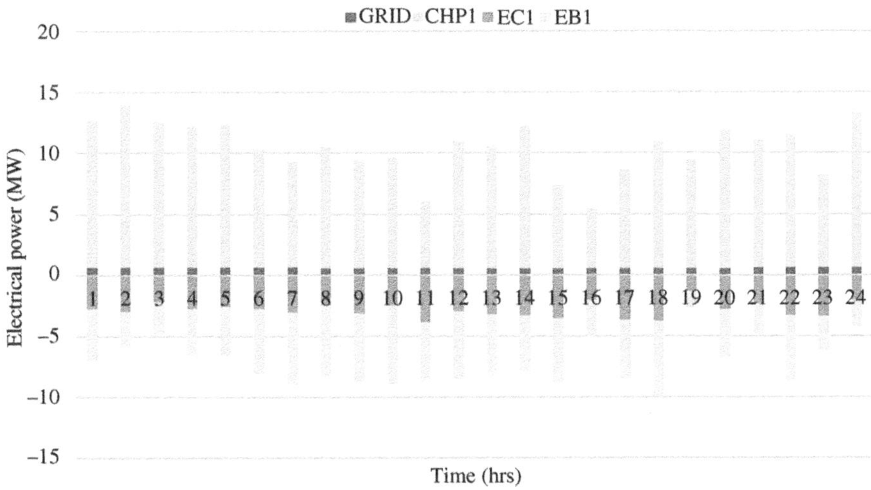

Figure 3.24 Electrical power scheduling in cooperative mode of EH1.

in Figure 3.24. RES and CHP provide most of EH1's electricity requirements. A modest amount of electrical power is bought from the grid throughout the day. The heat power produced by the CHP unit, HF, and EB, and consumed by the AC, is shown in Figure 3.25. EC and AC generate the cooling power in EH1, as illustrated in Figure 3.26.

The electrical power scheduling of EH2 components is represented in Figure 3.27. The CHP unit and RES in EH2 meet the electrical power demand, EB power consumption, and EHP power consumption during the scheduling period. Similar to EH1, a small amount of power is purchased from the upstream

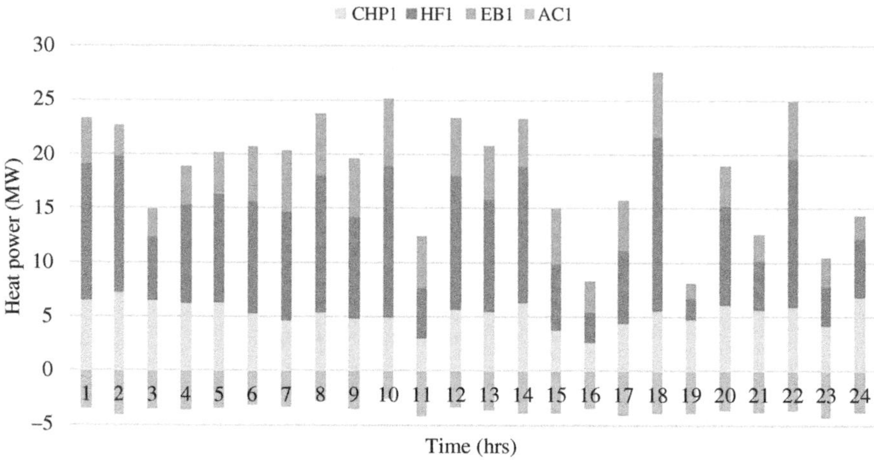

Figure 3.25 Heat power scheduling in cooperative mode of EH1.

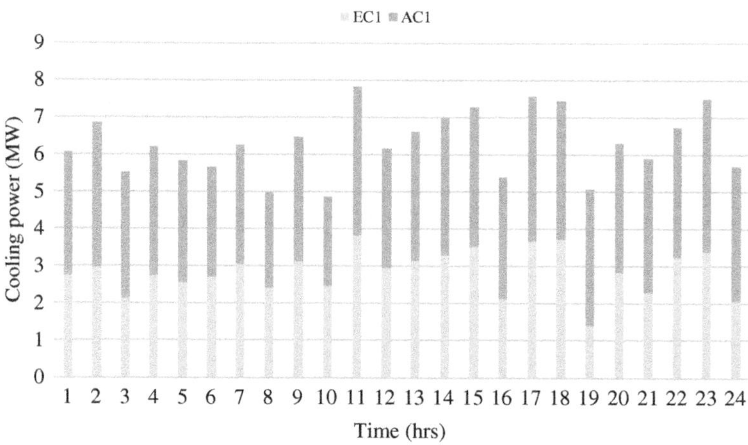

Figure 3.26 Cooling power scheduling in cooperative mode of EH1.

network during the day. In EH2, the CHP unit, EB, and EHP provide heating power while AC consumes it (Figure 3.28). All three units perform ideally to meet the heat demand based on their efficiency. EHP only operates during peak heating demand hours. Figure 3.29 shows the cooling power scheduling of EH2 components. Cooling demand in EH2 is met by EHP and AC. The majority of the cooling demand is fulfilled by AC. Figure 3.30 depicts the electrical power scheduling for EH3 components. The CHP unit and RES meet the majority of EH3's electrical power demand, as well as EB, EHP, and EC power consumption.

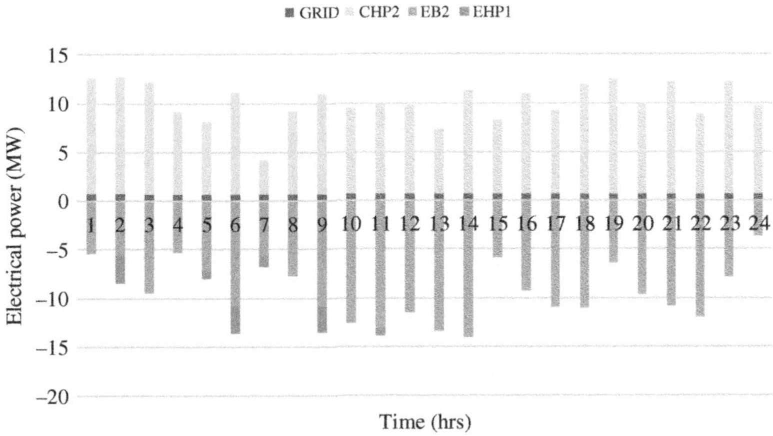

Figure 3.27 Electrical power scheduling in cooperative mode of EH2.

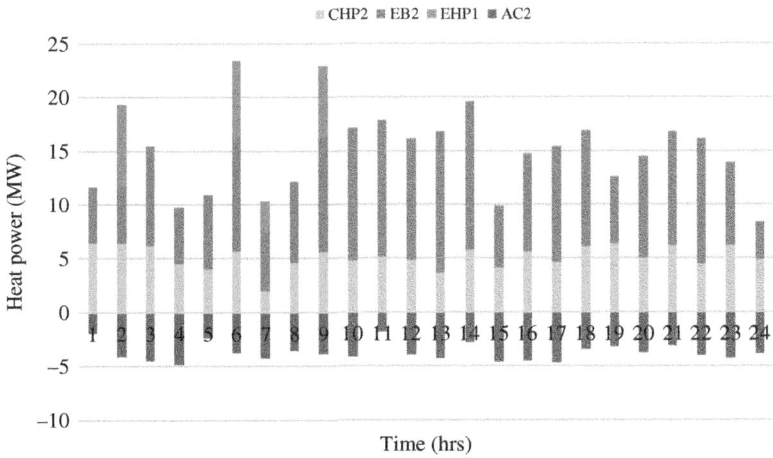

Figure 3.28 Heat power scheduling in cooperative mode of EH2.

Figure 3.31 illustrates the heat power generated by the CHP unit and EB and consumed by the AC. EHP, EC, and AC are the sources of cooling power in EH3 and operate in an optimal manner as presented in Figure 3.32. The amount of gas consumed by all three CHP units and HF in EH1, EH2, and EH3 is shown in Figure 3.33. Figures 3.34 and 3.35 show the charge–discharge state and SoC of BESS in EH2 and BESS and HES in EH3. Figure 3.34 indicates that BESS and HES in EH2 and EH3 charge between 1 and 3 a.m. due to off-peak demand hours. Discharges between 10 p.m. and 12 a.m. due to high demand and low RE generation. It can also be seen that BESS and HES are underutilized in

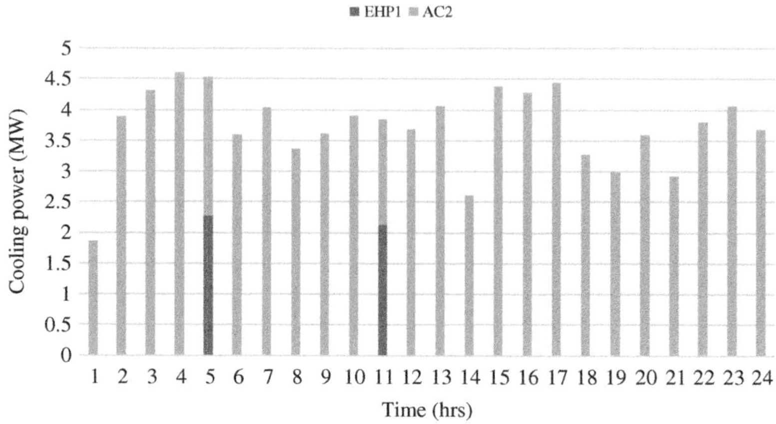

Figure 3.29 Cooling power scheduling in cooperative mode of EH2.

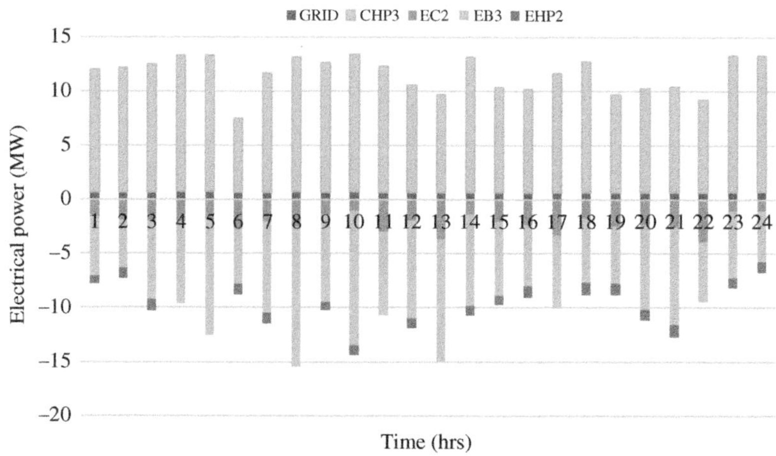

Figure 3.30 Electrical power scheduling in cooperative mode of EH3.

comparison to non-cooperative operations (Figures 3.21 and 3.22), owing to internal trading among EHs. Figure 3.36 represents the amount of electrical power traded from EH1 to EH2 (EH12), EH1 to EH3 (EH13), EH2 to EH1 (EH21), EH2 to EH3 (EH23), EH3 to EH1 (EH31), and EH3 to EH2 (EH32). The optimal heating and cooling power shared between EHs is depicted in Figures 3.37 and 3.38, respectively. In this case study, buying power is represented by negative values, and selling power is represented by positive values. Figure 3.39 shows the

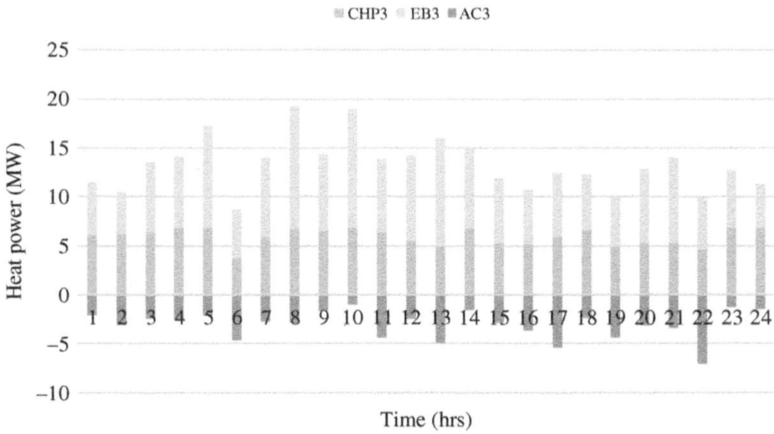

Figure 3.31 Heat power scheduling in cooperative mode of EH3.

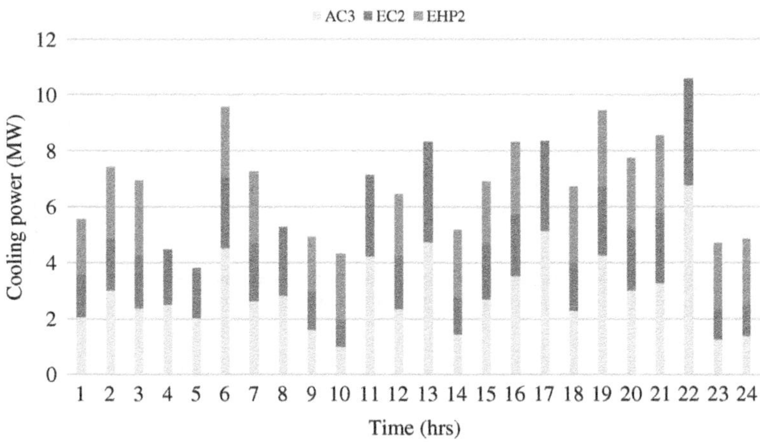

Figure 3.32 Cooling power scheduling in cooperative mode of EH3.

cumulative cost of all three EHs in cooperative mode of operation, which is much less compared to non-cooperative operation. Thus, in contrast to non-cooperative operation, this study shows that cooperative operation of networked EHs protects EH privacy while yielding economic benefits. These findings also suggest that EHs' cooperative interactions can significantly enhance their internal trade in proportion to their net energy. In the meantime, the decentralized framework's use guarantees the EHs' independent scheduling while protecting their personal data.

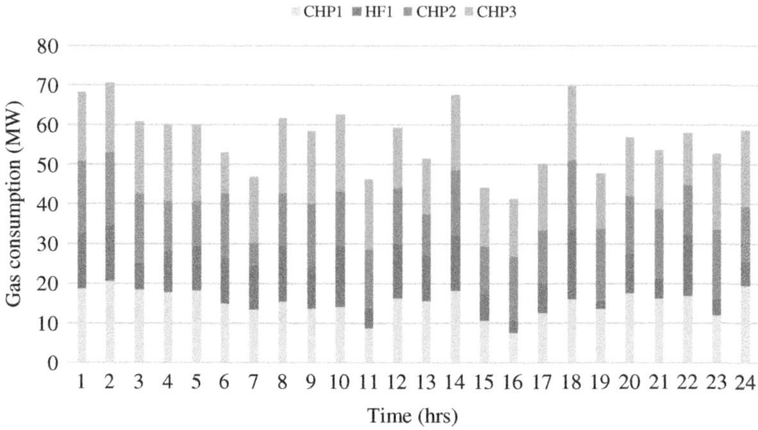

Figure 3.33 Gas consumption of EHs in cooperative mode.

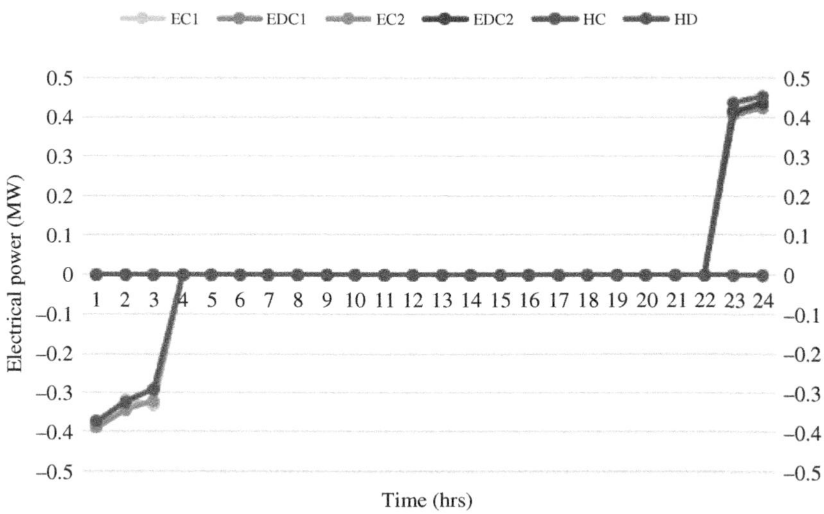

Figure 3.34 BESS and HES charge and discharge profile in cooperative mode.

3.5 Conclusion

This chapter proposes a comprehensive networked multi-energy EHs sharing model in centralized and decentralized modes of operation with a significant proportion of RES. The proposed framework addresses the volatile and intermittent nature of RES by incorporating various flexible devices such as CHP, EHP, EB, EC, AC, HF, BESS, and HES. The model is formulated as a multi-objective

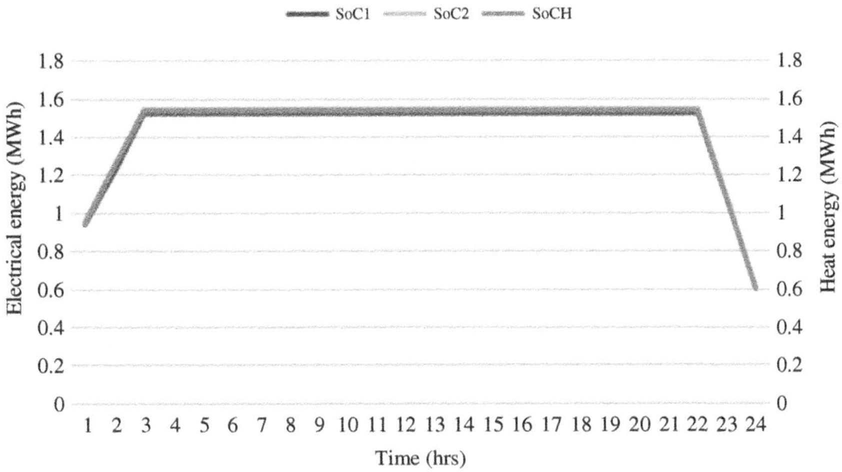

Figure 3.35 SoC profile of BESS and HES in cooperative mode.

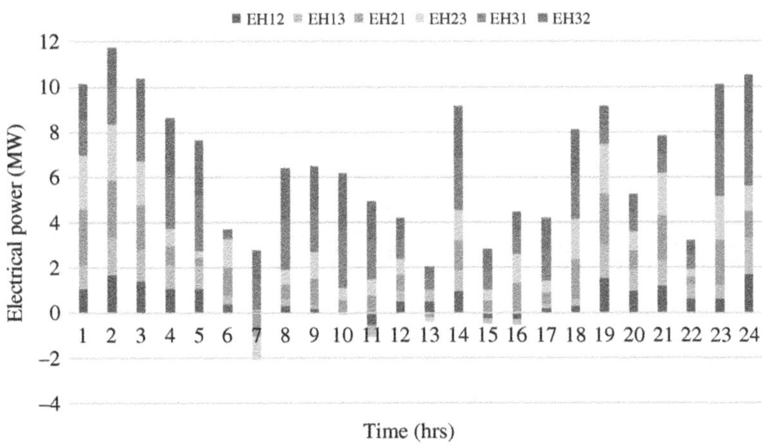

Figure 3.36 Electrical power exchange between EHs.

problem with two objective functions, namely, DSO profit maximization and EH's operational cost minimization, considering network constraints. The normalized weighted-sum strategy is used to solve multi-objective problems, resulting in compromised yet efficient solutions with objectives that are close to optimal values. To meet various operational requirements, MES operators could explore a variety of weight factor solutions. On the other side, the interaction among EHs is analyzed using game theory in a decentralized mode while the privacy and security of entities are preserved by sharing limited information. To share electricity,

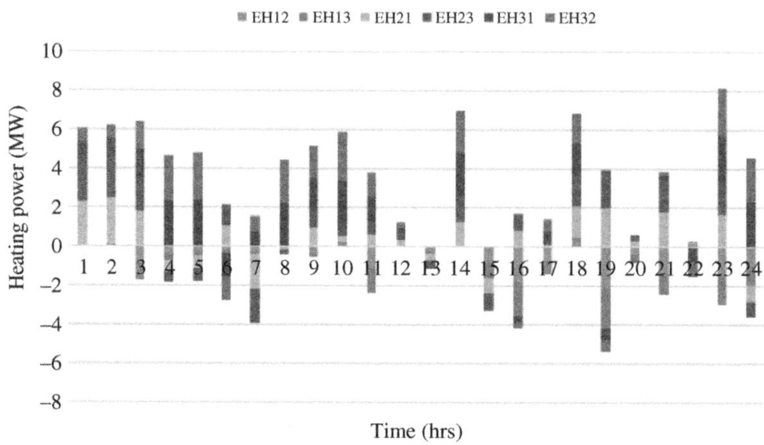

Figure 3.37 Heat power exchange between EHs.

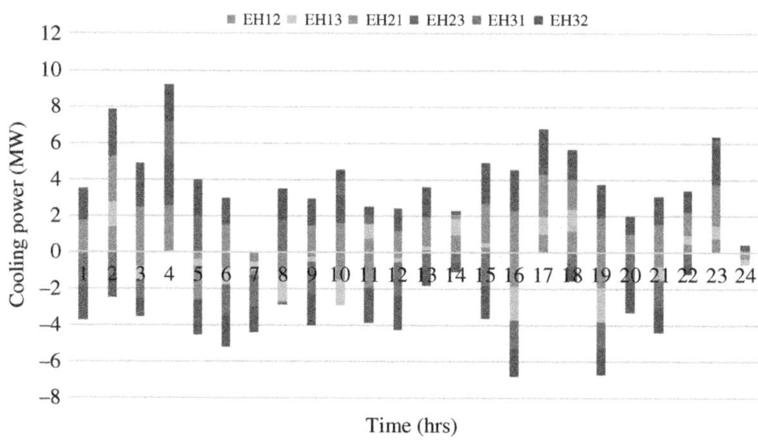

Figure 3.38 Cooling power exchange between EHs.

heating, and cooling power between EHs, a cooperative game is formulated that is typically connected to the gas network and power grid to minimize operational cost. By utilizing distinct supply and demand patterns, these cooperative EHs can rationally trade power (electrical, heating, and cooling) in order to reduce their own operating expenses. The optimization problem is solved fairly by using the bargaining game theory to ensure that all EHs are successfully motivated to cooperate with one another. In addition, the decentralized framework's use guarantees the EHs' independent scheduling while protecting their personal data. Based on numerical results, operating a multi-EH cooperatively provides greater

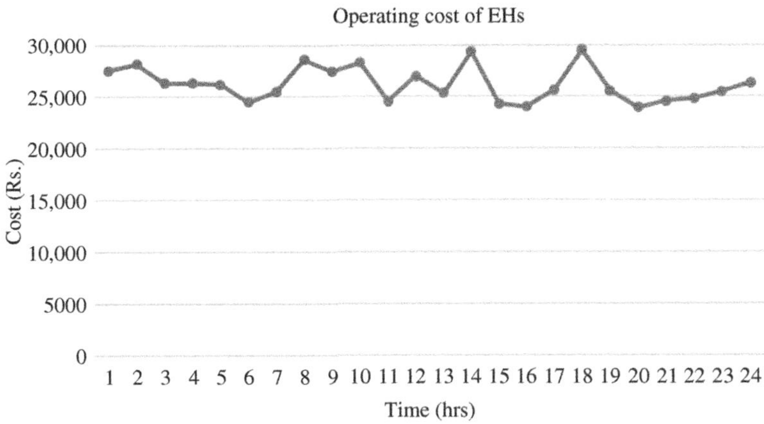

Figure 3.39 Operating cost of EHs in cooperative mode.

economic benefits than operating it non-cooperatively. Since the multi-energy networked EHs in the suggested model have the ability to meet needs for heating, cooling, and electricity, this improves the system's efficiency when compared to separate energy systems for heating, cooling, and electricity.

References

1 Mohammadi, M., Noorollahi, Y., Mohammadi-ivatioo, B. et al. (2017). Energy hub: from a model to a concept–a review. *Renewable and Sustainable Energy Reviews* 80: 1512–1527.
2 Mancarella, P. (2014). MES (multi-energy systems): An overview of concepts and evaluation models. *Energy* 65: 1–17.
3 Good, N. and Mancarella, P. (2019). Flexibility in multi-energy communities with electrical and thermal storage: a stochastic, robust approach for multi-service demand response. *IEEE Transactions on Smart Grid* 10 (1): 503–513.
4 Hu, J., Wang, Y., and Dong, L. (2024). Low carbon-oriented planning of shared energy storage station for multiple integrated energy systems considering energy-carbon flow and carbon emission reduction. *Energy* 290: 130139.
5 Wang, J., Zhong, H., Ma, Z. et al. (2017). Review and prospect of integrated demand response in the multi-energy system. *Applied Energy* 202: 772–782.
6 D. Sharma, N. M. Pindoriya, "An Overview of Optimization Techniques for Operation and Planning of Energy Hub Under Uncertainty," 2023 IEEE PES Innovative Smart Grid Technologies-Asia (ISGT Asia), 2023.

7 Chen, J. and Zhu, Q. (2017). A game-theoretic framework for resilient and distributed generation control of renewable energies in microgrids. *IEEE Trans. On Smart Grid* 8 (1): 285–295.

8 Xie, M., Ji, X., Hu, X., and Cheng, P. (2018). Autonomous optimized economic dispatch of active distribution system with multi-microgrids. *Energy* 153: 479–489.

9 Jalali, M., Zare, K., and Seyedi, H. (2017). Strategic decision-making of distribution network operator with multi-microgrids considering demand response program. *Energy* 141 (1): 1059–1071.

10 Nasiri, N., Yazdankhah, A.S., Mirzaei, M.A., and Loni, A. (2020). A bi-level market-clearing for coordinated regional-local multi- carrier systems in presence of energy storage technologies. *Sustainable Cities and Society* 63: 102439.

11 Tiwari, S., Singh, J.G., and Garg, A. (2024). A static robust energy management approach for modelling low emission multi-vectored energy hub including emission markets and power-to-gas units. *Energy* 294: 130827.

12 Wouters, C., Fraga, E.S., and James, A.M. (2015). An energy integrated, multi-microgrid, MILP (mixed-integer linear programming) approach for residential distributed energy system planning- a south Australian case-study. *Energy* 85: 30–44.

13 Fang, X., Ma, S., Yang, Q., and Zhang, J. (2019). Cooperative energy dispatch for multiple autonomous microgrids with distributed renewable sources and storages. *Smart Power Distrib. Syst. Control. Commun. Optim.* 99: 48–57.

14 Bui, V.H., Hussain, A., and Kim, H.M. (2018). A multiagent-based hierarchical energy management strategy for multi-microgrids considering adjustable power and demand response. *IEEE Trans. on Smart Grid* 9 (2): 1323–1333.

15 Zhao, T., Pan, X., Yao, S. et al. (2020). Strategic bidding of hybrid AC/DC microgrid embedded energy hubs: a two-stage chance constrained stochastic programming approach. *IEEE Trans. on Sustainable Energy* 11 (1): 116–125.

16 Guo, Q., Liang, X., Xie, D., and Jermsittiparsert, K. (2021). Efficient integration of demand response and plug-in electrical vehicle in microgrid: environmental and economic assessment. *Journal of Cleaner Production.*

17 G. Darivianakis, “A stochastic optimization approach to cooperative building energy management via an energy hub,” in Proc. 54th IEEE Conf. Decision Control, Osaka, Japan, pp. 7814–7819, Dec. 2015.

18 Hao, R., Ai, Q., Zhu, Y., and Wu, H. (2017). Hierarchical optimal dispatch based on energy hub for regional integrated energy system. *Elect. Power Autom. Equip.* 37 (6): 171–178.

19 Wei, J., Wang, J., Gao, H., and Gao, X. (2017). Optimal operation of micro integrated energy systems considering electrical and heat load classification and scheduling. In: Proc. IEEE Conf. Energy Internet Energy Syst. Integr., Beijing, China, 1–6.

20 Naja, A., Falaghi, H., Contreras, J., and Ramezani, M. (2017). A stochastic bilevel model for the energy hub manager problem. *IEEE Trans. on Smart Grid* 8 (5): 2394–2404.

21 Ma, T., Wu, J., and Hao, L. (2018). Energy flow modeling and optimal operation analysis of the micro energy grid based on energy hub. *Power Syst. Technol.* 133: 292–306.

22 Sheikhi, A., Bahrami, S., and Ranjbar, A.M. (2015). An autonomous demand response program for electricity and natural gas networks in smart energy hubs. *Energy* 89: 490–499.

23 Bahrami, S. and Sheikhi, A. (2016). From demand response in smart grid toward integrated demand response in smart energy hub. *IEEE Trans. on Smart Grid* 7 (2): 650–658.

24 Sheikhi, A., Rayati, M., Bahrami, S., and Ranjbar, A.M. (2015). Integrated demand side management game in smart energy hubs. *IEEE Trans. on Smart Grid* 6 (2): 675–683.

25 Kamyab, F. and Bahrami, S. (2016). Efficient operation of energy hubs in time of-use and dynamic pricing electricity market. *Energy* 106 (1): 343–355.

26 Bahrami, S., Toulabi, M., Ranjbar, S. et al. (2017). A decentralized energy management framework for energy hubs in dynamic pricing markets. *IEEE Trans. on Smart Grid* 9: 6780–6792.

27 Wu, F., Liu, X., Sun, Y. et al. (2017). Game optimization strategy for multiple parks based on combined cooling heating and power. *Automat. Electr. Power Syst.* 42: 68–75.

28 Maschler, M., Solan, E., and Zamir, S. (2013). *Game Theory*. New York, NY, USA: Cambridge University. Press.

29 Siqin, Z., Niu, D., Li, M. et al. (2022). Distributionally robust dispatching of multi-community integrated energy system considering energy sharing and profit allocation. *Applied Energy* 321: 119–202.

30 Wang, X., Jia, H., Wang, Z. et al. (2024). A real-time peer-to- peer energy trading for prosumers utilizing time-varying building virtual energy storage. *International Journal of Electrical Power & Energy Systems* 155: 109547.

31 Tiwari, S. and Singh, J.G. (2024). A cooperation based transactive energy management for networked energy hubs considering improved payoff allocation mechanism. *Sustainable Energy Technologies and Assessments* 65: 103777.

32 Martini, L. (2015). Trends of smart grids development as fostered by European research coordination: the contribution by the EERA JP on smart grids and the ELECTRA IRP. In: *Proc. IEEE 5th Int. Conf. Power Eng., Energy Elect*, 23–30. Drives, Riga, Latvia.

33 Ai, Q. and Hao, R. (2018). Key technologies and challenges for multi-energy complementarity and optimization of integrated energy system. *Autom. Electr. Power Syst.* 42 (4).

34 Ghappani, S.A. and Karimi, A. (2023). Optimal operation framework of an energy hub with combined heat, hydrogen, and power (CHHP) system based on ammonia. *Energy*.

35 Pezhmani, Y., Oskouei, M.Z., Rezaei, N., and Mehrjerdi, H. (2023). Optimal operation of active distribution networks hosting hybrid hydrogen-electricity refueling stations considering water demand under a stochastic-IGDT approach. *IET Gener. Transm. and Distrib.*

36 Arora, J.S. (2017). Chapter 18- multi-objective optimum design concepts and methods- Introduction to optimum design. In: *Introduction to Optimum Design*, 4the. Boston: Academic Press.

37 Zenginis., I., Vardakas., J., Echave., C. et al. (2017). Cooperation in microgrids through power exchange: an optimal sizing and operation approach. *Applied Energy* 203: 972–981.

38 Zhang, D., Samsatli, N.J., Hawkes, A.D. et al. (2013). Fair electricity transfer price and unit capacity selection for microgrids. *Energy Economics* 36 (3): 581–593.

39 Mhanna, S., Verbi, G., and Chapman, A.C. (2016). A faithful distributed mechanism for demand response aggregation. *IEEE Trans. on Smart Grid* 7 (3): 1743–1753.

40 Ehsanfar, A. and Heydari, B. (2018). An incentive-compatible scheme for electricity cooperatives: an axiomatic approach. *IEEE Trans. on Smart Grid* 9 (2): 1416–1424.

41 Tiwari, S. and Singh, J.G. (2022). Optimal energy management of multi-carrier networked energy hubs considering efficient integration of demand response and electrical vehicles: a cooperative energy management framework. *Journal of Energy Storage* 51: 104479.

42 Liu, N., Tan, L., Sun, H. et al. (2022). Bilevel heat-electricity energy sharing for integrated energy systems with energy hubs and prosumers. *IEEE Transactions on Industrial Informatics* 18 (6): 3754–3765.

43 Baran, M.E. and Wu, F.F. (1989). Network reconfiguration in distribution systems for loss reduction and load balancing. *IEEE Trans. on Power Del.* 4 (2): 1401–1407.

44 D. Sharma and N. M. Pindoriya, "An Emission and Uncertainty Aware Optimal Dispatch of Multi-Energy Hub," 2024 *IEEE PES Innovative Smart Grid Technologies - Asia (ISGT Asia),* Bengaluru, India, 2024, pp. 1–6.

45 Abdulnasser, G. et al. (2023). Optimizing the operation and coordination of multi-carrier energy systems in smart microgrids using a stochastic approach. *IEEE Access* 11: 58470–58490.

4

Impact of Hydrogen and Power-to-Gas Technology on MV-EHs

Subhendu Sekhar Sahoo[1], Ankit Garg[2], Khaleequr Rehman Niazi[2], and Shubham Tiwari[3]

[1] *Department of Electrical Engineering, GIST, Bhubaneswar, Odisha, India*
[2] *Department of Electrical Engineering, MNIT, Jaipur, Rajasthan, India*
[3] *Department of AFE, IIASA, Laxenburg, Vienna, Austria*

4.1 Introduction

The surroundings of green energy facilities are beginning to incorporate hydrogen and power-to-gas (P2G) technologies as they transition toward more durable and resilient multi-vector energy hubs (MV-EHs). This chapter explores the vital significance of these developments, emphasizing how they can improve the sustainability, dependability, and effectiveness of MV-EHs. Cleaner and more environmentally friendly forms of energy are becoming increasingly prevalent because of the global energy revolution. Hydrogen has emerged as a viable answer as nations and companies work to cut the production of greenhouse gases and fight climate change. Since hydrogen emits no direct emissions when used as a fuel or energetic carrier, it is a desirable choice for decarbonizing several different sectors and industries, including conveyance, industrial operations, and the production of electricity [1]. The distribution and administration of electrical energy from high-voltage lines of transmission to end users is greatly aided by MV-EHs, which normally run anywhere from 1 to 36 kV. To guarantee dependable power distribution in residential, commercial, and manufacturing facilities, these systems are essential. Nevertheless, the inclusion of unstable sources of renewable energy, such as wind and solar energy, is posing more and more issues for MV-EHs. These sources require complex management to ensure grid reliability and stability [2] (Figure 4.1).

Both P2G and hydrogen-based solutions are helpful in this scenario. The reliability of MV-EH could be significantly enhanced by the usage of gasoline as a sustainable energy source. Electrolysis is also a technique that separates moisture

Energy Management Strategies for Multi-Vectored Energy Hubs to Achieve Low Carbon Societies, First Edition.
Edited by Shubham Tiwari, Jai Govind Singh, Sivaraman Palanisamy, Sharmeela Chenniappan, Rupendra Kumar Pachauri, and Sanjeevikumar Padmanaban.

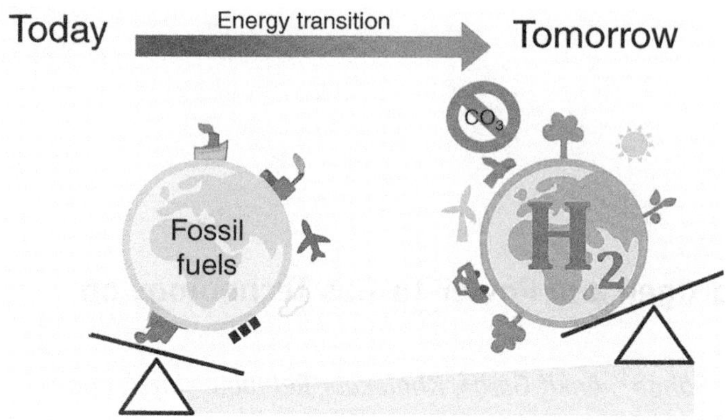

Figure 4.1 Hydrogen storage technology on multi-vector energy hub. *Source:* [3] / with permission of Elsevier.

through hydrogen and oxygen using electrical electricity. The process is widely applied to produce hydrogen by using excess energy from renewable sources. This technique, when coupled with P2G, the internet, allows excess green power to be converted into the element hydrogen, which is subsequently stored and used as needed. This reduces the challenges caused by the unpredictable nature of renewable energy sources in addition to aiding in the maintenance of the demand/supply stability. A set of processes known as P2G technology converts excess electricity into gases that are fuel. These techniques typically begin with the desalination of water, which yields hydrogen. Hydrogen is then combined with carbon monoxide to create methane, a process known as methanation. The methane it contains can be utilized to generate energy, injected throughout the existing natural gas systems, or employed as fuel for several different uses. P2G systems and MV-EHs can work together to capture additional green power that could otherwise end up in landfills, improving the overall efficiency of electricity and reducing reliance on energy from petroleum [4]. Efficient storage and administration of hydrogen is a major difficulty in the integration of P2G technologies and hydrogen into MV-EHs. Regarding energy density, cost, and safety, there are several storage options available, including as compressed hydrogen, liquid hydrogen, and metal hydrides, each with their own benefits and drawbacks. Hydrogen solution implementation in MV-EHs may be economically and practically feasible depending on the storage mechanism selected. To ensure that these storage solutions match the operational needs of MV-EHs while increasing efficiency and lowering costs, advanced modeling approaches are used to simulate and tune them [2]. The profitable operation of MV-EHs is mostly dependent on hydrogen rather than merely storage and conversion technologies. Hydrogen can increase the general endurance and adaptability of power networks by serving as a hedge against fluctuations in the output that comes from sources that are sustainable. When solar energy output is limited, stored nitrogen can be burned or converted into additional energy using fuel cells to guarantee a consistent and constant supply of electricity. In addition, by providing additional benefits like energy augmentation and frequency management, gasoline can enhance grid reliability and promote the adoption of distributed energy resources (DERs).

Incorporating P2G and hydrogen technology with MV-EHs has major economic implications. The development and application of these technologies may open new business opportunities and promote economic growth in the green energy sector. Nevertheless, the structure required for creating hydrogen storage spaces, and transfer may need significant upfront costs to set in motion and maintain. Therefore, to reduce prices, increase efficiency, and the general economic feasibility of these advancements, ongoing study, and improvement are necessary [5]. A revolutionary chance to advance sustainable energy systems exists at the interface of P2G and hydrogen-based technologies with low-voltage electrical infrastructure. These technologies provide a path toward more effective, dependable, and ecologically friendly MV-EHs by tackling the difficulties associated with the integration, archiving, and the conversion from renewable energy. Relentless advancements and investments in hydrogen and P2G technology will be essential to attaining their full potential and building a more resilient and environmentally conscious energy economy as the energy business continues to change [6].

4.2 Objectives

This chapter's main goal is to give a thorough analysis of how P2G and hydrogen development affect MV-EHs. The chapter aims to clarify how these technologies can improve the toughness, efficiency, and the sustainability of MV-EHs by thoroughly examining them. The chapter's initial goal is to examine the state of hydrogen storage technologies today and how they relate to MV-EHs. This entails examining and evaluating the viability of integrating several hydrogen storage techniques, such as compression as well, liquefaction, and solid-state storage spaces into MV-EHs. The intention is to shed light on how these storage options might be enhanced to facilitate the dependable and effective use of hydrogen in electrical infrastructure. The chapter's second goal is to examine the foundations and developments of P2G technology. This entails a thorough analysis of the relevant procedures, such as the synthesis of hydrogen through electrolysis plus the creation of synthetic methane from hydrogen by methanation. The chapter aims to show how P2G capabilities can be utilized to deal with issues with grid stability and renewable energy integration in MV-EHs by analyzing the technology elements and incorporating system methodologies. Evaluating the contribution of hydrogen toward the sustainable functioning of MV-EHs is another important goal. This entails examining the advantages of hydrogen for the surroundings, its potential economic effects, and its useful applications in improving the general efficiency and dependability of electrical installations. The goal of this chapter remains to demonstrate how hydrogen can help with the shift to low-carbon energy sources and reduce the emissions of greenhouse gases. Lastly, the chapter aims to give a general review of modeling and simulation methods that are employed in MV-EHs to assess P2G and hydrogen storage systems. The chapter intends to provide useful insights into how such technologies can be properly modelled and adjusted to improve their interoperability and efficiency in real-world situations by going over various ways and tackling frequent difficulties. The overall goals of the chapter include providing a comprehensive overview of P2G and hydrogen production methods and their implications for MV-EHs, and how they might propel the development of environmentally friendly electricity systems. Scope: This section will provide a thorough analysis of the functions and effects of P2G and hydrogen storage devices within the energy business. It will also address various important elements of these technologies and how they are integrated into MV-EHs. Hydrogen Storage Technologies: Initially the chapter will discuss different ways to store hydrogen, such as liquid hydrogen, solid-state preservation, and pressurized hydrogen. It will go over each preservation method's technical features, benefits, and drawbacks, with an emphasis on cost, capacity, and safety [6]. A review of current developments in technology for storage and how they affect the incorporation of hydrogen into MV-EHs will also be included in the curriculum. The goal of this part is to give readers a thorough grasp of how various storage options can affect the dependability and effectiveness of MV-EHs. P2G Technologies: The concepts and implementations of P2G technologies—which transform excess electrical power and produce protons and additional gases—will be covered in this chapter. It will go over the many types of electrolyzers and their efficiency and the electrolysis process used to produce oxygen. The chapter will also look at the mechanisms involved in producing synthetic methane by combining hydrogen and carbon dioxide. System elements, integration tactics,and P2G technologies' contribution to the supply and demand balance of renewable energy will all be covered in the conversation. The ways that P2G can improve the adaptability and sustainability of MV-EHs will be discussed in this section [7]. Integration with MV-EHs: The section will investigate the best ways to include P2G and hydrogen preservation into MV-EHs. Analyzing how these technologies affect operational adaptability, energy conservation, and grid stability is part of this. The scope will include realistic considerations for integrating the hydrogen into the current electrical structures and difficulties and potential solutions pertaining to regular system consumption, procedure, and design. Effective methods will be highlighted, while productive mergers will be demonstrated via instructional videos and actual-life instances. Modeling and Simulation: The section will conclude with a discussion of modeling and simulation methods for assessing P2G and hydrogen retention in MV-EHs. It will cover several methods for modeling these technologies, pinpoint shared difficulties, and investigate optimization techniques. Insights into how precise modeling can aid in the efficient deployment and monitoring of P2G and hydrogen techniques in real-world situations are the focus of this portion of the paper. The chapter will provide a comprehensive analysis of these elements, offering a comprehensive perspective on how P2G and hydrogen technologies might revolutionize MV-EHs and help us move toward a more environmentally friendly energy future.

4.3 Hydrogen Storage Technology

The use of hydrogen just like an alternate source of energy depends heavily on hydrogen storage, especially when it comes to MV-EHs. Since hydrogen has a low density when it is in its usual state, effective storage techniques are necessary to allow for its extensive application in energy systems [3]. Additionally, there are various important hydrogen storage techniques, including compressed hydrogen, liquid hydrogen, and metal hydrides. Each has unique properties that make it suitable for a variety of applications and operational requirements. The most popular and basic type of hydrogen storage is called compressed hydrogen, and this system works by compressing hydrogen gas to high challenges, such as usually between 350 and 700 bar. This makes it possible to store comparatively large amounts of hydrogen in cylindrical tanks. This technique is extensively utilized because of its simplicity, especially in small-scale renewable energy sources and fuel cell vehicles. But to compress hydrogen, a substantial amount of energy is needed—typically up to 10% of the hydrogen's energy content—and high-pressure tanks need to be made of sturdy materials with safety features to avoid leaks and explosions. Compressed carbon dioxide is preferred due to its established facilities and ease of accessibility, even with these obstacles [8]. To store liquid hydrogen, it must be cooled to cryogenic temperatures—below −253°C—where it compresses into a liquid. Compared to condensed hydrogen, liquid hydrogen has a substantially higher quantity of energy per volume, which makes it an improved space-efficient form of storage. Because of this, it can be applied in areas like aero or large-scale storage of energy where there is a limited amount of space. Nevertheless, boil-off losses can happen over lengthy periods, and permanent energy input is needed to retain hydrogen in its liquid condition at temperatures considered cryogenic. The energy-intensive procedure increases expenses and complexity, making liquid storage for hydrogen impractical for several uses. By utilizing substances that take hydrogen fuel into the form of crystals and release it whenever heated, metal hydroxides present a revolutionary method of storing hydrogen. Since hydrogen is stored at far lower pressures in these solid-state memory systems than in high-pressure or refrigerated storage, they offer a safer option. Although metal hydrides can store a considerable quantity of hydrogen by weight, they frequently need heat to release the element, which can be inefficient in terms of energy use [9]. Regarding large-scale use, considerable optimization is required for the weight of storage facilities and the price of metals themselves. Regarding the amount of energy stored, safety, cost, and practicality, every one of these types of storage techniques offers benefits and drawbacks. The infrastructure that is accessible and the unique requirements of the application determine which hydrogen storage technique is best. The full development of hydrogen in MV-EHs and other areas will require continuous breakthroughs in storage technology as the hydrogen sector grows. Technological Developments: The viability of hydrogen as a major energy carrier is being greatly increased by recent developments in hydrogen storage devices, especially for incorporating it into MV-EHs. It is imperative that these advancements tackle the issues of cost, safety, and effectiveness linked to hydrogen storage to facilitate the gas's extensive application in electrical power systems. The development of high-pressure compression hydrogen preservation is among the biggest breakthroughs. For preserving hydrogen at temperatures as high as 700 bar, novel composite substances are under development, which include carbon fiber-reinforced tanks. Compared to conventional steel tanks, these vehicles are stronger and lighter due to their superior strength-to-weight ratios. Furthermore, improvements in valve and sealing technology have raised the security and dependability of high-pressure storing. Notwithstanding these advancements, there are still issues to be resolved, such as the high energy cost of compression and worries about the material's long-term resilience to numerous stress cycles [10]. Notable advancements have also been made in liquid hydrogen storage, mainly in the creation of sophisticated insulating materials that limit heat transfer and lower boil-off losses. Improved cryogenic storage device designs, like vacuum-insulated canisters and multilayered insulation, have allowed for more effective gaseous hydrogen preservation. To enhance the capacity of storage and minimize energy losses, engineers are also investigating hybrid storage techniques that mix compressed and liquid hydrogen techniques. The energy-intensive liquefaction procedure is still very difficult to do, though, because it is expensive and difficult to keep hydrogen at freezing temperatures. A lot of progress has also been made in the field of metal hydrides and other solid-state hydrogen preservation technologies. Scientists are trying to create substances that are smaller, more effective, and more easily able to take in and let out hydrogen. Metallic hydrides' ability to store hydrogen while requiring less energy to release it has been demonstrated by experimental metal alloys and nanostructured substances. For instance, complex hydrides that are also chemical materials for storing hydrogen are currently being investigated for their ability to multi vectored energy hubs (MV-EHs) store hydrogen at reduced temperatures and pressures, whereas magnesium-based hydroxides are being investigated for their enormous hydrogen capability. These structures still have issues with cost, deterioration of the materials, and sluggish rates of absorption of hydrogen and adsorption. Furthermore, chemical hydrogen storage solutions, such as organic liquids, ammonia, and formic acid, are becoming more and more popular. These substances can chemically store hydrogen and release it when necessary. These systems have the benefit of storing hydrogen at room temperature and pressure; the difficulty is in effectively accelerating the exhalation of hydrogen from the storage medium [11]. Despite significant advancements in hydrogen storage technology, several obstacles still need to be overcome. Adoption is nevertheless hampered by high costs, energy inefficiency, material deterioration, and safety issues. Since better hydrogen storage options will be essential to the larger energy shift and the eventual incorporation of gasoline into MV-EHs alongside alternative renewable energies, continued studies and developments are essential to eliminating these obstacles [12]. Applications: MV-EHs could undergo a radical change thanks to hydrogen storage technology, which would increase sustainability, resilience, and flexibility. With the increasing integration of intermittent forms of renewable energy such as wind and sun power, battery storage for hydrogen can assist in maintaining grid stability, rebalancing demand and supply of energy, and guaranteeing consistent power delivery in. When used with MV-EHs, each type of hydrogen storage—compressed, liquid, and solid-state—offers special benefits. Integrating compressed hydrogen vaults into MV-EHs is one of the easiest approaches. Through the process of electrolysis, surplus electricity from energy sources that are renewable can be converted into hydrogen, which is subsequently held in high-pressure containers. Hydrogen can be utilized as a storage medium for excess electricity produced by renewable sources. Fuel cell technologies or combustible turbines can be used to convert the conserved hydrogen into usable electricity at times of high energy demand or low production from renewable sources, giving MV-EHs a dependable backup power supply. Compressed hydrogen's high-pressure storage capability makes it perfect for use in the microgrids and distributed energy sources, where quick deployment and ample room are crucial [13]. Larger-scale solutions within MV-EHs, including centralized storage structures supporting huge manufacturing sectors or utility-scale renewable energy assignments, are better suited for hydrogen in liquid state storage. Liquid hydrogen holding allows for the storage of large amounts of energy in a relatively tiny region because of its increased energy density. This is especially helpful for MV-EHs that supply high-energy businesses or areas where the generation of renewable energy varies seasonally. The ease of storing and transporting liquid hydrogen enables the deployment of storage for hydrogen hubs that can serve numerous MV-EHs over a larger region. Sustaining the cryogenic conditions necessary for hydrogen in liquid state storage, which might result in energy losses, is still difficult. For MV-EHs that place a high priority on long-term energy availability and safety, solid-state hydrogen stores and metal hydrides present an alluring alternative. Those methods lessen the possibility of leaks or explosions by storing hydrogen inside the crystal structure of metals. Long-duration reserve of energy can be added to MV-EHs through the integration of solid-state storage for hydrogen devices, guaranteeing energy availability during protracted times of low renewable energy supply. Because of this, they are ideal for MV-EHs in isolated locations or near vital infrastructure, where dependability and security are of utmost importance. Because of their expensive prices and sluggish rates of hydrogen release, mineral hydrides are not yet commonly utilized in MV-EHs; however, future developments may make them more feasible [14]. There are numerous important advantages to MV-EHs that incorporate hydrogen storage systems. By making load balancing possible, they help MV-EHs better control the unpredictability of green energy sources. The capacity of hydrogen batteries to supply additional power during blackouts or unstable times improves grid resilience as well. Hydrogen is a versatile energetic carrier within MV-EHs because of its versatility, which enables it to be employed for a wide range of applications, including industrial operations and the production of electricity. Hydrogen storage technology incorporation into MV-EHs can greatly improve grid environmental responsibility, range of motion, and reliability. MV-EHs can better control variations in supply and demand for electricity by storing solar power in the structure of hydrogen, resulting in a more stable and dependable energy infrastructure.

4.4 Power-to-Gas (P2G) Technologies

An inventive way to transform excess electrical energy—especially from renewable sources—into gaseous substances like methane and hydrogen is using P2G devices. P2G's main goal is to make it possible to store extra renewable energy and then utilize it later when consumption is higher or production is lower. This will improve grid reliability and efficacy in systems such as MV-EHs. Methanation and electrification are the two main techniques underlying P2G technology. The first and most important stage in the P2G procedure is electrification. Employing a source of electricity, water (H_2O) is split into hydrogen (H_3), and oxygen (O_3). An electrolyzer, which is particularly used for this procedure, is generally

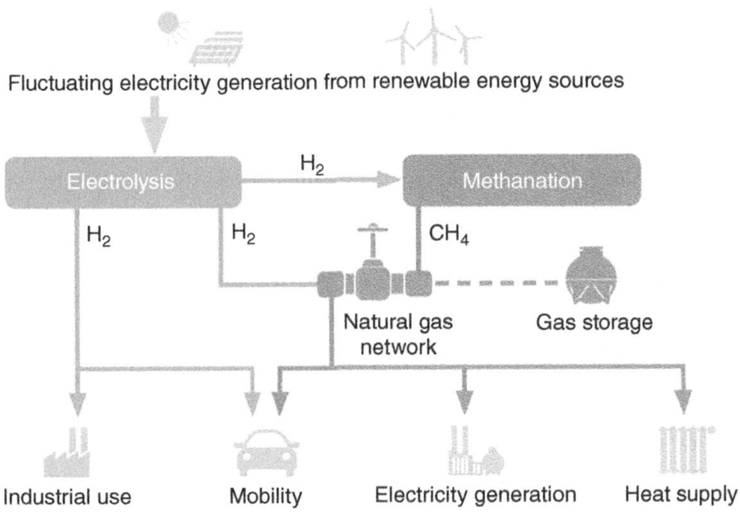

Figure 4.2 Power-to-gas (P2G) technologies. *Source:* Adapted from [11].

classified into three types: solid oxide, proton exchange membrane (PEM), and alkaline. The temperature ranges, prices, and operational efficiency of each variety vary. Because of their effectiveness and flexibility in handling variable energy inputs from renewable energies like sunlight and wind energy, alkaline and PEM electrolyzers becoming more frequently utilized. Electrolysis produces hydrogen, which can be preserved and utilized as a fuel or electricity carrier in MV-EHs or other industries like public transportation [15] (Figure 4.2).

The second step of the P2G mechanism is methanation, in which carbon dioxide (CO_2) and hydrogen generated from electrolysis are mixed to create synthesized methane (CH_4) using a process called the Sabatier process. The methane that results can be saved, utilized to generate energy, or heated. It can also be immediately pumped into the current gas pipeline grid. Methanation makes it possible for P2G to be integrated into the larger natural gas network and offers a practical means of long-term energy storage. P2G technologies are crucial for improving the sustainability and dependability of MV-EHs because they provide a versatile and scalable method of turning energy from renewable sources into accessible gas.

4.4.1 System Components

The authority to gas (P2G) innovations relies on a number of key sections that work in tandem to change extra electrical electricity into liquid fuels like gasoline or created methane. The sections include the electrolyzer, gas compressors, methanation reactors, CO_2 recording system, and storage units, every year performing a significant function in the general P2G management. The electrolytic device is the P2G technique's central constituent. Through the process of electrolysis, it divides particles of water into hydrogen (H_3) and oxygen (O_3), carrying out the vital function of transforming electrical energy toward hydrogen. An external lighting source, such as one derived from sunlight or wind energy, is necessary for the electrolyzer. The efficacy and the equipment are affected by the type of electrolyzer that is used—solid oxide, PEM, or alkaline [16]. After hydrogen is generated, the gas compressor is utilized. Since hydrogen has a low density, decompression is necessary for effective storage or additional treatment. The water-based gas's pressure is raised by the gas compressor. This was allowing it to be transported to other system components or stored in high-pressure drums. The methanation furnace is used if producing synthesized methane represents the intended result. This reactor helps to manufacture methane (CH_4) by combining hydrogen and collected carbon dioxide (CO_2) in the Sabatier reaction. Inhibitors in the reactor accelerate the chemical reactions and guarantee that hydrogen and CO_2 are efficiently converted to methane. These catalysts are usually nickel-based. One of the most important reactants in the methanation manipulation, carbon dioxide, must be obtained via the CO_2 capture system. By reusing discarded CO_2, it absorbs CO_2 from emissions from industries, biogas plants, or the natural environment directly. This increases the sustainability of the P2G equipment [17]. Lastly, storage facilities are essential for storing the generated methane or hydrogen. Although methane can be organized in gas storage sites or injected into the gas supply system for later use, hydrogen is frequently kept in high-pressure storage vessels or in a liquid state. These parts work together to effectively convert and, store solar energy while utilizing it in P2G systems. As a result, they are essential for improving grid flexibility and enabling environmentally friendly energy alternatives in MV-EHs.

4.4.2 Integration with Power Systems

P2G methods, especially in MV-EHs, provide a flexible way to incorporate renewable energy into current power networks. P2G can solve the issues of sustainable energy interruption and enhance grid stability by turning excess green power into hydrogen or manufactured methane. Several crucial tactics can be used to integrate P2G into MV-EHs [18]. P2G first makes the storage of energy possible, which is essential for energy hubs with a large percentage of energy generated from renewable sources. P2G has the capability to electrolyze surplus electricity produced by energy generated from renewable resources, such as wind and solar power, into gasoline when consumption is low. When consumption of energy is high, such hydrogen is capable of being stored and subsequently transformed back into electricity utilizing gasoline turbines or fuel cells, assisting in balancing the energy supply and demands in MV-EHs. Because of its flexibility, the power supply is more dependable and less dependent on energy from fossil fuels. Furthermore, by using the methanation method to produce synthetic methane, P2G can aid in the carbon reductions of the petroleum-based energy infrastructure [19]. This methane can take the place of conventional gas like natural gas in manufacturing, production of electricity, and heating once it is integrated into the current instinctual gas infrastructures. By using this green gas to run gas-powered generators, however, MV-EHs can lower greenhouse gases while preserving grid stability. Additionally, P2G technologies can be utilized in MV-EHs as tools for grid balancing, which helps control voltage and frequency variations brought on by the unpredictability of renewable energy sources. P2G can balance the grid in real-time, enhancing the general strength of MV-EHs, either by producing available for dispatch generation from stored protons or methane or by offering a configurable load (that is, during the electrolytic processes). Lastly, P2G incorporation with MV-EHs strengthens the connection between the gas, heating, and the generation sectors through sector coupling. This multipurpose position gives grid operators greater flexibility to administer supplies of energy across diverse energy subdomains while enabling excess clean power to be used in many sectors, optimizing its utilization and minimizing waste [20]. P2G technology gives MV-EHs improved energy storage, segment coupling, grid stability, and decarbonization routes, which makes them a crucial part of the power structures of the future that are safe.

4.5 Role of Hydrogen in Sustainable MV-EHs

4.5.1 Environmental Impact

In moderate-voltage power systems, deuterium is essential for cutting pollution and improving resilience (MV-EHs). By substituting fossil fuels in a variety of software, the element hydrogen, an alternative energy carrier, can dramatically reduce the ecological impacts of electrical grids and aid in the carbon neutrality of the electricity sector (Figure 4.3).

When utilized as a fuel, hydrogen has zero emissions, which is one of its biggest environmental advantages. Moreover, hydrogen makes it easier to use green energy sources. Hydrogen could be produced through electrolysis with excess electricity when the production of green electricity surpasses the need. This method maximizes the usage of clean energy by not only storing extra clean electricity but also preventing the reduction of the use of renewable energy [21]. Hydrogen serves as a link between erratic sources of sustainable energy and a steady supply of power in this manner, enabling MV-EHs to run more responsibly. Additionally, hydrogen helps decarbonize industries that are difficult to electrify,

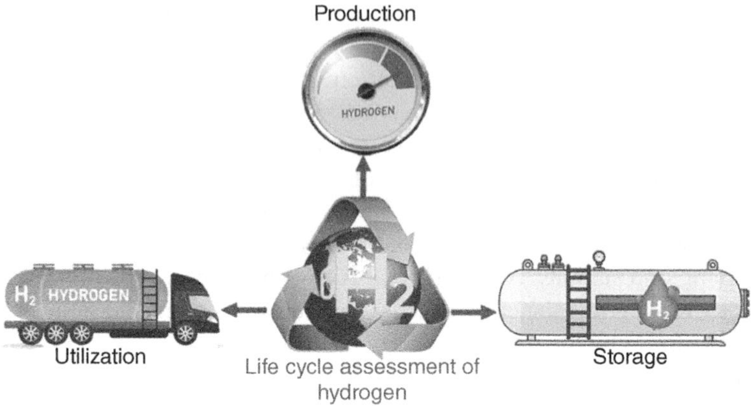

Figure 4.3 Role of hydrogen in sustainable. *Source:* [15] / with permission of Elsevier.

like transportation and heavy manufacturing. MV-EHs contribute to extending the advantages of renewable energy outside the grid by providing helium to industries that have historically relied on fossil fuels, hence lowering the total emissions [22, 23]. Furthermore, hydrogen provides a mechanism to reduce the carbon footprint of the fossil fuel grid by integrating with the current infrastructure using P2G technology. By combining hydrogen with petroleum products or turning protons into synthesis methane, MV-EHs can use what they have now to progressively move forward to a future without fossil fuels. Because it lowers emissions, makes it possible to use renewable energy sources more effectively, and eases the change to a carbon-free energy framework, hydrogen's function in MV-EHs encourages conservation.

4.5.2 Economic Considerations

Based on a financial viewpoint, the addition of P2G and hydrogen generation technologies into MV-EHs brings both potential and obstacles. Although hydrogen has a lot to offer in the long run in terms of reliability of electricity and long-term sustainability, the initial expenses and ongoing financial repercussions need to be carefully considered. The price of electroplating constitutes one of the main financial obstacles to incorporating hydrogen. The machinery needed to create hydrogen from water, known as an electrolyzer, is still somewhat expensive to operate and manufacture, especially when done on a large scale. Furthermore, the cost of creating hydrogen is highly influenced by the selling price of electricity due to the requiring energy process known as electrolysis. The cost of producing hydrogen may go down in areas where energy from renewable sources dominates the power mix, but for the time being, this is still a major obstacle to its widespread use in MV-EHs. The price of developing fresh construction is an additional economic consideration. Scaling up battery storage, delivery systems, refuelling stations, etc. to fulfill MV-EH demands will require large investments. Furthermore, adding hydrogen or manufactured methane through P2G to current gas grids through conversion or retrofitting causes additional financial burden. Nonetheless, attempts are being made to combine natural gas and hydrogen, which would facilitate the shift and lessen the requirement for urgent, extensive infrastructure improvements [24]. There are additionally several financial benefits to incorporating hydrogen into MV-EHs notwithstanding these expenses. Long-term energy preservation with hydrogen possesses the possibility of helping reduce the costs that come with intermittently generated electricity, and that means the requirement for costly backup sources of electricity. Furthermore, hydrogen can lower energy expenditures and boost security of electricity by maintaining the grid and lowering dependency on international fossil fuels. These are significant advantages for areas that are dependent on imported supplies of fossil resources. In addition, P2G techniques should become less expensive as they develop and grow, just like wind and solar electricity technologies have. The commercial viability of gasoline and P2G innovations might additionally be accelerated by incentive programs, assistance, and carbon pricing schemes. Although the beginning expenses of integrating P2G and hydrogen into MV-EHs are substantial, they are an acceptable expenditure for a future with renewable energy due to the long-term benefits to the economy, especially in terms of resource security, reliability in the grid, and lower emissions.

4.5.3 Case Study and Examples

Examples of MV-EHs in the real world that incorporate P2G and hydrogen technology show how these solutions can improve energy resilience and sustainability. These case studies shed light on the efficient integration of hydrogen into electric power plants to promote grid stability, reductions in emissions, and the integration of green energy sources, which is among the biggest P2G facilities worldwide. The facility employs excess renewable energy, mostly from prevailing wind to electrolyze hydrogen. After that, the hydrogen will be utilized as energy for fuel cells, stored, or injected into the nearby gas pipeline grid. Energiepark Mainz ensures that renewable energy is not wasted while aiding in system stabilization by turning surplus clean energy into hydrogen. This system demonstrates how P2G technology can manage changing clean energy inputs inside MV and enable massive energy storage MV-EHs. Another effective P2G connectivity effort is the Hy Balance initiative in Denmark. The Hobro-based project manages excess generated electricity by producing hydrogen. The generated hydrogen is employed in industrial, journeys, and network stabilization.

The system shows how oxygen could possibly be used to reduce carbon footprints in other industries in addition to producing power, improving the energy supply's overall effectiveness and environmental sustainability. A more modest instance is that of the Orkney Islands in Scotland, since the MV-EH system there incorporates hydrogen generation to counteract the excess supply of energy generated from tidal and wind sources. This research shows how hydrogen can be used to manage energy swings and maintain energy security in remote areas with significant renewable concentration. The following instances demonstrate the flexibility of P2G and hydrogen approaches for tackling the issues of climate change, grid balancing, and intermittent renewable energy. The gas hydrogen plays an increasingly important role in building a robust and environmentally conscious energy destiny since increasingly MV-EH systems implement these forms of technology.

4.6 Conclusion

The current section has looked at how P2G and hydrogen technologies can improve the sustainability and effectiveness of MV-EHs. Key findings show how important such innovations are for decarbonizing energy networks, including energy from nature, and to improve grid stability. First, there are several options for storing extra renewable energy using hydrogen preservation innovations, such as metal hydrides, liquid hydrogen, and compression hydrogen. Although each approach has unique benefits and drawbacks, taken as a whole, they offer useful alternatives for controlling variations in energy quantity and demand throughout MV-EHs. Improvements in energy storage technology, including lower costs and increased effectiveness, are making protons a more attractive alternative for massive quantities of energy storage. In addition, P2G techniques allow excess generated electricity to be converted into synthetic methanol or gasoline using the techniques of electrolysis and methanation. This conversion procedure improves the entire adaptability and dependability of the system by supporting storage of electricity and integrating clean energy sources into the current infrastructures. The capacity to incorporate hydrogen into the grid or combine it with gas from the earth illustrates P2G's potential to lower emission levels of greenhouse gases and maximize the use of resources that are renewable. Recognizing and optimizing such technologies requires a strong foundation in modeling and simulation. A range of methodologies, such as techno-economic modeling, optimization-based modeling, and physical system modeling, offer valuable perspectives on the efficiency and financial sustainability of P2G and hydrogen infrastructures. With the use of sophisticated simulation methodologies and real-time monitoring tools, problems like the complicated nature of chemical techniques, changeable sustainable inputs, and incorporation with current facilities are all being handled. The energy industry and long-term prosperity will be impacted more broadly by the merging of P2G and hydrogen technology. The development of ecologically friendly and sustainable energy systems is made possible in large part by these technologies. Hydrogen and P2G technologies facilitate the efficient storage and application of renewable energy sources, thereby mitigating dependence on non-renewable fossil fuels, curbing releases of greenhouse gases, and advancing energy autonomy. This change propels the move toward a future with alternatives to fossil fuels and supports international climate targets. Furthermore, the effective application of these devices can boost economic expansion by generating new sectors and employment possibilities in the infrastructures, storage, and industrialization of hydrogen. It also provides a way to integrate cutting-edge technologies with outdated energy infrastructure.

References

1 Arpino, F., Canale, C., Cortellessa, G. et al. (2024). Green hydrogen for energy storage and natural gas system decarbonization: an Italian case study. *International Journal of Hydrogen Energy* 49: 586–600.

2 Piazza, G., Delfino, F., Bergero, S. et al. (2023). Economic and environmental optimal design of a multi-vector energy hub feeding a local energy community. *Applied Energy* 347: 121259.

3 Zhang, G., Ge, Y., Ye, Z., and Al-Bahrani, M. (2023). Multi-objective planning of energy hub on economic aspects and resources with heat and power sources, energizable, electric vehicle and hydrogen storage system due to uncertainties and demand response. *Journal of Energy Storage* 57: 106160.

4 Liu, Q., Li, W., Zhao, Z., and Jian, G. (2024). Optimal operation of coordinated multi-carrier energy hubs for integrated electricity and gas networks. *Energy* 288: 129800.

5 Breyer, C., Lopez, G., Bogdanov, D., and Laaksonen, P. (2024). The role of electricity-based hydrogen in the emerging power-to-X economy. *International Journal of Hydrogen Energy* 49: 351–359.

6 Parolin, F., Colbertaldo, P., and Campanari, S. (2024). Optimal design of hydrogen delivery infrastructure for multi-sector end uses at regional scale. *International Journal of Hydrogen Energy* 79: 839–849.

7 Papadimitriou, C., Di Somma, M., Charalambous, C. et al. (2023). A comprehensive review of the design and operation optimization of energy hubs and their interaction with the markets and external networks. *Energies* 16 (10): 4018.

8 Samanta, S., Roy, D., Roy, S. et al. (2023). Techno-economic analysis of a fuel-cell driven integrated energy hub for decarbonising transportation. *Renewable and Sustainable Energy Reviews* 179: 113278.
9 Tiwari, S., Singh, J.G., and Garg, A. (2024). A static robust energy management approach for modelling low emission multi-vectored energy hub including emission markets and power-to-gas units. *Energy* 294: 130827.
10 Wang, X., Liu, Y., Liu, C., and Liu, J. (2020). Coordinating energy management for multiple energy hubs: from a transaction perspective. *International Journal of Electrical Power & Energy Systems* 121: 106060.
11 Zhu, F., Fu, J., Zhao, P., and Xie, D. (2020). Robust energy hub optimization with cross-vector demand response. *International Transactions on Electrical Energy Systems* 30 (10): e12559.
12 Seyednouri, S., Safari, A., Quteishat, A. et al. (2024). Stochastic energy management of a multi-microgrid system with battery/supercapacitor energy storages considering demand response and transactive energy. *Renewable Energy Focus* 48: 100531.
13 Tiwari, S. and Singh, J.G. (2024). A cooperation based transactive energy management for networked energy hubs considering improved payoff allocation mechanism. *Sustainable Energy Technologies and Assessments* 65: 103777.
14 Son, Y.-G., Oh, B.-C., Acquah, M.A. et al. (2021). Multi energy system with an associated energy hub: a review. *IEEE Access* 9: 127753–127766.
15 Hosseini, S.H.R., Allahham, A., Walker, S.L., and Taylor, P. (2020). Optimal planning and operation of multi-vector energy networks: a systematic review. *Renewable and Sustainable Energy Reviews* 133: 110216.
16 Arsoon, M.M. and Moghaddas-Tafreshi, S.M. (2022). Resilience-oriented proactive peer to peer multiple energy carriers swapping framework for the partial networked energy hubs. *Sustainable Energy Technologies and Assessments* 53: 102576.
17 Mobasseri, A., Tostado-Véliz, M., Ghadimi, A.A. et al. (2022). Multi-energy microgrid optimal operation with integrated power to gas technology considering uncertainties. *Journal of Cleaner Production* 333: 130174.
18 George-Williams, H., Wade, N., and Carpenter, R. (2022). A probabilistic framework for the techno-economic assessment of smart energy hubs for electric vehicle charging. *Renewable and Sustainable Energy Reviews* 162: 112386.
19 Li, C., Wang, N., Wang, Z. et al. (2022). Energy hub-based optimal planning framework for user-level integrated energy systems: considering synergistic effects under multiple uncertainties. *Applied Energy* 307: 118099.
20 Zhu, D., Yang, B., Liu, Q. et al. (2020). Energy trading in microgrids for synergies among electricity, hydrogen and heat networks. *Applied Energy* 272: 115225.

21 Sun, Y., Zhang, B., Ge, L. et al. (2020). Day-ahead optimization schedule for gas-electric integrated energy system based on second-order cone programming. *CSEE Journal of Power and Energy Systems* 6 (1): 142–151.

22 Garg, A. and Niazi, K. (2023). A robust energy management strategy for multi carrier energy system. *2023 7th International Conference on Computer Applications in Electrical Engineering-Recent Advances (CERA)*, 1–5. IEEE.

23 Ekhtiari, A., Flynn, D., and Syron, E. (2020). Investigation of the multi-point injection of green hydrogen from curtailed renewable power into a gas network. *Energies* 13 (22): 6047.

24 Petkov, I. and Gabrielli, P. (2020). Power-to-hydrogen as seasonal energy storage: an uncertainty analysis for optimal design of low-carbon multi-energy systems. *Applied Energy* 274: 115197.

5

Modeling and Analysis of MV-EHs with Advanced Energy Storage Units

Rengamani Shenbagalakshmi[1], *Jaganathan Subramaniyan*[2], *Govindasamy Ramprakash*[2], *and Veerappan Vengatesan*[2]

[1]*Department of Electrical & Electronics Engineering, Vel Tech Multi Tech Dr. Rangarajan, Dr. Sakunthala Engineering College, Avadi, Chennai, Tamilnadu, India*
[2]*Department of Electrical and Electronics Engineering, SRM TRP Engineering College, Trichy, Tamilnadu, India*

5.1 Introduction

The global consumption of energy has been rising dramatically, leading to the rapid depletion of readily available resources such as oil and coal. Existing systems are increasingly incapable of meeting expanding energy demands, highlighting the necessity for effective and affordable strategies to manage this challenge. Electric vehicles (EVs), energy storage systems (ESS), and distributed generation based on renewable energy sources (RES) like solar and wind have all contributed to the complexity and interconnection of today's electrical power grid networks. In recent decades, decarbonizing the power industry has become a central focus of energy transformation efforts due to growing environmental awareness. RES, including wind, solar, hydro, biomass, and geothermal, offer solutions to minimize energy scarcity and emissions. However, the integration of RES and EVs into the electrical system without adequate ESS poses challenges due to their variability and intermittency, complicating power network operations [1, 2].

ESSs are crucial for balancing supply and demand mismatches, ensuring reliable and consistent power generation from RES. They provide benefits such as peak reduction, supplemental services, and enhanced reliability. Hybrid ESSs, which combine various ESSs, further improve system stability and dependability. To meet global climate targets and reduce fossil fuel dependence, the proportion of power generated from RES must be significantly increased. The current electrical grid faces challenges from high RES penetration, as weather conditions affect RES output, making it difficult to provide consistent power

Energy Management Strategies for Multi-Vectored Energy Hubs to Achieve Low Carbon Societies, First Edition.
Edited by Shubham Tiwari, Jai Govind Singh, Sivaraman Palanisamy, Sharmeela Chenniappan, Rupendra Kumar Pachauri, and Sanjeevikumar Padmanaban.

services [3]. Interconversion technologies, like power-to-gas (P2G) systems, address these challenges by converting surplus renewable electricity into chemical fuels, enhancing energy storage capacity, and connecting gas and electricity networks [4–6].

Integrating traditionally separate energy infrastructures, such as gas, heat, and transportation networks, can provide flexibility to the electricity network and mitigate issues from high RES penetration. Multi-vector energy systems (MES) facilitate this integration by coupling various energy carriers to create an efficient and resilient energy system. The Energy hub (EH) approach, a key methodology for designing MES, improves planning by converting, conditioning, and storing multiple energy carriers, thereby enhancing system efficiency and resource utilization [7, 8]. The organization of this chapter is as follows: Section 5.2 presents the evolution of energy hubs, their components, benefits, and classification; Section 5.3 gives the overview of multi-vector energy hubs (MV-EHs); Section 5.4 illustrates the advanced energy storage units in MV-EHs with their overview, merits and demerits, and challenges faced while incorporating them in the MV-EHs; Section 5.5 describes the mathematical modeling of MV-EHs; and finally, conclusions and references are given.

5.2 Evolution of Energy Hubs, Their Components, Benefits, and Classification

5.2.1 Energy Hubs: Basic Definition and Structure

Various studies have offered distinct models and definitions of the EH notion up to this point [9–14]. A collection of interfaces connecting different energy systems—each with its own conversion devices, networks for transmission and distribution, and customers—makes up an energy hub. The main idea of this definition is that, in accordance with the physics concept of energy conservation, every system that is referred to as a "generating unit" can be thought of as a power-converting device. However, generating units are not addressed in the definition that is being presented. Nonetheless, it is evident that an energy hub with numerous energy carrier inputs and outputs can transform and store various energy types. These characteristics also make the term "multi-carrier energy system" applicable to an energy hub.

As per the reference [15], an energy hub is a simulated box that is depicted in Figure 5.1 and is used to convert load energy demand into energy carriers.

The energy storage units, conversion units, and input and output units make up the EH. It might therefore be referred to as the expansion or generalization of power system network nodes [16, 17]. A rough drawing of an EH structure

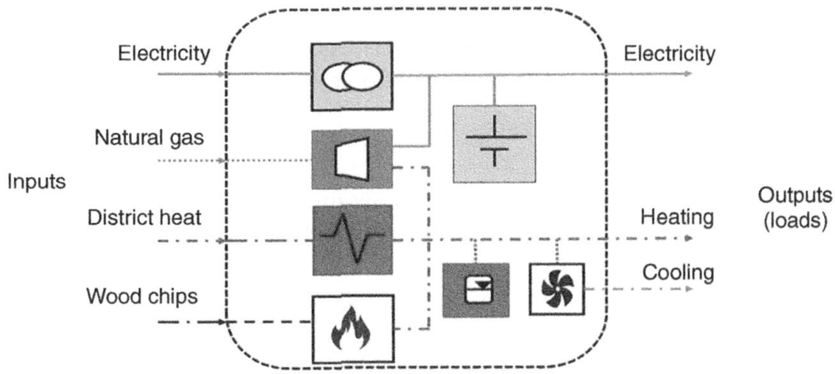

Figure 5.1 Image of an energy hub.

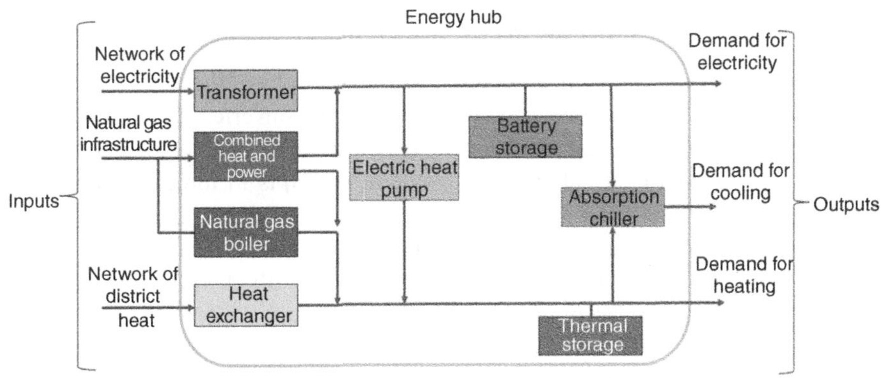

Figure 5.2 Energy hub detailed structure. *Source:* Adapted from [18].

with many energy carriers is shown in Figure 5.2. The structure's inputs include district heat networks, electricity, and natural gas. It also comprises storage devices like thermal storage units (TS) and battery storage units (BS), as well as energy-converting elements like heat exchangers, electric heat pumps, combined heat and power (CHPs), transformers (TR), and natural gas boilers (NGB). Lastly, it has outputs that must be met, like heat, cool, and electricity [18, 19].

In order to deliver specific energy services at the output, the energy carriers that the EH consumes at the input are transformed into different energy vectors [18]. Moreover, petroleum-based products and coal can be utilized as inputs. Municipal garbage, biomass, and hydrogen are further input energy carrier choices that could be investigated. If thoroughly explored, chemical reactants, waste, water, and air, in addition to the previously listed energy carriers, could all be employed as energy [20, 21]. The EH does not limit the size of the system it models in its development process. It enables the inclusion of any random number selection of energy products and carriers with sufficient resilience for system modeling [22].

In theory, EH falls under one of the four categories listed below:

Energy vectors at the input including electricity, hydrogen, water, gas, solar and wind energy, and fossil fuels.

Converters: Chillers, combined heat and power units, heat pumps, boilers, fuel cells, and electrolyzers (P2G) are examples of energy resources that are converted with the use of converters.

Systems for storing or preserving excess energy are known as ESSs; examples include flywheels, batteries, hot water tanks, hydrogen tanks, and heat storage.

Output: The energy requirements of the hub for end users (gas, water, hydrogen, heat, and cooling) [23].

5.2.2 The Background of the EH Methodology

The concept of an energy hub, which is essentially the integration of various energy sources with pertinent interface devices and converters in the operation and expansion plans of energy systems, was initially introduced in Ref. [24]. In fact, this concept came about as a result of attempts to forecast the energy landscape in the context of the growing use of distributed generation units and the developments made in the technologies of storage for various energy systems. After that, additional research on the energy hub idea was progressively conducted in order to accomplish a number of goals. PSP research analysis classifies the framework described in Ref. [25] as an early study that investigates the use of the energy hub notion in cogeneration or tri-generation system design. Accordingly, among the other subjects covered by the pertinent early research are the assessment of fuel cell systems [26] and the integration of energy sources into vehicle models within the energy hub paradigm [27]. However, the concept of multi-carrier energy systems has also been used by a number of other studies.

The EH concept was initially applied in published works by the Swiss municipal utility Regionalwerke AG Baden, according to an examination of an article that was published in an *IEEE* journal in 2007 [28]. With synthetic natural gas (SNG) as the output and electricity, heating, wood chips, steaming, and nitrogen as the inputs, this utility has tried to build an energy recycling unit. The fundamental component of that utility's energy plan can be regarded as providing biomass fuel to a central system that supplies both thermal and electrical needs, as shown in Figure 5.3.

The generation of SNG, which is converted into electrical power by the CHP incorporated into the main unit, can either directly or indirectly satisfy the

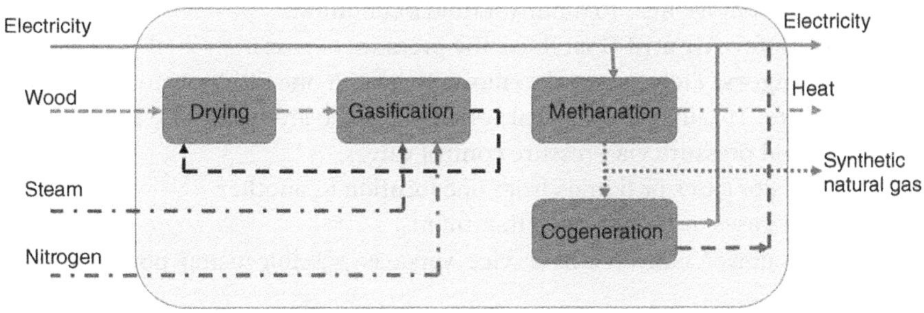

Figure 5.3 Overview of energy hubs.

electrical demands in the planned unit's output, as illustrated in Figure 5.3. Similar to this, the proposed system has two options for addressing thermal load simultaneously: (1) passive supply through the SNG, which is heated by the CHP, and (2) direct supply through the local district heating network, which absorbs waste heat from the gasification and methanation processes. Consequently, one could argue that multi-carrier energy systems were already in place prior to the introduction of the EH idea. It is crucial to remember that the EH idea can only work if specific connections between different energy forms are considered. The proposal can be implemented without the planned system of the regional utility Regionalwerke AG Baden.

5.2.3 Elements of Energy Hubs

Three primary parts make up an energy hub: ESSs, converters, and direct connections. While it is outside the purview of this section to examine the apparatus and infrastructure in these categories in depth, it is covered at great length in previous studies [29–32]. Here, we concentrate on classifying energy hub converters into two groups: switching converters and modifying converters, according to their input and output architectures. An overview of this suggested category for energy hub converters may be found below.

5.2.3.1 Adapting Converters

In order to tailor energy carriers' characteristics and quality to a given customer's request, converter modification is necessary. These converters, which are frequently used for conditioning, alter the input carriers to get the output meeting the required criteria. Only the characteristics or attributes of energy change from input to output; the energy type does not change. Here are some instances of altering converters:

Condensers: Remove heat to modify a fluid's condition.
Compressors for Air and Gas: Raise the pressure of gases.
Heat Exchangers: They are used to move heat from one fluid to another.
Transformers: Modify the electrical systems' voltage levels.
Control fluid pressure via pressure control valves.
Pumps: Transfer gases or liquids from one location to another.
Filters: Clear gases or liquids of contaminants.
DC to AC power conversion or vice versa is possible using power-electric inverters.

5.2.3.2 Converters for Switching

In order to achieve the desired kind, amount, and qualities of the required energy at the output, switching converters change the form of the energy carriers that are input. Switching converters alters the kind of energy carrier, as opposed to altering the converters. This group consists of

Electric Motors: Electrical energy can be transformed into mechanical energy using electric motors.
Generators: Produce electricity from mechanical energy.
Fuel Cells: Produce electricity from the chemical energy of fuels.
Solar Panels: Produce power from the sun.
Utilize wind turbines to generate electricity from wind energy.
Thermal Systems: These systems transform thermal energy into mechanical or electrical energy.
Gas Systems: Produce heat or power from gas energy.
Energy hubs can be better built to effectively manage and optimize different sources of energy to satisfy the various demands of customers by comprehending and classifying these converters. Understanding how various technologies inside an energy hub cooperate to regulate energy flows may be done in an organized manner, thanks to this division of energy hub converters into modifying and switching converters. Switching converters change energy carriers into new forms to suit particular purposes, whereas moderating converters alter the characteristics of energy carriers without altering their type. When creating energy hubs that are effective, adaptable, and able to combine different energy sources and technologies, this distinction is essential.

5.2.4 Benefits of Energy Hubs

Through the integration of various energy carriers and the optimization of their interactions to improve efficiency, sustainability, and resilience, energy hubs provide a revolutionary approach to energy management. Energy hubs provide a host of advantages in an ever-more-complex and dynamic energy market, meeting both present and future needs. The main benefits of energy hubs are covered in detail in this extensive discussion, with special attention to how they affect scalability, data utilization, market participation, resilience, efficiency, flexibility, cost savings, carbon emissions reduction, and integration of electric vehicles.

5.2.4.1 Management of Incorporated Energy

The efficient control of multiple energy sources, such as heat, gas, electricity, and even hydrogen, is made possible by energy hubs. Energy flows can be optimized with the help of this integrated strategy, guaranteeing that energy is used where it is most valuable and efficient. For instance, extra energy from renewable sources can be stored in batteries for later use or transformed into heat for district heating systems. Energy hubs reduce waste and enhance system performance by coordinating the usage of various energy carriers.

5.2.4.2 Enhanced Effectiveness

Energy hubs have the ability to improve energy system efficiency, which is one of their biggest advantages. Conventional energy systems frequently function as isolated units with little interaction between various energy sources. Inefficiencies include the loss of extra renewable energy during times of low demand, which may result from this isolation. On the other hand, energy hubs make it easier to convert, store, and distribute energy in a way that maximizes efficiency. For example, CHP systems can simultaneously produce electricity and useful heat, significantly increasing fuel utilization, as opposed to separate heating and electrical power generation.

5.2.4.3 Improved Adaptability

Unmatched flexibility in adapting to shifting supply and demand conditions is made possible by the integration of multiple energy carriers in energy hubs. In order to balance the erratic nature of nonconventional energy sources, including wind and solar, this flexibility is very important. Energy hubs can store extra energy in batteries or transform it into other forms, like heat or hydrogen, during times when renewable energy production is at its peak. In contrast, these energy forms that have been stored can be used directly to meet demand or transformed back into electricity during times when the renewable energy supply is low. This dynamic flexibility contributes to grid stabilization and a stable energy supply.

5.2.4.4 Savings on Costs

Energy hubs can result in significant cost savings by maximizing the use of already-existing infrastructure and increasing efficiency. Energy hubs cut operational expenses by minimizing energy loss and maximizing the use of available local resources. Energy storage and conversion also enable more efficient use of RES, lowering dependency on pricey fossil fuels and minimizing price fluctuation. By maximizing the efficiency of existing assets, energy hubs can also postpone or completely remove the need for expensive investments in new generation and transmission infrastructure.

5.2.4.5 Diminished Emissions of Carbon

Energy hubs' optimization powers are essential for cutting carbon emissions and moving the energy system closer to a low-carbon one. Energy hubs can lessen reliance on fossil fuels by integrating and optimizing RES. When renewable output is scarce, excess electricity from solar or wind power can be used to create hydrogen through electrolysis or store it as heat in thermal storage devices. This lowers greenhouse gas emissions while simultaneously increasing the usage of renewable energy, which makes it a more dependable and practical choice.

5.2.4.6 Adaptability and Dependability

Energy hubs provide various channels for energy delivery and storage, which increases the resilience and reliability of energy systems. Because of this redundancy, the total energy system is less vulnerable to disruptions like equipment failures or natural disasters. For instance, energy saved in batteries or as hydrogen can be used to sustain vital services in the event that the electrical grid fails. Furthermore, local generation and storage capacities strengthen communities' resilience to shocks and aid in their recovery, thereby augmenting energy security.

5.2.4.7 Local Production and Storage of Energy

To lessen reliance on centralized power plants and increase local energy security, energy hubs facilitate the integration of local energy generation and storage. Energy hubs could be equipped with sources of renewable energy like biomass generators, wind turbines, and rooftop solar panels to offer clean and sustainable energy to the local community. This ability is further improved by local storage options, such as thermal storage and batteries, which guarantee that energy produced locally can be kept and utilized when needed. In addition to boosting resilience, this decentralization of energy generation and storage gives communities more authority over their energy requirements.

5.2.4.8 Assistance with Electric Cars (EVs)

Energy systems face opportunities as well as challenges with the growing popularity of electric vehicles. Energy hubs enable vehicle-to-grid (V2G) capabilities, demand management, and effective charging options, all of which help to integrate EVs. When there is a spike in demand, EVs can function as mobile energy storage devices, storing extra renewable energy and reintroducing it to the grid. Energy is exchanged both ways, which improves grid stability and maximizes the usage of renewable resources. Energy hubs can also schedule EV charging to coincide with times when there is less demand on the grid or when renewable energy production is at its highest.

5.2.4.9 Reliability in Scale

Energy hubs can be easily scaled to meet different needs and situations because of their inherent scalability. They can be put into practice at various sizes, ranging from big urban or regional networks to tiny community-level systems. The scalability of energy hubs guarantees their applicability in diverse scenarios, encompassing both densely populated cities and rural areas. A major metropolis might create a sophisticated energy hub that controls gas, heat, and electricity for several areas, while a tiny village might install an energy hub that combines nearby biomass generators and solar panels.

5.2.4.10 Information and Tracking

Real-time data gathering and monitoring are incorporated into advanced energy hubs, which are crucial for improving decision-making and optimizing energy flows. Energy generation, use, and storage can be continuously monitored thanks to smart meters, sensors, and data analytics software. Predictive maintenance is made possible by this real-time data, which guarantees that equipment is maintained before malfunctions arise. Further enhancing the effectiveness and dependability of energy systems are improved demand forecasting and resource allocation made possible by data-driven insights. Energy hubs may continually optimize their operations and adjust to changing conditions by utilizing data.

5.2.4.11 Engagement in the Energy Market

Energy hubs improve prosumers' and consumers' access to the energy markets. Energy producers and consumers, or prosumers, have two options: they can take part in demand response programs or sell extra energy back to the grid. By offering the required management systems and infrastructure, energy hubs enable these connections. An energy hub, for instance, can combine the output of several small-scale renewable energy producers to form a virtual power plant that can engage in wholesale energy markets. Prosumers profit monetarily from this aggregation, which also helps integrate RES and maintain grid stability.

5.2.4.12 Support for Regulation and Policy

Policy and regulatory frameworks that promote clean energy and lower carbon footprints can be supported by energy hubs. Through the implementation of effective and sustainable energy management strategies, energy hubs offer concrete illustrations of how integrated energy systems can fulfill policy objectives. The effectiveness of energy hubs can be used by legislators to support and create laws that promote the implementation of comparable systems. Energy hubs can also offer insightful data and information that helps shape policy, guaranteeing that laws are grounded in actual performance and results.

Energy hubs offer a plethora of extensive advantages, including but not limited to greater productivity, greater adaptability, lower costs, decreased carbon emissions, enhanced resilience, localized energy production and storage, assistance for electric vehicles, scalability, data utilization, market involvement, and policy support. Energy hubs will be essential to developing resilient energy systems that are both efficient and sustainable and that satisfy the demands of contemporary civilization as the energy landscape changes. Communities and countries can take steps toward a more connected, dependable, and ecologically responsible energy future by adopting the idea of energy hubs.

5.3 Multi-Vector Energy Hubs

The literature that is being produced at the moment contains a wide variety of definitions for MES. The energy sectors—such as gases, electricity, water, the cooling process, temperatures, transportation, etc.—will communicate progressively at different points to allow the provision of additional products and lucrative revenue sources that lead to a more sophisticated, flexible, and carbon-emission-free energy system, according to Ref. [33]. An MES is an innovative integration that describes this interaction. As stated in Ref. [34], an MES is a system where different energy components—such as gases, temperatures, power, cooling processes, transportation, etc.—interact optimally at different points to produce favorable methodological, financial, and environmental efficacy. This is significantly dissimilar from traditional energy systems, in which each sector is managed independently. In order to decarbonize various aspects of the energy system while offering a system with greater adaptability through the inclusion of resilient and diverse energy sources, Ref. [35] stated their own version of MES as the technique of integrating and implementing expanded energy systems and services in a profitable manner.

An integrated energy system (MES) often unifies the planning, scheduling, and operation of integrated energy systems across various channels and topographical areas to offer a more stable and profitable energy service with minimal to no environmental impact [36, 37]. A comprehensive depiction of a potential future MES is provided in Figure 5.4.

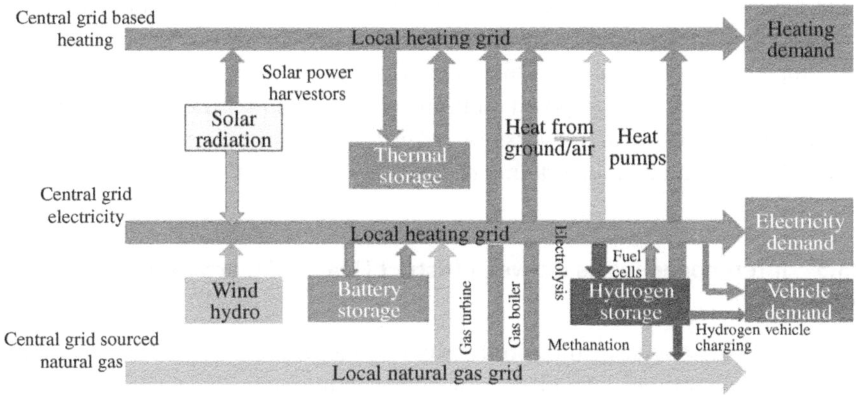

Figure 5.4 Multi-vector energy hub.

5.3.1 Different Types of Interactions and Interdependencies Among Energy Vectors

The scope of potential for integrating various energy infrastructures has expanded with the use of the MES technique in energy system management. The MES technique is acknowledged as the answer for next-generation energy systems, according to a number of recently conducted studies. Only with this approach will it be possible to transition to future energy systems that prioritize the shift to zero-carbon emissions: improving the incorporation of RES to reduce the electric sector's excessive reliance on fossil fuels, decarbonizing the gas and transportation sectors through P2G technologies, and connecting the gas heating and electricity networks through CHP. The following categories can be used to group the different types of energy vector exchanges that take place in an MES:

- Networks of power and natural gas.
- Networks for district electricity and heating.
- District heating, natural gas, and electricity networks.

5.3.2 Interdependencies Between Natural Gas and Electricity Networks

The increasing usage of modern machinery like CHP and gas-fired generators today makes it possible for natural gas and electricity networks to interact. These gadgets make it possible for the two networks to couple. The erratic nature of environmentally friendly generation in the electrical grid is compensated for by natural gas. Reference [38] examined how increasing the integration of sustainable energy sources into natural gas and electrical networks can enhance system flexibility.

It has been noted that the electrical network is significantly more flexible when power generation from non-conventional sources of energy is converted into H_2 and supplementary forms of vectors of energies using contemporary techniques. The authors in Ref. [39] proposed an innovative solution for the incorporation of natural gas and electrical networks in order to address the complications associated with the prediction uncertainty of wind power generation.

5.3.3 Interdependencies Between District Heat and Electricity Networks

Either the district stage or the community stage is where the district's electricity and heat collaborations take place. Cogeneration methods, like CHP, are used to yield electricity and heat concurrently, resulting in this interaction [40]. Furthermore, as the globe moves closer to net zero, electrifying the heating system is an excellent way to reduce carbon emissions by switching from the existing practice of using gas boilers, which produce emissions, to the utilization of electricity heating techniques like heat pumps. Reference [41] looked into the design and operation of cogeneration, paying close attention to how the district's electrical and heat networks were integrated. The study showed that using CHP to simultaneously generate heat and electricity resulted in a substantial decrease in greenhouse gas emissions and overall cost savings. The authors in Ref. [42] examined how demand response might be used to investigate potential connections between district heating and electrical networks, as well as the potential savings on household energy expenses through the application of rational and cost-driven control concepts. The outcomes of the simulation demonstrated that the suggested intelligent control strategy outperformed the traditional strategy in terms of reduced variable energy expenses.

5.3.4 Interdependencies Between Natural Gas, District Heating, and Electricity Networks

Systems for cogeneration, such as CHP, serve as a bridge between the networks for district heating, electricity, and natural gas. CHP units use natural gas as their input to simultaneously produce heat and electricity, supplying the heating network or end consumers with heating and the electricity network.

Reference [43] employed a Sankey diagram to illustrate the many technological penetrations, including CHP and electric heat pumps, that are utilized to describe the flow of energy that occurs in interconnected electricity, heating, and natural gas networks. It also examined how various technologies affected each network's steady-state operational characteristics.

5.3.5 Advantages of MV-EHs

There are several benefits to MES over traditional independent energy systems. As shown in Table 5.1, these are categorized according to their advantages in terms of technology, economy, and the environment.

5.3.6 Challenges in MV-EHs

5.3.6.1 Technical Difficulties

i) **Integration Difficulty:**
   - **System Collaboration:** Making certain that various energy systems and technologies work together.
   - **Coordination:** Handling how several energy vectors interact and depend on one another.

ii) **Data Management:**
   - **Data Variety and Volume:** Managing a wide range of data types and volumes from several sources.
   - **Real-Time Monitoring and Control:** Putting mechanisms in place for processing data and making decisions in real time.

iii) **Infrastructure Needs:**
   - **Modernizing Present Infrastructure:** Adapting present infrastructure to facilitate multi-vector integration.
   - **Grid Stability:** Keeping several interconnected systems stable and reliable

iv) **Energy Storage:**
   - **Technological Development:** Improving storage technologies to efficiently manage various energy vectors.
   - **Cost and Efficiency:** Efficient and affordable energy storage options should be balanced.

v) **Cybersecurity:**
   - **Threat Detection:** Recognizing and addressing cybersecurity risks in tandem with interconnected systems.
   - **Protection Measures:** To safeguard infrastructure and data, strong cybersecurity measures should be implemented.

5.3.6.2 The Financial Challenges

i) **Costs of Investment:**
   - **Initial Capital:** The development of infrastructure and technology necessitates a large initial expenditure.
   - **Funding Sources:** Getting money and investments for initiatives using multiple vectors.

Table 5.1 Advantages of multi-vector energy hubs.

Category	Advantages	Description
Technical	Improved efficiency in energy use	Minimized waste and maximized usage of resources
	Enhanced dependability and adaptability	Disaster resilience and redundancy
	Grid stability and management with renewable integration	Improved use of sporadic renewable energy sources
		Regulating frequency and load balancing
	Industry pairing	Integration across sectors and electrification
	Technologies for energy storage	Many storage choices and optimization
	Innovation and the progress of technology	Propels R&D, storage innovation, and smart grids
	City energy infrastructure	Development of smart cities and urban sustainability
	Energy strategy planning	Combining scenario analysis and integrated planning
Economic	Savings on costs	Improved capital utilization and decreased operating expenses
	Opportunities in the market	Fresh market prospects for energy services and trading
	Benefits of regulation and policy	Encourages broad energy policy and adaptable rules
	Possibilities for education and training	Transitioning from one workforce to another and developing skills
	Extended strategic advantages	Infrastructure that is future-proof and supports sustainable growth
Environmental	Decreased emissions	Reduced pollution and emissions of greenhouse gases
	Managed resources in a sustainable way	Economical use of renewable and local resources
	Social and community advantages	Increased access to energy and resilience in communities
	Customer involvement and options	More options for energy and consumer empowerment
	International cooperation	International collaboration and the creation of standards

ii) **Dynamics of the Market:**
- **Regulatory Uncertainty:** Handling divergent laws and guidelines in several areas.
- **Market Structure:** Adjusting to multi-vector integration's effects on the dynamics and structure of the market.

iii) **Cost Allocation:**
- **Cost Sharing:** Establishing a just price for shared services and infrastructure among stakeholders.
- **Economic Viability:** Preserving multi-vector systems' competitiveness and economic sustainability.

iv) **Revenue Models:**
- **New Business Models:** Crafting fresh business plans and revenue streams for energy systems with several vectors.
- **Incentive Mechanisms:** Developing rewards to promote multi-vector system adoption and investment.

5.3.6.3 Social and Environmental Challenges

i) **Durability:**
- **Sustainable Resource Management:** Ensuring sustainable resource management while reducing environmental effects.
- **Life Cycle Impact:** Evaluating and reducing multi-vector systems' environmental impact over their whole life cycle.

ii) **Acceptance in Society:**
- **Public Knowledge:** Increasing the general public's knowledge and acceptance of MES.
- **Community Engagement:** Including stakeholders and communities in the process of planning and carrying out.

iii) **Alignment of Regulations and Policies:**
- **Policy Coherence:** coordinating laws and policies among various industries and legal systems.
- **Standardization:** Creating and implementing multi-vector integration best practices and standards.

iv) **Accessibility and Equity:**
- **Electricity Access:** Guaranteeing that all communities, particularly underprivileged and marginalized groups, have fair access to electricity.
- **Affordability:** Maintaining consumer-friendly energy prices while multi-vector systems are implemented.

Significant technical, financial, environmental, and societal obstacles confront MV-EHs. Coordinated efforts in technological advancement, development of policies, economic reconfiguration, and community engagement are needed to overcome these obstacles. In order to successfully deploy and maintain MES, which have the potential to improve sustainability, dependability, and efficiency in the energy industry, these problems must be resolved.

5.4 Role of Advanced Energy Storage Technologies in MV-EHs

The efficiency and dependability of MV-EHs, which combine several energy sources including gas, heat, and electricity, depend heavily on advanced energy storage technology. Modern storage technologies are used by these centers to improve sustainability, lower costs, and balance supply and demand. Lithium-ion (Li-ion) batteries are one of the key technologies; they are well-known for having a high energy density, efficiency, and scalability, which makes them perfect for storing extra renewable electricity and supplying it during the high-demand periods. Systems for thermal energy storage (TES) use substances like water or molten salts to store heat or cold. TES increases energy efficiency by capturing excess heat from solar thermal systems or industrial processes and releasing it when needed for heating or electricity production. The storage of hydrogen is becoming more crucial. When surplus renewable electricity is utilized for electrolysis, the resulting hydrogen can be stored, converted back into electricity later, or used as heat. This process works well with the current gas infrastructure.

Flywheels are quick-response and high-power output energy storage devices that can be used to manage short-term power fluctuations and stabilize grid frequency. Supercapacitors are good for managing transient supply and demand imbalances because they store energy electrostatically and deliver fast power bursts. Through enhanced resilience, flexibility, and sustainability, these solutions support transitioning to a low-carbon energy system for MV-EHs. There are several categories into which the various types of storage for energy can be further separated. The bulk of energy storage types in this case are categorized as follows: magnetic energy storage, chemical and hydrogen energy storage, flywheel energy storage, compressed air energy storage (CAES), pumped energy storage, TES, and electrochemical and battery energy storage.

5.4.1 Flywheel Energy Storage

Energy can be mechanically stored in the form of kinetic energy storage, also known as flywheel energy storage, which allows machines to run smoothly and produce high power and energy densities. Kinetic energy is transferred into and out of flywheels by an electric machine operating as a generator or motor, based on the charging and discharging mode. Permanent magnet machines are

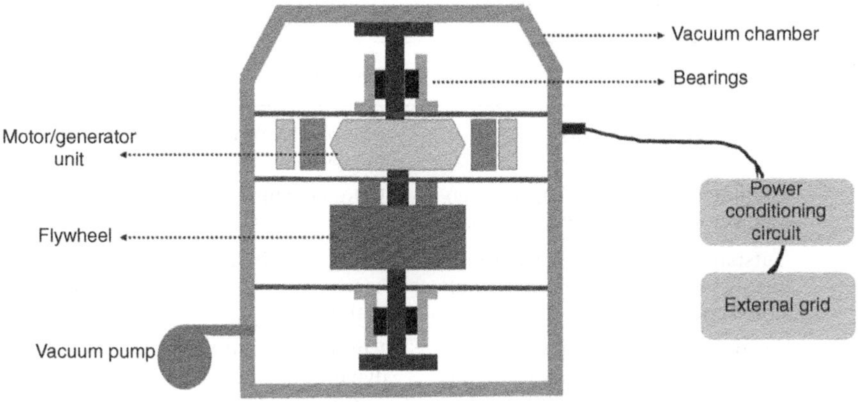

Figure 5.5 Flywheel energy storage technique.

often used to power flywheels due to their high efficiencies, low rotor losses, and high power densities [44]. There are several constraints for other electrical machines in terms of potential use speed, idling losses, noise, vibration, and pricing. These machines include, for example, variable-reluctance, induction, and bearingless models. The flywheel Energy storage system is illustrated in Figure 5.5. A flywheel's revolving mass receives charging energy, which is then converted to kinetic energy and stored. Electricity can be produced on demand from this stored energy. Magnetic bearings, which function in a vacuum to reduce losses due to friction over long-term preservation and safety concerns, support the rotating mass [45]. To benefit from the advantages of each kind, the rotating bearing system might be either mechanical, magnetic, or a hybrid of the two. Due to their lack of friction, magnetic bearings do not need to be lubricated; nonetheless, some of them have intricate control systems and need energy to run. Superconductive magnetic bearings (SMBs) can be used at high speeds, but they need energy to run a cryogenic cooling mechanism. In flywheels, it is desirable to achieve high rotational velocity and high power density, as the energy stored is only linearly related to mass but proportionate to the square of velocity. Material science and systems engineering provide the essential enabling technologies [46]. The flywheel's rotor and its housing are made of steel, alloys (such as titanium or aluminum alloys), and, more recently, tough materials like composites. Up to 10,000 rpm may currently be reached, and rotor materials and design are the subject of extensive research [41]. Compared to chemical batteries, the utilization of composite materials allows for higher rotational speeds and power densities. In cars where a lot of peak power is required during acceleration and a lot of power is made available for storage quickly during braking, high power density is preferred. Flywheel energy storage devices are also known for their high energy

Table 5.2 Merits and demerits of flywheel storage systems.

Sl. No	Merits	Demerits
1	Rapid charging capacity	Comparatively low density of energy to the battery system
2	Lifetime is mainly insensitive to the quantity of charge/discharge cycles	Extremely futile losses
3	Outstanding density of power that is mostly unaffected by the amount of stored energy	Energy required to power magnetic bearings or the requirement for bearing maintenance
4	The usage of motors and various industrial applications results in low maintenance requirements, easy state of charge determination, and extensive operational expertise	Failure may result from unforeseen dynamic stresses or external disturbances

and power density, extended operating life, tremendous cycle life (several tens of thousands), high round-trip efficiency, and low environmental effect [47]. The drawbacks of flywheel ESSs include reduced power density, expense, vibrations and sound, upkeep requirements, and security issues when compared to batteries and supercapacitors [48, 49]. The literature [49] suggests an innovative concept for a combinational structure wherein supercapacitors are housed inside the disc of a rotary flywheel system in order to increase their power density. This makes it possible to store enormous amounts of energy and exchange pulsed power.

Due to their high substantial discharge of up to 15% per hour and high energy installation costs, which vary from USD 1500 to USD 6000/kWh, FES systems are unsuitable for medium- or extended-term storage demands. The cost of installing a flywheel system should decrease to approximately USD 1000 to USD 3900/kWh once cycles and calendar lifetimes start to rise. Table 5.2 discusses the benefits and drawbacks of flywheel energy storage technologies.

5.4.1.1 Significant Progress to Improve the Energy Storage Performance of Flywheels

Important advancements consist of:

1) **New Materials:** Increasing energy density is the goal of developments in high-strength, low-density materials. But given the size and weight of the necessary containment vessels, these enhancements need to be financially feasible. Though cost is still a major consideration, research into efficient and lightweight containment systems could increase energy density even more.

2) **Bearings with Superconductivity:** For flywheel systems, reducing friction losses is essential since it permits lower self-discharge rates and faster rotational speeds. Bearings made of high-temperature superconducting materials can greatly improve performance by reducing cooling expenses and friction loss. One such device that stabilizes the grid is a 300 kW/100 kWh flywheel with superconducting magnetic bearings that has been in use in Japan since 2015.
3) **Electric Machines:** Cutting prospective system costs and reducing reliance on rare earth resources are possible with novel designs for electrical machinery (motors/generators) that employ fewer permanent magnets. Promising are switched reluctance machines, which run on reluctance torque instead of permanent magnets. Because these machines do not have any physical friction elements, they can operate at incredibly high speeds with little maintenance. However, given the increased complexity of the system, sophisticated control techniques are required to guarantee reliability.

   Improved materials, electric machines, and superconducting bearings are essential for raising flywheel ESSs' performance, affordability, and sustainability.

5.4.1.2 Challenges in Integrating Flywheels into MV-EHs

In spite of significant advancements in flywheel storage techniques, there are significant challenges in integrating them into MV-EHs. They are summarized in Table 5.3.

5.4.2 CAES Technology

A CAES system stores air that is compressed, or potentially elastic as energy in a reservoir.

Large-scale CAES facilities require air reservoirs with a high volume. Depleted gas fields or ancient natural salt deposits can be prepared for use as appropriate storage caverns for compressed air. Where an appropriate and pre-existing cavern is available, costs are substantially reduced. Building a dedicated cavern to store compressed air results in a significant rise in energy storage expenses. Although they are typically too costly to be deemed economically viable, metal vessels as reservoirs are technically possible. Similar to traditional gas turbines, CAES systems operate on the same principles, but instead of having simultaneous compression and expansion phases, they are decoupled. In order to charge a CAES, excess or off-peak electricity is supplied to a motor that drives a sequence of air compressors for storing energy in a storage tank. The air gets warmer throughout this process. An air cooler, also known as a radiator, removes this heat in a traditional (diabatic) CAES system and releases it into the atmosphere. Underground

Table 5.3 Challenges in integrating flywheels into MV-EHs.

Sl. No.	Challenges	Description
1	High starting prices	Superconducting bearings, sophisticated materials, and novel electric machines all demand a substantial upfront expenditure.
2	Financial sustainability	It is difficult to weigh the economic benefits of larger, heavier containment vessels against gains in energy density and performance.
3	Integration of complex systems	Control systems that are complex and cutting-edge technology are needed to ensure a smooth integration with other energy vectors such as heat, gas, and electricity.
4	Scale and density of energy	It is still very difficult to increase energy density without dramatically increasing costs or structural complexity.
5	Technological development	Superconducting bearings and switching reluctance machines are two examples of components that are still in the early stages of research and require more testing and development.
6	The environmental circumstances	The performance and lifetime of flywheel systems depend on their ability to withstand changes in temperature and humidity.

caverns, mostly made of salt, are where compressed air is usually kept at pressures ranging from 4.0 megapascals (MPa) to 8.0 megapascals [50]. Usually, the stored air powers a gas-fired turbine generator to discharge the CAES system during periods of high energy demand. When the expanded compressed air leaves the reservoir, it cools down; hence, heating is required to improve the power quality of the turbine/generator unit.

To power the turbine system, compressed air and fuel—natural gas, for example—are combined in a combustion chamber. To increase efficiency, combustion exhaust gases are frequently recovered. To generate heat throughout the expansion phase, the traditional CAES design burns fossil fuels in the turbine chambers, which has the disadvantage of producing CO_2. More recently, a concept called advanced adiabatic compressed energy storage (AA-CAES) systems was created to overcome this problem.

A thermal storage system (TES) in the AA-CAES concept holds the heat that would normally be released into the surrounding atmosphere during the compression stage. During expansion-mode operation, this heat is transferred back to the

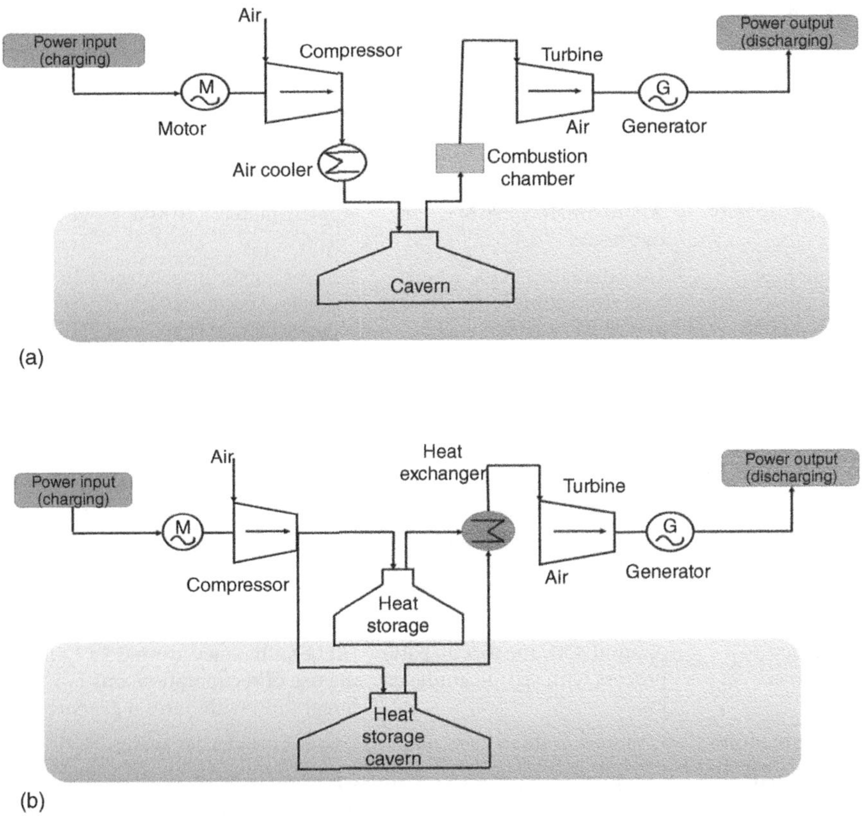

Figure 5.6 (a) Diabatic system of compressed air energy storage technique. (b) Adiabatic system of compressed air energy storage technique.

air being discharged from the reservoir via heat exchangers. This makes it possible for AA-CAES systems to generate electricity from compressed air without the need for burning and without producing any associated emissions. These two systems are schematically compared in Figure 5.6.

The charging of the storage system in CAES systems produces a considerable quantity of heat. On the other hand, during discharge, the pressurized air rapidly cools down and may freeze, potentially causing damage to the system. As a result, CAES systems are typically situated near gas-fired power stations, utilizing the waste heat from these facilities to introduce it into the purified air inside the CAES system. As of 2016, there are just two large-scale plants linked to the grid: a 290 MW plant in Huntorf, Germany, and one in McIntosh, Alabama, United States. This is in spite of the growing interest in CAES technology. After years of development, a 270 MW CAES project was reportedly abandoned in Iowa, United States, due to the risk of bankruptcy.

Table 5.4 Key features of Huntorf and McIntosh CAES plants.

Feature	Huntorf CAES plant	McIntosh CAES plant
Location	Huntorf, Germany	McIntosh, Alabama, USA
Operational since	1978	1991
Storage capacity	Approximately 290 MW for two hours	Approximately 110 MW for 26 hours
Technology	Uses diabatic CAES, where heat from compression is lost and must be added back during expansion	Utilizes a hybrid approach with a combined cycle, incorporating recuperation to reuse some of the compression heat
Energy source	Primarily relies on off-peak electricity to compress air, which is stored in underground salt caverns	Similar approach, but also integrates natural gas for heating the air during expansion, increasing efficiency
Key feature	The world's first commercialized CAES plant, proving the viability and dependability of huge-scale energy storage	Advanced use of recuperators to improve efficiency and reduce fuel consumption
Efficiency	Around 42% due to a diabatic process with significant heat loss	Higher efficiency around 54% due to the use of recuperation and integration with natural gas turbines
Storage medium	Underground salt caverns	Underground salt caverns
Operational flexibility	Can provide rapid response to grid demand fluctuations, enhancing grid stability	Offers extended discharge duration, making it suitable for longer periods of power support
Environmental impact	Involves CO_2 emissions due to the use of natural gas for reheating the air during expansion	Improved environmental profile through better efficiency and reduced fuel use compared to Huntorf

Source: Adapted from [51].

The salient features of these two systems are displayed in Table 5.4.

Table 5.4 summarizes the important aspects of the Huntorf and McIntosh CAES plants, highlighting their capacities, technologies, efficiencies, and operational characteristics.

5.4.2.1 Challenges Faced by Compressed Air Storage Systems in MV-EHs

There are various obstacles to overcome when integrating CAES into MV-EHs:

1) **High Initial Costs:** Infrastructure, such as compressors, turbines, and subterranean storage facilities, requires significant investment.

2) **Energy Efficiency:** Because heat is lost during air compression and expansion, CAES systems sometimes have poorer energy efficiency, necessitating the use of sophisticated thermal management techniques.
3) **Space Requirements:** In order to store air, sizable subterranean caverns or reservoirs are required, which restricts site selection and raises project complexity.
4) **Complex Integration:** Sophisticated control systems and flawless interoperability are necessary to coordinate CAES with other energy vectors (gas, heat, and electricity).
5) **Environmental Concerns:** Combustion-related air quality problems and underground storage hazards are examples of possible environmental effects.

5.4.3 Pumped Hydro Storage (PHS)

PHS is, by far, the most extensively used vast-scale ESS. This advanced technology was first made available for purchase in the 1890s. By the end of 2016, more than 150 GW of PHS electricity had been installed and put into service [52]; some estimates suggest even larger capabilities [53]. PHS uses a water pump to transfer water between two reservoirs that are spaced apart in order to store energy as gravitational potential energy. Using external power, water is circulated by means of a penstock through the lowest point toward the top water reservoir when there is less need for electricity. This completes the ESS charging process. Table 5.5 demonstrates the merits and demerits of the pumped hydro storage techniques.

Table 5.5 The merits and demerits of the pumped hydro storage techniques.

Sl. No.	Merits	Demerits
1	High technological maturity and a wealth of operational knowledge in an established technology	Geographical constraints, as a suitable location requiring extensive land utilization, are required
2	Incredibly low discharge	Reduced energy density with a big footprint
3	Reasonable efficiency over a circular trip	Extensive construction duration, high initial investment prices, and a lengthy time to return investment
4	Both lengthy storage durations and high-volume storage are achievable, minimal installation costs for energy, excellent flexibility in starting and stopping, long lifespan and inexpensive storage	Concerns regarding the environment

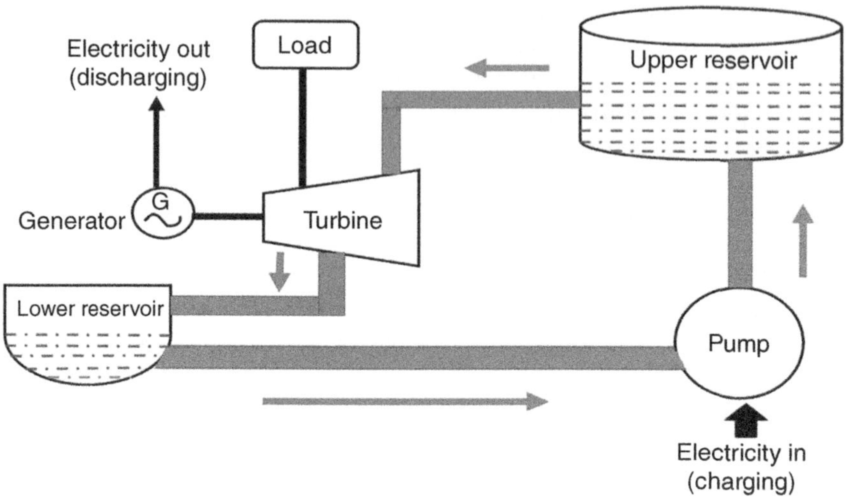

Figure 5.7 Pumped energy storage technique.

The pump and turbine unit are coupled to a reversible electric generator/motor system. When there is a high demand for electricity, the liquid flow turns around, and the liquid that has accumulated in the upper reservoir is pumped to the lower reservoir via the energy-generating turbine system. Following that, the electricity produced is added to the system (Figure 5.7). The quantity of water stored in the reservoir with greater elevation and the elevation difference between tanks are directly proportional to the PHS plant's energy storage capacity. Large rivers or lakes are commonly used as lower reserves to save construction costs.

Technically speaking, there are still more design options available. As lower reservoirs, flooding underground mines or other cavities can be utilized. These projects are frequently called "underground" or "subsurface" PHS plants. Since there are fewer eligible above-surface places and they may have less of an environmental impact, the notion has attracted attention again, even if no projects have been completed thus far, especially in Europe [54, 55]. The Okinawa Yanbaru Seawater Pumped Storage Power Station in Japan was the sole additional plant in the entire globe that employed seawater for electricity generation, even though it is no longer in service for commercial purposes. There have been plans for other projects, most notably in Chile. The Pacific Ocean serves as the lower reservoir for the 30 MW Japanese project, while a human-made top reservoir is used [56, 57]. PHS facilities have traditionally handled medium- or long-term storage, with discharge periods ranging from a few hours to many days.

PHS plants typically possess an overall efficiency ranging from 74% to 84%. And they are projected to last for a very long time—between 40 and 60 years—although substantial renovations are required. In electrical systems that have multiple base-load energy facilities, the majority of PHS plants were previously utilized to equalize the difference between generation and load during periods of high and low demand. Cheap off-peak electricity is typically stored in reservoirs and sold during the peak demand hours of the morning and evening. This established business model faces challenges due to the growing use of Variable Renewable Energy in electrical systems, particularly solar PV, which has the ability to level off daytime peaks. PHS needs to adapt and learn new operational principles in order to generate profitable revenue streams from modifying its offerings to provide additional options while maintaining systems operation, a marketplace that is predicted to grow as VRE usage rises. When producing, PHS was once able to control power, but not when pumping.

Nonetheless, PHS engineers are progressively considering the implementation of a PHS system with variable speed that allows power control both in the pumping and generating stages. Additionally, this setup is more efficient than the conventional setup. A ternary system will allow for concurrent production and pumping to offer more precise control over frequency [58–60].

5.4.4 Batteries and Electrochemical Systems for Energy Storage

Batteries and capacitors are two different devices with different uses and ways to store electrical energy. High energy densities and voltages are available from batteries, a dependable energy storage technology. Lithium-ion (Li-ion), flow, lead-acid (Pb-acid), lead-carbon, nickel-cadmium (NiCd), sodium-sulfur (NaS), and nickel-cadmium (NiCd) are some of the several types of batteries. There are three types of capacitors: electrostatic, electrolytic, and electrochemical. Capacitors use electrochemistry to store and deliver energy. Because of their porous electrode designs, electrochemical capacitors, also known as supercapacitors (UCs), have the largest capacitance per unit volume among them. Innovations in electrode materials and electrolytes, with the goal of enhancing cost, energy density, power density, cycle life, and safety, have propelled significant breakthroughs in battery technology. Thorough assessments of numerous battery technologies, including lithium-ion battery and supercapacitor performance measures, can be found in research by the authors in Ref. [61]. In their investigation of methods for improving Li-ion battery performance in Ref. [62], they offer techniques such as cutting the size of the active material, creating composites, doping, adjusting the shape of the particles, using coatings, and changing the electrolyte.

Lithium batteries are becoming more and more significant due to their elevated electrical power and energy density. Li-air batteries, in particular, have potential energy densities that are on par with petrol, which makes them ideal for use in electric vehicles. For commercial viability, however, issues including cathode design, electrolyte optimization, and comprehension of intricate charge–discharge interactions must be resolved. The promise of Li-air and Li-sulfur (Li-S) batteries over conventional Li-ion batteries is highlighted by research by authors in Refs. [63, 64], but they also stress the need for additional chemical research before these technologies can be commercially viable.

Because sodium is abundant and has similar insertion chemistries to Li-ion batteries, sodium-ion (Na-ion) batteries offer a more affordable option. Significant developments in cathode, anode, and electrolyte materials are required to match Li-ion technology's performance. In their assessment of Na-ion battery materials, [65] include sodium "β"-alumina and gel polymer electrolytes as possible electrolytes and recommend phosphates and fluorophosphates as viable cathode materials. Furthermore, because of their abundance of materials, affordability, and environmental friendliness, aluminum-ion batteries have been proposed as a viable replacement for Li-ion batteries. Nevertheless, before they can be commercialized, they must improve in terms of steadiness and energy storage capacity.

Electrochemical capacitors are known for their extended life cycles and excellent storage efficiency. They can withstand hundreds of thousands of cycles without experiencing a noticeable reduction in capacity. Nevertheless, they have voltage restrictions and self-discharge issues. Electrochemical capacitors' energy and power densities are increasing due to developments in nanoscale research. Recent advancements in supercapacitor technologies are reviewed by the authors in Refs. [66, 67], with an emphasis on better electrode materials and device topologies. Pseudocapacitors, hybrids, and double-layer capacitors are the three types of electrochemical capacitors; each has a different electrode material and charge storage method.

Because of their independent energy and power densities, flow batteries, including zinc bromide (ZnBr), vanadium redox (VRB), and polysulfide bromide (PSB), are appropriate for utility-scale renewable energy storage. High prices and technical difficulties with materials and cell design still exist, though. To increase flow battery performance, research focuses on mass transfer, electrolyte wetting, and electrode surface areas. As a more affordable option, the literature [68] provides a metal-free flow battery based on quinone. For better performance metrics, hybrid flow batteries (HFB), which combine characteristics of flow and conventional batteries, are also being investigated.

To put it briefly, the goals of current research and development in electrochemical energy storage technologies are to solve current problems, enhance functionality, and bring these technologies to market. In order to advance battery and capacitor technology to fulfill future energy storage demands, improvements in electrode materials, electrolytes, and cell designs are essential.

Table 5.6 Merits and demerits of battery energy storage systems.

Sl. No	Merits	Demerits
1	Independent power and energy modification to specific applications is made possible by external electrolyte storage	Handling chemicals with the possibility of an acid solution leaking
2	A comparatively elevated conversion efficiency was attained	Maintenance expenses could go up if sensors, pumps, and flow control devices are required
3	High durability and cycle life, together with long-term performance supported by the electrodes' capacity to maintain their morphological integrity	Elevated expenses for specific active components or essential system components, like electrolyte or membrane storage tanks

5.4.4.1 Merits and Demerits of Battery ESSs

Table 5.6 shows the benefits and drawbacks of battery-powered energy storage technologies.

5.4.4.2 Challenges in Integrating Battery Energy Storage in MV-EHs

MV-EH integration of battery storage involves a number of operational, technological, financial, legal, regulatory, and social problems.

i) **Technical Challenges:** Ensuring system interoperability, creating standardized communication protocols, regulating battery performance and lifespan, optimizing energy management, and preserving grid stability are some of the technical issues.
ii) **Economic Challenges:** High initial expenses, finding revenue streams, negotiating incentives, and doing cost–benefit evaluations to guarantee long-term savings and economic viability are the main economic hurdles.
iii) **Regulatory and Policy Challenges:** Understanding and adhering to intricate rules, adhering to grid codes, and creating creative business models and supportive policies that explicitly outline ownership and responsibility structures are all part of the regulatory and policy issues.
iv) **Challenges in Operation:** Coordinating MES, efficiently controlling loads, preserving dependable operation through stringent maintenance routines, and guaranteeing effective data integration and cybersecurity are some of the operational issues.
v) **Sustainable Challenges:** The ecological and social consequences of batteries are discussed, along with life cycle assessments, improving public perception, and involving communities to win support.

5.4.5 Thermal Energy Storage Technology

The act of retaining heating or cooling in a substance, whether it be naturally existing, like the ground, or intentionally created, like water tanks, is known as TES, or TES. TES modes include sensible, latent, and thermochemical storage. The traditional method, referred to as sensible heat storage, uses appliances that raise the medium's temperature, such as hot water storage and subsurface TES. Latent heat storage is a new technology that gives benefits like compactness, improved thermal properties, and isothermal phase-change operations by changing the phase of the storage medium. Examples include eutectic salts, fatty acids, esters, paraffin waxes, and other common phase change materials (PCMs).

When the source of energy varies, whether temporarily or seasonally, TES systems store the energy. Ground heat systems and other forms of seasonal storage hold heat for months at a time, conserving summer heat for use in the winter. Similar to storing solar energy throughout the day for use at night, short-term storage maintains heat for a few hours to several days. By storing thermal energy for use during peak times and low-rate periods, these devices can lower power prices. Enhancing thermal characteristics, creating new storage media, and developing modeling techniques are the main areas of concentration for TES technology research. Using circulating media, such as air or water, for ground thermal storage is becoming more and more popular. The commercialization of TES is hindered by various factors such as a lack of awareness, policy concerns, high costs, and technical feasibility.

5.4.6 Magnetic Energy Storage Technology

Superconducting magnetic energy storage (SMES) can be accomplished by using a massive superconducting coil, which has almost no resistance to electricity at temperatures close to zero degrees Celsius and is capable of holding electric energy in the magnetic field formed by DC current passing through it. The superconducting coil is kept at cryogenic temperatures by using nitrogen or liquid helium containers. Superconductors offer almost no resistance to electron flow, so that there are almost no energy losses in the coil.

The cooling system that maintains the cryogenic temperature at that level is responsible for certain energy losses. SMES coils include an almost instantaneous power discharge capacity as well as an infinitely efficient cycle of charging and discharging. The main SMES design elements that affect storage performance are the structure, energy capacity, coil arrangement, and operating temperature. High-temperature superconductor devices are still in the research and development stage due to their high cost, but low-temperature ones are widely accessible.

Applications for SMES include automatic generation control, frequency regulation, load leveling, power quality improvement, transmission capability enhancement, load stability, voltage stability, and uninterruptible power supplies. The literature given in Ref. [69] reviews configurations of voltage-source-converter-based, thyristor-based, and S. Koohi-Fayegh and M.A. Rosen current-source-converter-based SMES. They suggest categorizing the price of SMES technologies based on how much it costs for energy storage capacity and power handling capabilities (i.e., conductor, coil structural components, cryogenic vessel, refrigeration, protection, and control equipment).

They contend that there is a significant cost variance in the latter and that concentrating research on it might significantly lower the cost of SMES as a whole. SMES systems are reviewed for power systems by the authors in Ref. [70]. In contrast to a number of studies conducted in labs or using simulations on computers, they highlight the importance of creating practical SMES technologies for electrical systems. Furthermore, they advise that efficient control methods be created in order to integrate small grades of SMES systems at various locations and boost their power capacities.

5.4.7 Chemical and Hydrogen Energy Storage

An energy-dense, reversible chemical process could be a viable option for energy storage. Systems for storing chemical energy can occasionally be categorized based on the type of energy they use. Electrical energy is stored electrochemically, and heat energy is stored thermochemically. In the process of producing hydrogen for energy storage, either directly (e.g., via photoconversion) or electrolytically, it is held for a while before being oxidized or undergoing another chemical reaction to recover the energy input. Energy does not originate from hydrogen; rather, hydrogen is the product of a chemical reaction. For numerous decades, electricity has served as the main energy source for numerous energy-related technologies in civilization.

The properties of hydrogen energy are complementary to those of electrical power. A "hydrogen economy" that encompasses the creation, delivery, storage, and use of hydrogen electrical systems has been suggested by some [71]. The author of the literature [72] discusses the hydrogen economy, its relevance to the environment and climate, how it improves the system's energy quality, how it helps decarbonize fossil fuel-fueled power plants, and how it applies to portable electronics, both stationary and mobile, as well as the innovative electrochemical energy converter fuel cell. This section describes methods for storing hydrogen and the processes involved in creating hydrogen for energy storage.

A popular technique for producing hydrogen is water splitting. Fossil fuels, RES, and other energy sources can supply the energy needed for this operation.

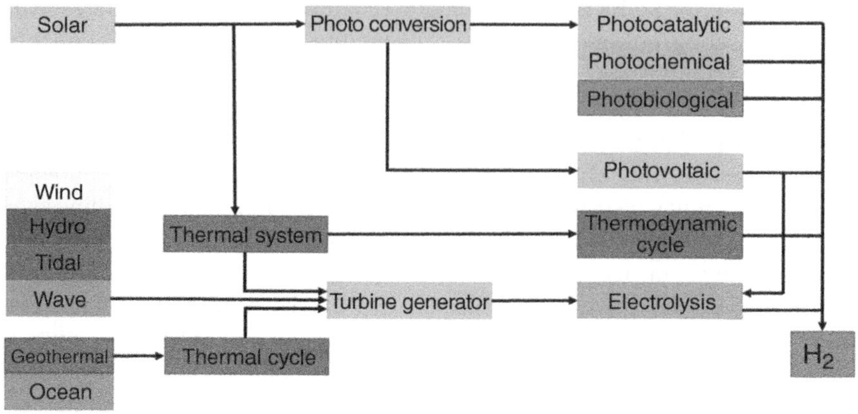

Figure 5.8 Hydrogen production.

Because renewable energy is frequently sporadic, it must be stored before use. The processes that the authors in Ref. [73] discuss using to produce hydrogen from solar energy include an assembly of a solar-powered device and an electrolysis system, an intensive thermal solar system with a propeller and an electrolysis device, a solar energy-centered structure and a thermochemical water-splitting cycle, and direct photoconversion (photocatalytic, photoelectrochemical, and photobiological water splitting). The Hydrogen production process is demonstrated in Figure 5.8.

Other RES such as wind, hydropower, geothermal, ocean thermal energy conversion, anaerobic digestion of biomass, and biowaste are also covered in this study [73]. Some people link photosynthetic and photovoltaic processes because they both use sunlight to harvest energy. However, in the case of natural photosynthesis, biomass or chemical fuels are produced, whereas in the case of photovoltaic, non-stored electrical current is. These two processes also function differently. In order to manufacture hydrogen, Blankenship et al. [74] compare current photovoltaic-driven electrolysis technologies with natural photosynthesis and demonstrate that the latter is more efficient annually.

A discussion is held regarding potential applications of recent advancements in synthetic biology to enhance the solar energy conversion efficiency of natural photosynthesis. Hydrogen storage presents significant challenges, particularly for automotive applications.

A car running on hydrogen instead of petrol or diesel needs a substantially larger fuel tank since hydrogen has a lower energy density per unit of volume than the other fuels. In addition, compared to methane and propane, hydrogen is the lightest element and is more difficult to liquefy. Its small molecular size and low density make it susceptible to leaking from confinement vessels.

It is possible to store pure hydrogen in three different forms: mixed phase (hydrogen slush), compressed gas, or cryogenic liquid. The liquefaction of hydrogen and pressurization of a cryogenic liquid to achieve higher densities in a fluid or mixed-phase condition requires a substantial amount of energy and sophisticated equipment. Various porous networks, including zeolites, metal–organic frameworks, clathrate hydrates, various carbon materials (e.g., nanotubes, fullerenes, and graphene), and typical organic polymers, absorb and release hydrogen during adsorptive storage. It is feasible to repeat this process multiple times without the solid disintegrating or the gas being lost. Hydrogen is kept in chemical storage in solid or liquid phases by forming bonds chemically with additional elements in a substance that is abundant in hydrogen.

Ammonia-like complexes, amines, amides, and metal and non-metal hydrides are examples of solid-phase systems. Alcohols, formic acid, and N-ethylperhydrocarbazole are examples of liquid carriers. Two major obstacles to using ammonia as a hydrogen storage intermediate are energy-efficient ammonia breakdown and the subsequent separation and purification of the hydrogen product, according to the authors in Ref. [75], who investigate materials and technologies for hydrogen purification from decomposed ammonia gas streams. They demonstrate how dense-metal membranes free of flaws can produce products with acceptable levels of purity. A review of several features of hydrogen storage techniques has been conducted in Refs. [76–78].

They comprise many techniques for storing hydrogen, such as cryogenic liquid and high-pressure storage [78, 79], adsorptive storage on large-surface-area adsorbents, chemical storage in metal hydrides, complex hydrides, and intermetallic compounds, and storage in boranes. The filling phase of compressed hydrogen storage is thermodynamically modeled and analyzed by the authors in Ref. [78] using the second law of thermodynamics.

Similar to batteries, fuel cells are low-power density devices that transform chemical energy into electrical energy. They demonstrate approximately 70–80% energy efficiency, but certain power plants (such as combined cycle units) can reach up to 60% efficiency.

Fuel, such as hydrogen and oxygen, is used in fuel cells. Their power densities can be increased by combining them with supercapacitors. Reference [80, 81] highlights research to take advantage of the special properties of graphene to create supported catalysts with higher electrocatalytic activity, better durability, and high-performance electrode designs in fuel cells.

5.5 Mathematical Model of MV-EHs

Over a given time horizon, the MV-EH model seeks to minimize total costs, which include those for power, heat, gas, operations, and emissions. Each cost component is a function of time, and the objective function is defined as the sum of these costs. The volumes of heat, gas, and electricity used, as well as emissions and operational factors, are important factors. In order to guarantee system efficiency and balance, the model includes a number of constraints. Constraints on the energy balance guarantee that each energy vector's supply and demand are met. Energy storage devices behave in a way that is captured by storage dynamics. The system is maintained within its capacity by generation and conversion limitations. While emission restrictions guarantee adherence to environmental norms, operational constraints guarantee compliance with technological and regulatory requirements. With the use of this thorough model, MV-EHs may be optimized, offering insights into sustainable energy management and performance enhancement.

The objective is to reduce the overall cost, which takes into account the cost of operations, the cost of energy purchases, and perhaps the cost of emissions. We can express the objective function as follows:

$$\min_{x(t)} \sum_{t=1}^{T} \mathrm{C_{el}}(t)\,\mathrm{E_{el}}(t) + \mathrm{C_{ht}}(t)\,\mathrm{H_{ht}}(t) + \mathrm{C_{gs}}(t)\,\mathrm{G_{gs}}(t) + \mathrm{C_{op}}(t)\,\mathrm{O_{op}}(\mathrm{t}) \\ + \mathrm{C_{emm}}(t) \cdot \mathrm{Emm}(t)$$

where

$\mathrm{C_{el}}(t)$, $\mathrm{C_{ht}}(t)$, and $\mathrm{C_{gs}}(t)$ are the cost of electric power, heat, and gas at a given time t

$\mathrm{E_{el}}(t)$, $\mathrm{H_{ht}}(t)$, and $\mathrm{G_{gs}}(t)$ are the amounts of electricity, heat, and gas used at the time t

$\mathrm{C_{op}}(t)$ is the operational cost

$\mathrm{O_{op}}(t)$ is the other operational variable

$\mathrm{C_{emm}}(t)$ is the emission cost

$\mathrm{Emm}(t)$ is the emissions at time t

5.5.1 Modeling Approaches

MV-EHs are a unified system for controlling several energy carriers, including heat, gas, and electricity. Complex methodologies are used in the modeling of these hubs in order to maximize performance, improve efficiency, and guarantee dependability. The approaches and analytical procedures for modeling MV-EHs are examined in this article, with a particular emphasis on mathematical models, optimization algorithms, and simulation tools.

5.5.1.1 Mathematical Modeling

- **Deterministic Models:** These models imitate the operation of MV-EHs under presumptive settings by using fixed inputs. In order to optimize energy distribution and storage activities, linear programming (LP) and mixed-integer linear programming (MILP) are frequently utilized. Under certain conditions, these techniques offer concise, uncomplicated solutions.
- **Stochastic Models:** Stochastic models use probabilistic techniques to address the uncertainties in energy supply, demand, and prices. Stochastic programming and Monte Carlo simulations are two methods that aid in the creation of resilient models that can manage unpredictability and variability in the system.

5.5.1.2 Tools for Simulation

- **EnergyPlus:** This program is well-suited for simulating the thermal elements of MV-EHs since it is frequently used to simulate in-depth building energy consumption, including heating, cooling, and other energy flows.
- **HOMER (Hybrid Optimization of Multiple Energy Resources):** HOMER is a microgrid optimization tool that works with storage systems and RES. It can simulate many setups and determine which is the most economical and productive.
- **MATLAB/Simulink:** A flexible platform that makes thorough simulation, analysis, and modeling of dynamic systems possible. It is widely utilized for MV-EH electrical and thermal system simulation.

5.5.1.3 Hybrid Models

Hybrid models provide a balanced approach by combining deterministic and stochastic elements. To account for both the predictable and unpredictable components of energy management, they employ stochastic approaches for long-term planning and deterministic methods for precise, short-term optimizations.

5.5.2 Analytical Techniques

5.5.2.1 Optimization Algorithms

- **Genetic Algorithms (GA):** Inspired by natural selection, GAs are used for solving complex optimization problems in MV-EHs by iteratively improving solutions.
- **Particle Swarm Optimization (PSO):** This method optimizes problems by simulating the social behavior of swarms, effectively handling nonlinear and complex optimization challenges.

- **Linear and Nonlinear Programming:** These mathematical approaches solve optimization problems by finding the best outcome subject to a set of linear or nonlinear constraints, suitable for various operational scenarios in MV-EHs.

5.5.2.2 Performance Analysis

- **Efficiency Metrics:** Key metrics include round-trip efficiency (efficiency of the ESS), overall system efficiency, and specific efficiencies of different energy conversion processes.
- **Reliability Assessment:** Tools such as reliability block diagrams (RBD) and fault tree analysis (FTA) evaluate the dependability and robustness of MV-EHs, identifying potential failure points and improvement areas.

5.5.2.3 Economic and Environmental Analysis

- **Cost–Benefit Analysis:** Evaluates the economic viability of MV-EHs by comparing implementation and operational costs against the benefits, including energy savings and increased reliability.
- **Life Cycle Assessment (LCA):** Assesses the environmental impact of MV-EHs from production through disposal, aiming to identify sustainable practices and reduce the ecological footprint.

5.5.3 Challenges and Opportunities

5.5.3.1 Challenges

- **Complexity and Scalability:** Modeling large-scale MV-EHs involves managing numerous variables and interactions, making the process computationally intensive.
- **Data Availability and Accuracy:** High-resolution, reliable data are crucial for accurate modeling and analysis but can be challenging to obtain.
- **Integration of RES:** The intermittent nature of RES adds complexity to the optimization and reliability of MV-EHs.

5.5.3.2 Opportunities

- **Advanced Computational Techniques:** Utilizing machine learning and artificial intelligence for predictive modeling and optimization can enhance the performance and reliability of MV-EHs.
- **Development of Hybrid Storage Systems:** Integrating various storage technologies (e.g., batteries, thermal storage) can improve the flexibility and efficiency of energy hubs.

Table 5.7 Comparative evaluation of various energy storage systems.

Energy storage devices/parameters	Flywheel energy storage	Pumped hydro storage	CAES	Li-ion batteries	Lead-acid batteries	Supercapacitors	H_2 fuel storage
Overall setup cost in USD/kWh	1500–4000	500–2500	2000–4000	200–600	150–400	1000–4000	3000–6000
Energy density (Wh/kg)		0.5–1.5	30–60	100–250	30–50	5–10	100–150
Power density (W/kg)	500–2000	1–2	200–400	250–440	75–300	10,000–15,000	1000–3000
Cycle life (number of cycles)	20,000–100,000	10,000–60,000	10,000–25,000	1000–10,000	500–1800	500,000+	500–1500
Efficiency (%)	90–95	70–85	40–70	85–98		85–98	70–80
Response time (ms)	<10	1000–10,000	1000–2,000	<1	100–200	1000–10,000	1000–10,000
Lifespan (years)	15–20	30–60	20–40	8–15	3–5	10–15	10–20
Impact on the environment	Low	High	Medium	Medium	High	Low	Medium

Source: [82–87].

5.5.4 Policy and Incentive Design

Creating supportive policies and incentives can promote the adoption and development of MV-EHs, accelerating the transition to sustainable energy systems.

5.5.4.1 Future Research Directions

- **Integrated Modeling Frameworks:** Developing comprehensive frameworks that integrate multiple energy vectors and storage technologies, providing holistic solutions for MV-EHs.
- **Real-Time Optimization:** Enhancing real-time optimization techniques to better manage dynamic energy systems and the integration of RES.
- **Resilience and Adaptability:** Focusing on the resilience and adaptability of MV-EHs to cope with unforeseen events and changes in energy demand and supply.
- **Collaborative Platforms:** Encouraging collaboration between academia, industry, and policymakers to share data, tools, and best practices for advancing MV-EH technologies.

By addressing these challenges and leveraging the opportunities, modeling and analyzing MV-EHs can significantly contribute to the development of efficient, reliable, and sustainable energy systems.

The comparison of all the ESSs is illustrated in Table 5.7.

5.6 Conclusion

A more integrated, effective, and sustainable energy system is made possible by the modeling and analysis of MV-EHs with improved energy storage units. This represents a revolutionary approach to energy management. The transition from standalone energy systems to networked multi-vector systems that can maximize resource utilization and improve resilience has been brought to light by the development of energy hubs. In order to enable effective energy conversion and distribution, MV-EH components, including generation units, storage technologies, and distribution networks, collaborate.

Through an examination of the various forms of interactions and interdependencies across energy vectors, such as electricity, natural gas, and district heating networks, we can get insight into the complex dynamics underlying contemporary energy systems. In order to maximize energy flows and guarantee the dependability and stability of the network as a whole, these interdependencies are essential. Significant benefits from the integration of these networks include decreased environmental impact, increased flexibility, and higher energy efficiency. It does, however, also come with certain drawbacks, such as the requirement for sophisticated coordination and control mechanisms and complexity in system design and operation.

Modern energy storage technologies are essential to MV-EHs because they integrate RES, improve system dependability, and offer the flexibility needed to balance supply and demand. By allowing excess energy to be stored and released when needed, these technologies help to minimize volatility and guarantee a steady supply of energy. A strong framework for optimizing the operation of MV-EHs is provided by their mathematical modeling. These models allow for the cost-effective distribution of resources while respecting capacity and regulatory limits by integrating a range of cost considerations, constraints, and operational variables. This all-encompassing strategy opens the door for in-depth study and optimization, leading to the development of energy systems that are more resilient and sustainable.

References

1. Guo, Z., Wei, W., Chen, L. et al. (2021). Impact of energy storage on renewable energy utilization: a geometric description. *IEEE Transactions on Sustainable Energy* 12: 874–885.
2. König, A., Nicoletti, L., Schröder, D. et al. (2021). An overview of parameter and cost for battery electric vehicles. *World Electric Vehicle Journal* 12: 21.
3. Sioshansi, R., Denholm, P., Arteaga, J. et al. (2022). Energy-storage modeling: state-of-the-art and future research directions. *IEEE Transactions on Power Apparatus and Systems* 37: 860–875.
4. Reynolds, J., Ahmad, M.W., and Rezgui, Y. (2018). Holistic modelling techniques for the operational optimisation of multi-vector energy systems. *Energy and Buildings* 169: 397–416.
5. Wang, C., Dong, S., Xu, S. et al. (2019). Impact of power-to-gas cost characteristics on power-gas-heating integrated system scheduling. *IEEE Access* 7: 17654–17662.
6. Kholardi, F., Assili, M., Lasemi, M.A., Hajizadeh, A. (2018). Optimal management of energy hub with considering hydrogen network. *In Proceedings of the 2018 International Conference on Smart Energy Systems and Technologies (SEST)*, Seville, Spain (10–12 September 2018), pp. 1–6.
7. Rosa, R.N. (2017). The role of synthetic fuels for a carbon neutral economy. *C—Journal of Carbon Research* 3: 11.
8. Park, C., Sedundo, R., Knazkins, V., Korbakorba, P. (2016). Feasibility analysis of the power-to-gas concept in the future Swiss power system. *CIRED Workshop*, Zurich University of Applied Sciences, Zurich, Switzerland, pp. 1768–1772.

9 Krause, T., Andersson, G., Frohlich, K., and Vaccaro, A. (2011). Multiple-energy carriers: modeling of production, delivery, and consumption. *Proceedings of the IEEE* 99: 15–27.

10 Hajimiragha, A., Canizares, C., Fowler, M., Geidl, M., Andersson, G. (2007). Optimal energy flow of integrated energy systems with hydrogen economy considerations. *Bulk Power System Dynamics and Control-VII. Revitalizing Operational Reliability—September, 2007 iREP Symposium*, IEEE, pp. 1–11.

11 Galus, M.D., Koch, S., and Andersson, G. (2011). Provision of load frequency control by PHEVs, controllable loads, and a cogeneration unit. *IEEE Transactions on Industrial Electronics* 58: 4568–4582.

12 Chicco, G. and Mancarella, P. (2009). Matrix modelling of small-scale trigeneration systems and application to operational optimization. *Energy* 34: 261–273.

13 Yokoyama, R., Hasegawa, Y., and Ito, K. (2002). A MILP decomposition approach to large scale optimization in structural design of energy supply systems. *Energy Conversion and Management* 43: 771–790.

14 Voll, P., Klaffke, C., Hennen, M., and Bardow, A. (2013). Automated superstructure-based synthesis and optimization of distributed energy supply systems. *Energy* 50: 374–388.

15 Johnson, N., Krey, V., McCollum, D.L. et al. (2015). Stranded on a low carbon planet: implications of climate policy for the phase-out of coal-based power plants. *Technological Forecasting and Social Change* 90: 89–102.

16 Geidl, M. and Andersson, G. (2007). Optimal power flow of multiple energy carriers. *IEEE Transactions on Power Apparatus and Systems* 22: 145–155.

17 Mohammadi, M., Noorollahi, Y., Mohammadi-Ivatloo, B., and Yousefi, H. (2017). Energy hub: from a model to a concept–a review. *Renewable and Sustainable Energy Reviews* 80: 1512–1527.

18 Kampouropoulos, K. (2018). Multi-objective optimization of an energy hub using artificial intelligence; centre technologies: Barcelona, Spain, http://www.tdx.cat (accessed 2 May 2022).

19 Walker, S., Labeodan, T., Maassen, W., and Zeiler, W. (2017). A review study of the current research on energy hub for energy positive neighborhoods. *Energy Procedia* 122: 727–732.

20 Geidl, M. (2007). *Integrated Modeling and Optimization of Multi-Carrier Energy Systems*. Zurich, Switzerland: ETH.

21 Walker, S.B., Mukherjee, U., Fowler, M., and Elkamel, A. (2016). Benchmarking and selection of power-to-gas utilizing electrolytic hydrogen as an energy storage alternative. *International Journal of Hydrogen Energy* 41: 7717–7731.

22 Schulze, M., Friedrich, L., Gautschi, M. (2008). Modeling and optimization of renewables: Applying the energy hub approach. In *Proceedings of the 2008 IEEE International Conference on Sustainable Energy Technologies*, Singapore (24–27 November 2008), pp. 83–88.

23 Suslov, K., Piskunova, V., Gerasimov, D. et al. (2019). Development of the methodological basis of the simulation modelling of the multi-energy systems. *E3S Web Conferences* 124: 01049.

24 Favre-Perrod, P. (2005). A vision of future energy networks. *IEEE Power Engineering Society Inaugural Conference and Exposition in Africa* (11–15 July 2005), IEEE, pp. 13–17.

25 Chicco, G., Mancarella, P. (2006). A comprehensive approach to the characterization of trigeneration systems. *Proceeding VI World Energy System Conference*, Torino, Italy (10–12 July).

26 Hemmes, K., Zachariah-Wolf, J., Geidl, M., and Andersson, G. (2007). Towards multi-source multi-product energy systems. *International Journal of Hydrogen Energy* 32: 1332–1338.

27 Marti, B. (2006). Integrated analysis of energy and transportation systems. Master thesis. University of Zurich.

28 Geidl, M., Koeppel, G., Favre-Perrod, P. et al. (2007). Energy hubs for the future. *IEEE Power and Energy Magazine* 5: 24–30.

29 Geidl, M. (2007). Integrated modeling and optimization of multi-carrier energy systems. Master thesis. TU Graz.

30 Zeng, Q., Fang, J., Chen, Z., Li, J., Zhang, B. (2016). A multistage coordinative optimization for sitting and sizing P2G plants in an integrated electricity and natural gas system. *Energy Conference (ENERGYCON), 2016 IEEE International*, IEEE, Belgium (4–8 April 2016), pp. 1–6.

31 Clegg, S. and Mancarella, P. (2015). Integrated modeling and assessment of the operational impact of power-to-gas (P2G) on electrical and gas transmission networks. *IEEE Transactions on Sustainable Energy* 6: 1234–1244.

32 Jokar, S.M., Rahimpour, H.R., Momeni, H. et al. (2014). Simulation and feasibility analysis of structured packing replacement in absorption column of natural gas dehydration process: a case study for Farashband gas processing plant, Iran. *Journal of Natural Gas Science and Engineering* 18: 336–350.

33 Hussain, B., Thirkill, A. (2018). Multi-energy vector integration innovation opportunities; catapult energy systems: Birmingham, UK. https://es.catapult.org.uk (accessed 5 May 2022).

34 Mancarella, P. (2014). MES (multi-energy systems): an overview of concepts and evaluation models. *Energy* 65: 1–17.

35 Lowe, R., Chiu, L., Pye, S., Gallo Cassarino, T., Barrett, M., Scamman, D., Smith, A., Solano-Rodriguez, B. (2020). Lost generation: system resilience and flexibility. *In Proceedings of the Proceedings of the 2020 MIT "A+ B" Applied Energy Symposium*, Cambridge, MA, USA (1 August 2020).

36 O'Malley, M., Kroposki, B., Hannegan, B. et al. (2016). *Energy Systems Integration. Defining and Describing the Value Proposition*. Golden, CO, USA: National Renewable Energy Lab (NREL).

37 Carradore, L., Turri, R. (2009). Modeling and simulation of multi-vector energy systems. In *Proceedings of the 2009 IEEE Bucharest PowerTech,* Bucharest, Romania (28 June–2 July 2009), pp. 1–7.

38 Clegg, S. and Mancarella, P. (2016). Storing renewables in the gas network: modelling of power-to-gas seasonal storage flexibility in low-carbon power systems. *IET Generation Transmission and Distribution* 10: 566–575.

39 Qadrdan, M., Wu, J., Jenkins, N., and Ekanayake, J. (2013). Operating strategies for a GB integrated gas and electricity network considering the uncertainty in wind power forecasts. *IEEE Transactions on Sustainable Energy* 5: 128–138.

40 Abeysekera, M., Jenkins, N., Wu, J. (2016). Integrated energy systems: an overview of benefits, analysis, research gaps and opportunities. Hubnet, UK: www.hubnet.org.uk (accessed 10 May 2022).

41 Salgado, F. and Pedrero, P. (2008). Short-term operation planning on cogeneration systems: a survey. *Electric Power Systems Research* 78: 835–848.

42 Houwing, M., Negenborn, R.R., and De Schutter, B. (2010). Demand response with micro-CHP systems. *Proceedings of the IEEE* 99: 200–213.

43 Liu, X. and Mancarella, P. (2016). Modelling, assessment and Sankey diagrams of integrated electricity-heat-gas networks in multi-vector district energy systems. *Applied Energy* 167: 336–352.

44 Mousavi, S.M., Faraji, F., Majazi, A., and Al-Haddad, K. (2017). A comprehensive review of flywheel energy storage system technology. *Renewable and Sustainable Energy Reviews* 67: 477–490.

45 Bolund, B., Bernhoff, H., and Leijon, M. (2007). Flywheel energy and power storage systems. *Renewable and Sustainable Energy Reviews* 11: 235–258.

46 Baker, J. (2008). New technology and possible advances in energy storage. *Energy Policy* 36 (12): 4368–4373.

47 Amiryar, M.E. and Pullen, K.R. (2017). A review of flywheel energy storage system technologies and their applications. *Applied Sciences* 7 (3): 286.

48 Meishner, F. and Sauer, D.U. (2019). Wayside energy recovery systems in DC urban railway grids. *eTransportation* 1: 100001.

49 Toodeji, H. (2019). A developed flywheel energy storage with built-in rotating super capacitors. *Turkish Journal of Electrical Engineering and Computer Sciences* 27: 213–229.

50 Chen, H., Cong, T.N., Yang, W. et al. (2009). Progress in electrical energy storage system: a critical review. *Progress in Natural Science* 19 (3): 291–312. https://doi.org/10.1016/j.pnsc.2008.07.014.

51 IRENA (2017). *Electricity Storage and Renewables: Costs and Markets to 2030.* Abu Dhabi: International Renewable Energy Agency.

52 IHA (2017). Briefing. 2017 Key Trends in Hydropower, International Hydropower Association. https://www.hydropower.org/sites/default/files/

publications-docs/2017%20Key%20Trends%20in%20Hydropower_0.pdf (accessed 30 August 2017).
53 US DOE (2017). DOE Global Energy Storage Database. United States Department of Energy. Office of Electricity & Energy Reliability. http://www.energystorageexchange.org/projects (accessed 4 August 2017).
54 EERA (2016). Underground pumped hydro storage, Brussels, European Energy Research Alliance. https://eera-energystorage.eu/wp-content/uploads/2016/03/EERA_Factsheet_Underground-Pumped-Hydro-Energy-Storage_not-final.pdf.
55 Akinyele, D.O. and Rayudu, R.K. (2014). Review of energy storage technologies for sustainable power networks. *Sustainable Energy Technologies and Assessments* 8: 74–91. https://doi.org/10.1016/j.seta.2014.07.004.
56 Oshima, K., Kawai, J., Otsuka, S., Wada, T. and Imano, H. (1999). Development of pump-turbine for seawater pumped storage power plant, American Society of Civil Engineers, pp. 1–6 https://doi.org/10.1061/40440(1999)67 (accessed 31 August 2017).
57 Hiratsuka, A., Arai, T., and Yoshimura, T. (1993). Seawater pumped-storage power plant in Okinawa island, Japan. *Engineering Geology* 35 (3–4): 237–246. https://doi.org/10.1016/0013-7952(93)90012-2.
58 IHA (2017). *2017 Hydropower Status Report*. London: International Hydropower Association.
59 ANL (2013). *Modeling Ternary Pumped Storage Units*. Oak Ridge, Tennessee, USA: Argonne National Laboratory http://ceeesa.es.anl.gov/projects/psh/ANL_DIS-13_07_Modeling_Ternary_Units.pdf.
60 Ciocan, G.D., Teller, O., and Czerwinski, F. (2012). Variable speed pump-turbines technology. *UPB Scientific Bulletin. Series D: Mechanical Engineering* 74 (1): https://www.scientificbulletin.upb.ro/rev_docs_arhiva/full67d_700564.pdf.
61 Hall, P.J. and Bain, E.J. (2008). Energy-storage technologies and electricity generation. *Energy Policy* 36 (12): 4352–4355.
62 Nitta, N., Wu, F., Lee, J.T., and Yushin, G. (2015). Li-ion battery materials: present and future. *Materials Today* 18 (5): 252–264.
63 Bruce, P.G., Freunberger, S.A., Hardwick, L.J., and Tarascon, J. (2012). Li–O_2 and Li–S batteries with high energy storage. *Nature Materials* 11 (1): 19–29.
64 Thackeray, M.M., Wolverton, C., and Isaacs, E.D. (2012). Electrical energy storage for transportation—approaching the limits of, and going beyond, lithium-ion batteries. *Energy & Environmental Science* 5: 7854–7863.
65 Palomares, V., Serras, P., Villaluenga, I. et al. (2012). Na-ion batteries, recent advances and present challenges to become low cost energy storage systems. *Energy & Environmental Science* 5: 5884–5901.
66 Kötz, R. and Carlen, M. (2000). Principles and applications of electrochemical capacitors. *Electrochimica Acta* 45: 2483–2498.

67 Lu, P., Xue, D., Yang, H., and Liu, Y. (2013). Supercapacitor and nanoscale research towards electrochemical energy storage. *International Journal of Smart and Nano Materials* 4 (1): 2–26.

68 Huskinson, B., Marshak, M.P., Suh, C. et al. (2014). A metal-free organic–inorganic aqueous f low battery. *Nature* 505: 195–198.

69 Hasan Ali, M., Wu, B., and Dougal, R.A. (2010). An overview of SMES applications in power and energy systems. *IEEE Transactions on Sustainable Energy* 1 (1): 38–47.

70 Sutanto, D., Cheng, K.W.E. (2009). Superconducting magnetic energy storage systems for power system applications. *Proceedings of International Conference on Applied Superconductivity and Electromagnetic Devices, ASEMD, 2009* (25–27 September 2009).

71 Rosen, M.A. and Koohi-Fayegh, S. (2016). The prospects for hydrogen as an energy carrier: an overview of hydrogen energy and hydrogen energy systems. *Energy, Ecology and Environment* 1 (1): 10–29.

72 Winter, C.J. (2009). Hydrogen energy — abundant, efficient, clean: a debate over the energy-system-of-change. *International Journal of Hydrogen Energy* 34 (14): S1–S52. Supplement 1.

73 Abbasi, T. and Abbasi, S.A. (2011). 'Renewable' hydrogen: prospects and challenges. *Renewable and Sustainable Energy Reviews* 15 (6): 3034–3040.

74 Blankenship, R.E., Tiede, D.M., Barber, J. et al. (2011). Comparing photosynthetic and photovoltaic efficiencies and recognizing the potential for improvement. *Science* 332 (6031): 805–809.

75 Lamb, K., Dolan, M.D., and Kennedy, D.F. (2019). Ammonia for hydrogen storage: a review of catalytic ammonia decomposition and hydrogen separation and purification. *International Journal of Hydrogen Energy* 44 (7): 3580–3593.

76 Eberle, U., Felderhoff, M., and Schüth, F. (2009). Chemical and physical solutions for hydrogen storage. *Chemie International Edition* 48: 6608–6630.

77 Dalebrook, A.F., Gan, W., Grasemann, M. et al. (2013). Hydrogen storage: beyond conventional methods. *Chemical Communications* 49: 8735–8751.

78 Hosseini, M., Dincer, I., Naterer, G.F., and Rosen, M.A. (2012). Thermodynamic analysis of filling compressed gaseous hydrogen storage tanks. *International Journal of Hydrogen Energy* 37 (6): 5063–5071.

79 Sakintuna, B., Lamari-Darkrim, F., and Hirscher, M. (2007). Metal hydride materials for solid hydrogen storage: a review. *International Journal of Hydrogen Energy* 32 (9): 1121–1140.

80 Hou, J., Shao, Y., Ellis, M.W. et al. (2011). Graphene-based electrochemical energy conversion and storage: fuel cells, supercapacitors and lithium ion batteries. *Physical Chemistry Chemical Physics* 13: 15384–15402.

81 Koohi-Fayegh, S. and Rosen, M.A. (2020, 101047). A review of energy storage types, applications and recent developments. *Journal of Energy Storage* 27: 1–23.

82 Izzati, L., Ariffin, W., Zain, M. et al. (2021). Comparative study of energy storage systems (ESSs). *Journal of Physics: Conference Series* 1962: 012035. https://doi.org/10.1088/1742-6596/1962/1/012035.

83 Chen, H., Cong, T., Yang, W. et al. (2009). Progress in electrical energy storage systems: a critical review. *Progress in Natural Science* 19: 291–312. https://doi.org/10.1016/j.pnsc.2008.07.014.

84 International Renewable Energy Agency (IRENA) (2017). *Electricity Storage and Renewables: Costs and Markets to 2030*. IRENA.

85 International Hydropower Association (IHA). (2017). 2017 Hydropower Status Report. IHA.

86 Department of Energy (DOE) (2017). Global Energy Storage Database.

87 Akinyele, D. and Rayudu, R. (2014). Review of energy storage technologies for sustainable power networks. *Sustainable Energy Technologies and Assessments* 8: 74–91.

6

Market and Energy Trading Mechanism in MV-EHs

Rajasekharan Rajasree[1], *Dhandapani Lakshmi*[1], *Ravichandran Karthick Manoj*[1], *Kesavan Stalin*[2], *Sivaraman Palanisamy*[3], *and Sharmeela Chenniappan*[3]

[1] *Department of EEE, AMET University, Chennai, Tamilnadu, India*
[2] *Department of Marine, AMET University, Chennai, Tamilnadu, India*
[3] *Department of EEE, CEG Campus, Anna University, Chennai, Tamilnadu, India*

6.1 Introduction to Different Market Clearing Mechanisms in MEH

Market Equilibrium Models or Hypotheses (MEHs) are essential tools in economics that analyze how supply and demand interact to determine prices, quantities, and resource allocation in markets. They rest on the concept of market equilibrium, where the quantity demanded by consumers matches the quantity supplied by producers at a specific price, ensuring that the market clears without any surplus or shortage. MEHs can be categorized into partial equilibrium models, which focus on a single market in isolation, and general equilibrium models, which consider the interdependencies across multiple markets [1]. These models assume rational behavior, perfect competition, full information, and flexible prices, making them valuable for theoretical analysis, policy evaluation, and understanding market dynamics. However, they also have limitations, such as simplifying assumptions and a tendency to overlook dynamic and behavioral factors. Despite these limitations, MEHs are widely used in economic forecasting, policy design, and market analysis, providing a foundational framework for understanding how markets function.

Energy Management Strategies for Multi-Vectored Energy Hubs to Achieve Low Carbon Societies, First Edition.
Edited by Shubham Tiwari, Jai Govind Singh, Sivaraman Palanisamy, Sharmeela Chenniappan, Rupendra Kumar Pachauri, and Sanjeevikumar Padmanaban.

6.2 Concepts of Market Equilibrium Models

Market equilibrium models (MEMs) are built around the concept of market equilibrium, where the forces of supply and demand balance each other to determine the prices and quantities of goods or services. At the core of MEMs is the idea that markets naturally move toward an equilibrium point where the quantity demanded by consumers equals the quantity supplied by producers at a specific price. This equilibrium ensures that there are no shortages (where demand exceeds supply) or surpluses (where supply exceeds demand), leading to an efficient allocation of resources. The equilibrium price is where the supply and demand curves intersect, indicating the price at which both consumers and producers are satisfied [2].

To understand how MEMs reach this equilibrium, it's essential to consider the underlying assumptions of these models. MEMs typically assume that all market participants, both consumers and producers, behave rationally, meaning they make decisions aimed at maximizing their utility (for consumers) or profit (for producers). This rational behavior is modeled through supply and demand curves, which depict the relationship between the price of a good and the quantity demanded or supplied. MEMs also assume perfect competition, where numerous buyers and sellers exist, none of whom can influence the market price individually [3]. This competitive environment ensures that prices are determined purely by market forces rather than by the actions of any single participant.

Another crucial concept in MEMs is the assumption of full information and flexible prices. Full information means that all participants in the market have complete knowledge of prices, available quantities, and other relevant market conditions, allowing them to make informed decisions. Flexible prices refer to the ability of prices to adjust freely in response to changes in supply and demand, ensuring that the market can reach equilibrium efficiently [4]. These assumptions, while simplifying the complexities of real-world markets, provide a robust framework for analyzing market dynamics, predicting the effects of economic policies, and understanding how different factors influence market outcomes.

6.3 Mechanisms of Energy Trading in MEH

6.3.1 Market Structure and Participants

Energy trading operates within a multifaceted market structure that includes various stakeholders, such as producers, consumers, traders, and regulatory authorities. Producers generate energy from a range of sources, including fossil fuels and renewables, while consumers—ranging from industrial entities to residential users—demand energy for various purposes. Traders act as intermediaries, facilitating transactions to enhance market liquidity and efficiency. Regulatory authorities oversee market practices, ensuring fair competition and consumer protection. This diverse ecosystem shapes the dynamics of supply and demand, influencing pricing mechanisms and trading strategies.

6.3.2 Spot and Futures Markets

Energy trading is characterized by the existence of spot and futures markets, each serving distinct purposes. Spot markets facilitate immediate transactions, allowing energy to be bought and sold for prompt delivery at current prices. This real-time trading is crucial for balancing supply and demand on short notice. Conversely, futures markets enable participants to hedge against price volatility by agreeing on prices for future delivery [5]. These contracts provide certainty for both producers and consumers, allowing them to plan and manage financial risk associated with fluctuating energy prices.

6.3.3 Pricing Mechanisms and Instruments

Pricing in energy markets is determined by a combination of factors, including supply and demand dynamics, geopolitical influences, and weather patterns. Traders use various financial instruments, such as options, swaps, and contracts for difference (CFDs), to manage risk and capitalize on price movements. These instruments facilitate both speculative trading and risk management, enabling participants to navigate the complexities of energy pricing [6]. Understanding these mechanisms is crucial for market participants, as they help inform strategies for buying and selling energy effectively.

6.3.4 Environmental and Regulatory Considerations

The energy trading landscape is increasingly influenced by environmental regulations and sustainability initiatives. Governments and regulatory bodies are implementing policies aimed at reducing carbon emissions and promoting renewable energy sources. Mechanisms such as carbon trading systems and renewable energy certificates (RECs) incentivize cleaner energy practices, impacting trading behaviors and pricing structures. As regulatory frameworks evolve, market participants must adapt their strategies to align with sustainability goals, balancing economic interests with environmental responsibilities.

6.3.5 Technological Innovations and Market Integration

Technological advancements are transforming energy trading by enhancing market efficiency and enabling greater integration of renewable energy sources. Innovations such as blockchain technology and smart grid systems facilitate transparent and secure transactions, while real-time data analytics improve decision-making for traders. Additionally, increased connectivity among regional energy markets allows for more efficient cross-border trading, optimizing resource allocation. As technology continues to evolve, it shapes the future of energy trading, promoting a more resilient and sustainable energy ecosystem [7].

6.4 Types of Market Equilibrium in MEHs

The different types of market equilibrium explain the diverse ways in which supply and demand interact, reflecting the complexity of real-world economic scenarios. Each type provides valuable insights for analyzing market behavior and the effects of various factors on equilibrium outcomes.

6.4.1 Stable Equilibrium

In a stable equilibrium, any small deviation from the equilibrium price leads to automatic corrective actions that restore the balance between supply and demand. If prices rise above the equilibrium, a surplus occurs, prompting sellers to lower prices. Conversely, if prices fall, a shortage arises, leading buyers to compete for the limited supply, driving prices back up. This self-correcting nature ensures the market returns to equilibrium.

6.4.2 Unstable Equilibrium

An unstable equilibrium occurs when a small change in price or quantity leads to a larger divergence from the equilibrium point. In this scenario, if prices rise or fall slightly, the market may spiral into either a persistent surplus or a shortage, moving further away from equilibrium [8]. Such conditions often require external intervention to restore balance, as market forces alone may not suffice.

6.4.3 Dynamic Equilibrium

Dynamic equilibrium refers to a state where market conditions are in continuous flux due to changes in external factors, such as consumer preferences, technological advancements, or regulatory changes. While the market may not be in a static balance, it adjusts over time to reach a new equilibrium as conditions evolve. This type emphasizes the importance of time and adaptability in markets.

6.4.4 Partial Equilibrium

Partial equilibrium focuses on the equilibrium condition in a specific market while assuming other markets remain constant. This approach simplifies analysis by isolating one market to understand its dynamics without accounting for interactions with other markets. It is particularly useful for studying short-term impacts of policy changes or external shocks in a single market context.

6.4.5 General Equilibrium

General equilibrium encompasses the interactions among multiple markets simultaneously. It examines how changes in one market affect others, capturing the interconnectedness of economic systems. This comprehensive approach is essential for understanding broader economic policies and their implications across different sectors.

6.4.6 Long-Run Equilibrium

Long-run equilibrium considers a situation where all factors of production and costs are variable, allowing for adjustments in supply and demand over time [9]. In this state, firms can enter or exit the market, leading to an equilibrium price that reflects the average cost of production in the long term.

6.4.7 Short-Run Equilibrium

Short-run equilibrium occurs when at least one factor of production is fixed. In this state, firms may experience either supernormal profits or losses, prompting adjustments only in variable factors. This type highlights how immediate market responses can differ from long-term outcomes.

6.5 Graphical Representation of Market Equilibrium

The graphical representation of market equilibrium is a powerful tool for visualizing how supply and demand interact to determine the price and quantity of goods in a market. The basic components of this representation include the supply curve, the demand curve, and the equilibrium point.

Figure 6.1 represents the trading mechanisms and market equilibrium model in MV-EHs. Figure 6.2 illustrates the structure and architecture of Multi-Vector Energy Hubs (MV-EHs). Figure 6.3 represents the different components of Blockchain used in MV-EHs.

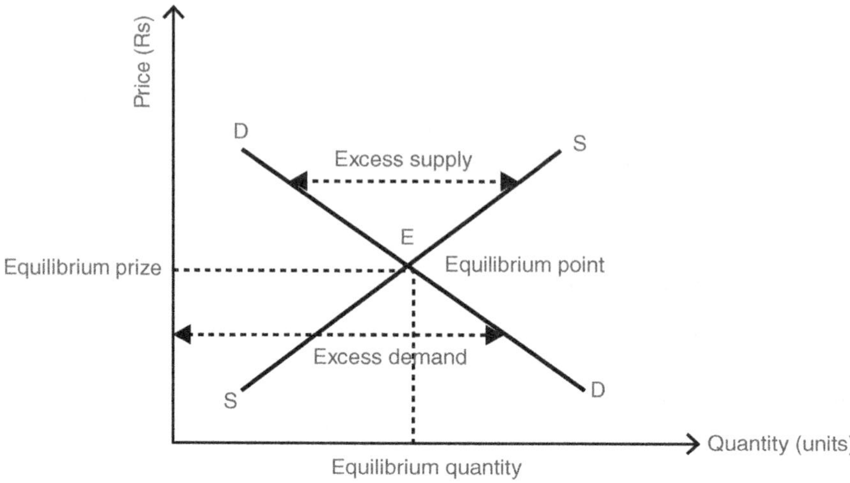

Figure 6.1 Graphical representation of market equilibrium in MEH.

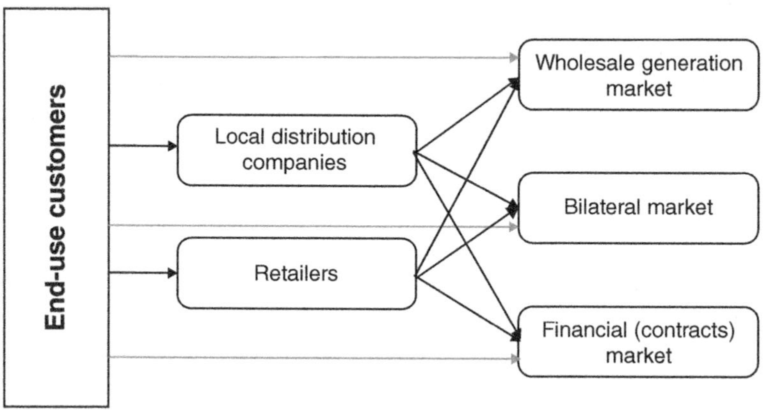

Figure 6.2 Energy market design.

6.5.1 Demand and Supply Curves

- Demand Curve: This curve typically slopes downward from left to right, reflecting the inverse relationship between price and quantity demanded. As prices decrease, consumers are willing to buy more of the good, leading to higher quantities demanded.
- Supply Curve: In contrast, the supply curve usually slopes upward from left to right, indicating a direct relationship between price and quantity supplied. As prices increase, producers are incentivized to supply more of the good, resulting in higher quantities supplied.

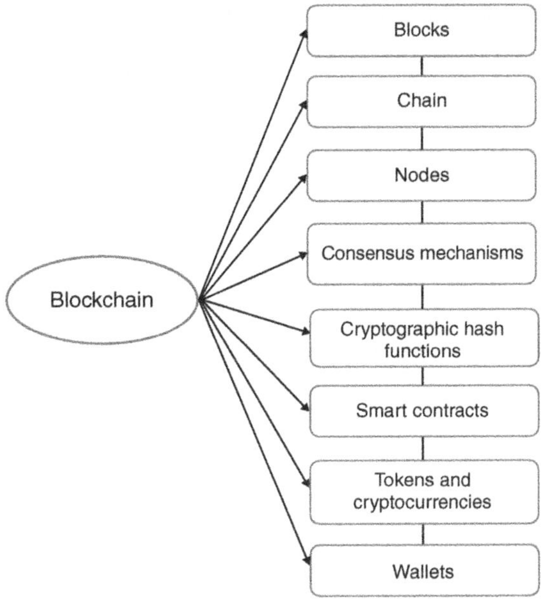

Figure 6.3 Components of blockchain.

6.5.2 Equilibrium Point

The intersection of the demand and supply curves represents the market equilibrium. At this point, the quantity demanded by consumers equals the quantity supplied by producers, establishing the equilibrium price and quantity [10]. This equilibrium point is essential for understanding how markets clear without excess supply or demand.

6.5.3 Shifts in Curves

When external factors influence either supply or demand, the curves can shift:

- Shift in Demand Curve: If consumer preferences change or income levels increase, the demand curve may shift to the right, resulting in a higher equilibrium price and quantity. Conversely, a leftward shift (due to decreased preferences or income) leads to a lower equilibrium price and quantity.
- Shift in Supply Curve: Factors like production costs, technology, or regulations can shift the supply curve. A rightward shift indicates increased supply (lower production costs), resulting in a lower equilibrium price and a higher quantity. A leftward shift (e.g., due to increased costs) leads to a higher equilibrium price and lower quantity.

6.5.4 Surpluses and Shortages

Graphical representation also helps illustrate market imbalances:

- Surplus: If the market price is set above the equilibrium price, quantity supplied exceeds quantity demanded, leading to a surplus. This excess supply will generally push prices down toward equilibrium.
- Shortage: Conversely, if the price is below equilibrium, quantity demanded exceeds quantity supplied, creating a shortage. This scarcity often drives prices up until the market reaches equilibrium.

6.6 Factors Affecting Market Equilibrium Models

The different factors of the market equilibrium interact in complex ways, affecting the overall market equilibrium and leading to shifts in prices and quantities. The following is the list of factors that affect market equilibrium:

- Supply and Demand Dynamics
  - Changes in Consumer Preferences: Shifts in tastes can increase or decrease demand.
  - Changes in Production Costs: Variations in the cost of inputs can affect supply.
  - Technological Advancements: Innovations can enhance supply efficiency.
- External Shocks
  - Natural Disasters: Events like hurricanes can disrupt supply chains.
  - Economic Events: Recessions or booms can alter consumer spending.
- Government Policies
  - Taxes and Subsidies: Tax increases can reduce supply, while subsidies can boost it.
  - Regulations: Policies can impose restrictions that impact supply.
- Market Competition
  - Number of Sellers: More competition generally leads to lower prices and higher supply.
  - Market Power: Monopolies can dictate prices and alter equilibrium.
- Consumer Income Levels
  - Changes in Income: Higher incomes typically increase demand for normal goods.
  - Inferior Goods: Demand for these decreases as income rises.
- Expectations
  - Future Price Expectations: Anticipating future price changes can influence current buying and selling behavior.

  - Economic Outlook: Optimism or pessimism about the economy can affect consumer confidence and spending.
- Substitutes and Complements
  - Availability of Substitutes: If the price of a substitute rises, demand for the original product may increase.
  - Complementary Goods: A decrease in the price of one good can increase demand for its complement.
- Global Influences
  - International Trade: Tariffs and trade agreements can impact supply and demand.
  - Global Economic Conditions: Economic fluctuations in other countries can influence domestic markets.
- Seasonal Factors
  - Seasonality of Demand: Certain products see fluctuations based on the time of year (e.g., holidays and weather).
- Demographic Changes
  - Population Growth: An increase in population can lead to higher demand for various goods and services.
  - Age Distribution: Changes in demographics can shift demand patterns (e.g., aging population).

6.7 Energy Market Designs

Energy market designs refer to the frameworks and structures that govern how energy is produced, distributed, and consumed within a market [11]. Effective energy market designs are essential for ensuring efficiency, reliability, and sustainability in energy systems.

6.7.1 Types of Energy Markets

- **Wholesale Markets:** These facilitate the buying and selling of large quantities of electricity among generators and distributors. Prices are determined based on supply and demand.
- **Retail Markets:** These involve the sale of electricity directly to consumers, allowing for competition among suppliers and various pricing options.
- **Capacity Markets:** These ensure that there is enough generation capacity available to meet peak demand, providing financial incentives for new and existing plants to maintain readiness.

6.7.2 Market Clearing Mechanisms

- **Auction Systems:** Competitive bidding processes where energy suppliers submit offers to sell power, and buyers submit bids to purchase. The market clears at a price where supply meets demand.
- **Clearing Prices:** These are determined based on the marginal cost of the last unit of energy needed to meet demand, reflecting the true cost of electricity at any given time.

6.7.3 Regulatory Framework

- **Role of Regulatory Bodies:** Organizations such as the Federal Energy Regulatory Commission (FERC) in the U.S. oversee market operations to ensure fairness, competition, and reliability.
- **Compliance and Standards:** Regulatory frameworks set standards for market participants, promoting transparency and protecting consumer interests.

6.7.4 Incentives for Renewable Energy

- **Feed-In Tariffs:** Guarantees fixed payments for energy producers who generate power from renewable sources, encouraging investment in renewables.
- **Renewable Portfolio Standards:** Mandates requiring utilities to obtain a certain percentage of their energy from renewable sources [12].

6.7.5 Demand Response Programs

These programs incentivize consumers to reduce their energy usage during peak demand times, helping to balance supply and demand. Technologies like smart meters facilitate this by providing real-time data.

6.7.6 Integration of Distributed Energy Resources

- **Decentralized Generation:** Involves local generation sources, such as solar panels or wind turbines, which can supply power directly to consumers or back to the grid.
- **Microgrids and Virtual Power Plants:** These allow for localized energy management, enhancing resilience and efficiency.

6.7.7 Market Interconnections

- **Regional Transmission Organizations (RTOs):** Facilitate coordination and integration of energy markets across different regions, optimizing resource use.
- **Cross-Border Trading:** Enables electricity exchanges between countries, enhancing energy security and efficiency.

6.7.8 Pricing Mechanisms

- **Locational Marginal Pricing (LMP):** Prices reflect the cost of delivering electricity at specific locations, accounting for transmission constraints and generation costs.
- **Time-of-Use Pricing:** Prices vary based on the time of day, encouraging consumers to shift usage to off-peak periods.

6.7.9 Environmental Considerations

- **Carbon Pricing:** Incorporates the cost of carbon emissions into market prices, encouraging cleaner energy sources.
- **Emissions Trading Systems:** Market-based approaches that cap total emissions and allow trading of emission permits, promoting cost-effective reductions.

6.7.10 Challenges and Barriers

- **Market Power and Manipulation:** Risks of monopolistic behavior can lead to inefficiencies and higher prices.
- **Infrastructure Limitations:** Aging grids and insufficient capacity can hinder the effective functioning of energy markets.

6.7.11 Future Trends in Energy Market Design

- **Decarbonization Strategies:** Aligning market designs with climate goals through policies that promote renewable energy and energy efficiency.
- **Blockchain and Digital Innovations:** These technologies have the potential to enhance transaction transparency and efficiency, as well as facilitate peer-to-peer (P2P) energy trading.

6.8 Blockchain Technologies

Blockchain technology is a decentralized digital ledger system that securely records transactions across a network of computers, ensuring transparency, security, and immutability. Each transaction is grouped into blocks, which are linked together in chronological order to form a chain [13]. This structure prevents tampering, as altering one block would require changing all subsequent blocks, making the data virtually immutable.

6.8.1 Key Components of Blockchain Technology

Blockchain technology consists of several key components that work together to ensure its functionality, security, and integrity. Here are the main components:

6.8.1.1 Blocks

Each block contains:

- **Data**: Information about transactions (e.g., sender, receiver, and amount).
- **Timestamp**: The time the block was created.
- **Nonce**: A unique number used in mining (in proof-of-work systems).
- **Hash**: A cryptographic hash of the block's contents and the hash of the previous block, linking it to the chain.

6.8.1.2 Chain

The chain is the sequential linking of blocks, where each block references the previous one. This structure maintains the order of transactions and enhances security through interconnectivity.

6.8.1.3 Nodes

Nodes are computers participating in the blockchain network. They maintain copies of the blockchain, validate transactions, and propagate information throughout the network [14]. Nodes can be categorized as:

- **Full Nodes**: Store the entire blockchain and validate transactions.
- **Light Nodes**: Store only a portion of the blockchain and rely on full nodes for verification.

6.8.1.4 Consensus Mechanisms

These are protocols used to achieve agreement among nodes on the validity of transactions. Common mechanisms include:

- **Proof of Work (PoW)**: Requires computational effort to solve complex puzzles.
- **Proof of Stake (PoS)**: Validators are chosen based on the amount of cryptocurrencies they hold and are willing to stake.

6.8.1.5 Cryptographic Hash Functions

These are algorithms that convert data into a fixed-size string of characters, making it secure and unique. They ensure the integrity of data within the blockchain, as even a minor change in the input will result in a completely different hash.

6.8.1.6 Smart Contracts

Self-executing contracts with the terms of the agreement directly written into code. They automate processes and enforce agreements without intermediaries.

6.8.1.7 Tokens and Cryptocurrencies

Digital assets that can represent value or utility within the blockchain ecosystem. They can be used for transactions, accessing services, or incentivizing participants.

6.8.1.8 Wallets

Digital tools that allow users to store, send, and receive cryptocurrencies. Wallets can be:

- **Hot Wallets**: Connected to the internet, making them more accessible but potentially less secure.
- **Cold Wallets**: Offline storage, providing enhanced security against hacks.

6.8.2 Types of Blockchain Technology

The different blockchain technologies have distinct purposes, making them versatile tools for a variety of applications across different industries.

6.8.2.1 Public Blockchain

Public blockchains are open to anyone, allowing unrestricted participation in reading, writing, and validating transactions. They are highly decentralized and transparent, with notable examples including Bitcoin and Ethereum [15]. These blockchains provide a secure environment but can face scalability challenges due to their wide accessibility.

6.8.2.2 Private Blockchain

Private blockchains restrict access to authorized users, making them ideal for organizations that require greater control and confidentiality. They are typically faster and more efficient than public blockchains, as seen in platforms like Hyperledger. These blockchains prioritize privacy but sacrifice some degree of transparency.

6.8.2.3 Consortium Blockchain

Consortium blockchains are governed by a group of organizations rather than a single entity. This structure facilitates collaboration among businesses while maintaining a level of decentralization. Examples include Quorum and Hyperledger Fabric, which are particularly suited for industries where multiple stakeholders need to work together securely.

6.8.2.4 Hybrid Blockchain

Hybrid blockchains merge the features of both public and private blockchains, allowing selective transparency and privacy [16]. This flexibility makes them suitable for various applications that require different access levels, as demonstrated by platforms like Dragonchain, which cater to both public and private needs.

6.8.2.5 Sidechains

Sidechains are separate blockchains linked to a main blockchain, enabling the transfer of assets between the two. They operate independently and allow for experimentation and development without impacting the main chain, as seen in solutions like the Liquid Network for Bitcoin.

6.8.2.6 Layer 2 Solutions

Layer 2 solutions are built on top of existing blockchains to enhance scalability and transaction speeds. They facilitate faster transactions while retaining the security of the main blockchain, with examples including the Lightning Network for Bitcoin and Optimistic Rollups for Ethereum.

6.8.3 Features of Blockchain Technology

- **Decentralization**: Unlike traditional databases controlled by a single entity, blockchains distribute data across multiple nodes, reducing the risk of failure or manipulation.
- **Immutability**: Each block is linked to the previous one through cryptographic hashes, creating a secure chain. Changing any block would require altering all subsequent blocks, which is extremely difficult.
- **Transparency**: All participants in the network can view the entire blockchain, promoting trust and accountability.
- **Security**: Transactions are encrypted and require consensus from multiple nodes, making them secure against fraud and hacking.

6.8.4 Benefits of Blockchain Technology

- **Transparency**: Transactions on a blockchain are visible to all participants, enhancing accountability and trust among parties.
- **Security**: Blockchain uses cryptographic techniques, making it highly secure against fraud and unauthorized access.
- **Decentralization**: It eliminates the need for a central authority, reducing the risk of a single point of failure and allowing for P2P interactions.
- **Immutability**: Once data is recorded on a blockchain, it cannot be easily altered or deleted, providing a reliable audit trail.
- **Efficiency**: Smart contracts can automate processes, reducing the need for intermediaries and speeding up transactions.
- **Cost Reduction**: By minimizing the need for middlemen and lowering transaction fees, blockchain can reduce costs for businesses.
- **Traceability**: Blockchain allows for easy tracking of assets through the supply chain, enhancing provenance and reducing fraud.

- **Accessibility**: It can provide financial services to unbanked populations, increasing financial inclusion.
- **Interoperability**: Different blockchain networks can interact with each other, enabling more extensive and versatile applications.
- **Innovation**: The technology fosters new business models and services, driving innovation across various industries.

6.8.5 Challenges and Limitations of Blockchain Technology

Blockchain technology offers many benefits, but it also faces several challenges and limitations that can impact its adoption and effectiveness. Here are some key challenges:

- **Scalability:** Many blockchain networks struggle to handle a high volume of transactions efficiently. As user demand increases, transaction speeds can slow down, leading to higher fees and delays, particularly in public blockchains like Bitcoin and Ethereum.
- **Energy Consumption:** PoW consensus mechanisms, used by many cryptocurrencies, require significant computational power and energy. This has raised environmental concerns and prompted a search for more sustainable alternatives.
- **Regulatory Uncertainty:** The legal status of blockchain technology and cryptocurrencies varies by jurisdiction, leading to uncertainty in regulatory frameworks. This can hinder investment and innovation in the space, as businesses navigate complex compliance requirements.
- **Complexity and Usability:** Understanding and implementing blockchain technology can be complex, requiring specialized knowledge. This complexity can deter businesses and individuals from adopting it, especially if user interfaces are not intuitive.
- **Interoperability Issues:** Different blockchain networks often lack compatibility, making it challenging to share data and transfer assets across platforms. This limits the overall potential of blockchain solutions in a multinetwork environment.
- **Security Vulnerabilities:** While blockchain itself is secure, applications built on top of it can have vulnerabilities. Additionally, social engineering attacks (like phishing) can target users and undermine security.
- **Cost of Implementation:** Establishing and maintaining a blockchain network can involve significant costs, including infrastructure setup, transaction fees, and ongoing operational expenses. For private blockchains, this can be particularly high.
- **Data Privacy Concerns:** While blockchain provides transparency, this can conflict with privacy needs. Public blockchains expose transaction details, which

may not be suitable for all applications, particularly those involving sensitive information.

- **Governance Challenges:** Deciding how to manage changes or upgrades to blockchain protocols can be contentious, leading to forks or splits within the community. Governance models can be unclear or ineffective, causing disputes among stakeholders.
- **Limited Adoption:** Despite its potential, widespread adoption of blockchain technology remains limited in many industries. This can be due to a lack of understanding, resistance to change, or the perceived risks associated with new technologies.

6.8.6 Applications of Blockchain Technology

Blockchain technology has a wide range of applications across various industries due to its inherent qualities of security, transparency, and decentralization. The following are some of the applications:

- **Cryptocurrencies:** The most well-known application, cryptocurrencies like Bitcoin and Ethereum use blockchain to enable secure P2P transactions without the need for intermediaries.
- **Supply Chain Management:** Blockchain enhances transparency and traceability in supply chains. It allows all stakeholders to track products from origin to consumer, reducing fraud and improving efficiency.
- **Healthcare:** In healthcare, blockchain can securely store patient records, ensuring privacy while allowing authorized access to medical data. This improves data integrity and streamlines the sharing of information between healthcare providers.
- **Voting Systems:** Blockchain can be utilized to create secure and transparent voting systems, ensuring that votes are accurately recorded and tamper-proof. This can increase trust in electoral processes.
- **Digital Identity Verification:** Blockchain provides secure solutions for identity verification, reducing the risk of identity theft and fraud. It allows individuals to control their personal data while providing proof of identity when needed.
- **Smart Contracts:** These self-executing contracts with the terms written into code automate processes and reduce the need for intermediaries. They can be used in various industries, from real estate transactions to insurance claims.
- **Financial Services:** Blockchain can streamline processes in banking and finance, such as cross-border payments, trade finance, and asset management. It can reduce transaction times and costs while increasing security.
- **Real Estate:** In real estate, blockchain can simplify property transactions by providing transparent ownership records and facilitating the transfer of property titles, reducing fraud and increasing efficiency.

- **Intellectual Property Rights:** Blockchain can protect intellectual property by providing a secure way to register and track ownership of digital assets, ensuring creators are compensated fairly.
- **Energy Sector:** Blockchain can facilitate P2P energy trading, allowing consumers to buy and sell excess energy generated from renewable sources directly to one another.

6.9 Role of Market Makers in MEHs

Market makers play a crucial role in maintaining liquidity and facilitating trade in financial markets, helping to achieve and maintain market equilibrium. In the context of a market equilibrium model, their role can be understood through the following key functions.

6.9.1 Providing Liquidity

Market makers continuously quote buy (bid) and sell (ask) prices for a security, ensuring that there is always a counterparty for both buyers and sellers. This helps to:

- Reduce price volatility by providing immediate liquidity.
- Allow traders to execute transactions without having to wait for other participants, which stabilizes the market.

6.9.2 Reducing Bid-Ask Spreads

By providing tight bid-ask spreads (the difference between the highest price a buyer is willing to pay and the lowest price a seller is willing to accept), market makers help minimize transaction costs for investors [17]. A narrower spread encourages trading and makes the market more efficient, which can contribute to achieving equilibrium.

6.9.3 Price Discovery

Market makers adjust their prices based on supply and demand conditions. If there are more buyers than sellers, the price will rise, and vice versa. By constantly adjusting prices, market makers help reflect the true underlying value of securities, facilitating efficient price discovery that leads toward market equilibrium.

6.9.4 Stabilizing Markets

Market makers can act as shock absorbers during periods of market volatility. By continuing to provide liquidity when there are significant imbalances in supply and demand, they help prevent excessive price swings, which can help the market gravitate toward equilibrium over time.

6.9.5 Reducing Information Asymmetry

Market makers reduce information asymmetry by providing regular, public price quotes. This transparency allows all market participants to access the same information, reducing the chances of mispricing and improving the efficiency of the market [18].

6.9.6 Risk Management

Market makers take on the risk of holding inventories of securities. By doing so, they mitigate the risk for other market participants, allowing buyers and sellers to trade with minimal concern about whether their orders will be filled. This risk management function helps maintain a smoother trading environment that supports equilibrium.

6.9.7 Facilitating Arbitrage

Market makers may also facilitate arbitrage opportunities by pricing assets in such a way that discrepancies between different markets or related securities are quickly resolved. This ensures that prices reflect fair value, further promoting equilibrium across markets.

6.10 Smart Contracts Between EHs

Smart contracts between **energy hubs (EHs)** can play a transformative role in decentralized energy systems. EHs are typically nodes in a network that manage the supply, demand, and storage of energy. These hubs can represent households, businesses, or industrial entities that both consume and produce energy, often referred to as **prosumers.**

6.10.1 Role of Smart Contracts Between Energy Hubs

Smart contracts are self-executing contracts with the terms of the agreement directly written into code. In the context of energy hubs, they can automatically facilitate, verify, and enforce energy transactions and interactions based on predefined conditions [19]. Here are some key roles and potential applications:

6.10.1.1 Energy Trading

Smart contracts enable direct, P2P energy trading between energy hubs. In decentralized energy grids, prosumers can generate surplus energy (e.g., from solar panels) and sell it to other hubs that need additional energy. Smart contracts automatically execute these trades based on the below conditions:

- Price agreements between hubs.
- Energy availability and demand.
- Energy quality or type (e.g., renewable vs. nonrenewable).

6.10.1.2 Dynamic Pricing

Smart contracts allow for dynamic pricing models based on real-time market conditions. In traditional energy systems, prices are often fixed or regulated. In a smart contract-enabled system, the energy prices are based on

- Time of day (peak vs. off-peak hours).
- Weather conditions affecting renewable energy production.
- Grid congestion or localized supply–demand mismatches.

This flexibility promotes more efficient energy use and encourages hubs to trade at optimal times.

6.10.1.3 Automated Energy Distribution

Smart contracts can manage energy distribution by ensuring that energy flows optimally between different hubs. When one hub generates excess energy, a smart contract can direct the energy to nearby hubs with the highest demand, minimizing transmission losses and ensuring efficient grid operation [20].

6.10.1.4 Microgrid Management

In decentralized energy systems, microgrids (small-scale power grids) often operate semi-independently. Smart contracts between energy hubs within a microgrid can handle:

- Balancing supply and demand within the microgrid.
- Automatically adjusting energy flows based on local production (e.g., from solar and wind).
- Prioritizing critical services (e.g., hospitals) during power shortages.

This enables microgrids to operate more autonomously while maintaining stability and reliability.

6.10.1.5 Energy Storage Management

Many energy hubs have integrated energy storage systems (e.g., batteries) to store excess energy for later use. Smart contracts can manage storage capacity by:

- Automatically storing energy during times of low demand or low prices.
- Selling stored energy back into the grid or to other hubs when demand is high.
- Optimizing storage use based on preset conditions (e.g., energy prices and demand forecasts).

6.10.1.6 Grid Balancing and Stability

Energy hubs and smart contracts can collectively maintain grid stability by:

- Adjusting energy production or consumption based on grid requirements.
- Preventing overloading of the grid by throttling energy flows during high-demand periods.
- Automatically responding to grid frequency changes or voltage fluctuations by adjusting hub output.

Smart contracts can coordinate these actions across multiple energy hubs without human intervention.

6.10.1.7 Carbon Credits and Sustainability Incentives

In some smart energy systems, energy hubs can earn carbon credits or other sustainability incentives for producing renewable energy or reducing consumption during peak periods. Smart contracts can:

- Automatically track energy generation from renewable sources.
- Transfer carbon credits between hubs.
- Ensure compliance with sustainability goals by enforcing green energy quotas.

6.10.1.8 Grid Services (Demand Response)

Smart contracts allow energy hubs to participate in demand response programs, where they adjust energy consumption based on grid needs. For example, during times of peak demand, a smart contract could automatically reduce energy use in noncritical areas of a building or sell energy back to the grid. These contracts can be triggered by:

- Real-time grid conditions.
- Energy prices reaching a certain threshold.
- Requests from grid operators for load shedding.

6.10.1.9 Dispute Resolution

In case of disputes (e.g., energy shortages and contractual disagreements), smart contracts can automate resolutions based on predefined rules. For example, if one hub fails to deliver the agreed amount of energy, penalties or compensations can be automatically enforced.

6.10.2 Benefits of Smart Contracts in Energy Hubs

- **Transparency**: All transactions are recorded on the blockchain, ensuring auditability and trust.
- **Efficiency**: By automating transactions and energy flows, smart contracts reduce the need for intermediaries and minimize delays.
- **Cost-Effectiveness**: Reducing intermediaries and enabling direct trading lowers transaction costs.
- **Decentralization**: Smart contracts empower energy hubs to operate independently, reducing reliance on centralized grid operators.
- **Scalability**: They enable the development of large-scale, decentralized energy networks by coordinating multiple hubs.

6.11 Algorithms for Energy Trading Among EHs

Energy trading among EHs can involve various algorithms to optimize the allocation and trading of energy resources. Some of the algorithms and methods used for energy trading are

6.11.1 Market-Based Algorithms

6.11.1.1 Auction Mechanisms

Market-based algorithms often utilize auction mechanisms where EHs submit bids to either buy or sell energy. The clearing process determines the prices based on supply and demand, ensuring efficient market operation. Variants like double auctions enable simultaneous bidding from buyers and sellers, fostering competitive pricing.

6.11.2 Game Theory Approaches

6.11.2.1 Nash Equilibrium

Game theory provides a framework for understanding strategic interactions among EHs. In competitive settings, Nash equilibrium can be analyzed to determine stable strategies where no EH can benefit from changing its actions unilaterally [21]. This helps in formulating robust trading strategies.

6.11.2.2 Cooperative Game Theory

In scenarios where collaboration is possible, cooperative game theory can be applied to form coalitions among EHs. This approach seeks to maximize the overall utility of the group, enabling better resource allocation and sharing of benefits.

6.11.3 Optimization Algorithms

6.11.3.1 Linear Programming (LP)

Linear programming techniques are effective for modeling energy trading as an optimization problem, aiming to minimize costs or maximize profits subject to various constraints. This approach is beneficial for straightforward scenarios with linear relationships.

6.11.3.2 Mixed-Integer Programming (MIP)

For more complex trading situations that involve discrete decisions (like capacity limits), MIP provides a robust optimization framework. MIP can handle multiple variables and constraints, making it suitable for diverse trading scenarios.

6.11.3.3 Dynamic Programming

Dynamic programming is useful for addressing multistage decision problems, where EHs can optimize their trading strategies over time, taking future uncertainties into account.

6.11.4 Machine Learning Techniques

6.11.4.1 Reinforcement Learning (RL)

Reinforcement learning allows EHs to learn optimal trading strategies through trial and error. By simulating various trading scenarios, RL models can adaptively improve decision-making based on past experiences and outcomes.

6.11.4.2 Neural Networks

Neural networks can be employed to predict price movements and demand patterns, helping EHs make informed trading decisions based on historical data.

6.11.5 Multiagent Systems

6.11.5.1 Distributed Algorithms

In a multiagent system setup, distributed algorithms enable EHs to operate autonomously while communicating with each other. This fosters a decentralized approach to trading, allowing for adaptive responses to market changes [22].

6.11.5.2 Consensus Algorithms

Consensus algorithms ensure that all EHs agree on trading terms and conditions, facilitating cooperation and reducing conflicts in decentralized energy trading.

6.11.6 Forecasting Models

6.11.6.1 Time Series Analysis

Time series analysis leverages historical data to forecast future energy demand and supply trends. Accurate forecasting is critical for strategic trading and efficient resource management.

6.11.6.2 Weather Forecasting Models

Incorporating weather data into trading models helps predict renewable energy generation, such as solar and wind, allowing EHs to optimize their trading strategies accordingly.

6.11.7 Blockchain and Smart Contracts

6.11.7.1 Decentralized Trading Platforms

Blockchain technology provides a secure and transparent framework for energy trading. Smart contracts can automate transactions based on predefined conditions, enhancing efficiency and trust in the trading process.

6.11.8 Heuristic Methods

6.11.8.1 Genetic Algorithms

Genetic algorithms simulate natural selection processes to explore optimal trading strategies. They are particularly useful for solving complex optimization problems with many variables.

6.11.8.2 Particle Swarm Optimization

This method employs a swarm-based approach to find optimal trading conditions by mimicking the social behavior of birds or fish, allowing EHs to converge on efficient trading solutions.

6.12 Regulatory Framework for MEHs

The regulatory framework for market equilibrium models in energy trading plays a crucial role in ensuring fair, efficient, and sustainable energy markets. Here's an overview of the key components and considerations within this framework.

6.12.1 Market Structure and Design

The regulatory framework defines the market structure, distinguishing between centralized (pool) and decentralized (bilateral) markets. Each structure impacts how market equilibrium is achieved. Regulations specify the roles and responsibilities of market participants, including producers, consumers, and aggregators, ensuring fair access and participation for all stakeholders.

6.12.2 Price Formation Mechanisms

Regulations govern the mechanisms of price formation, outlining auction designs, such as how bids are submitted and how prices are cleared. Price caps and floors may also be implemented to mitigate excessive volatility and prevent market manipulation, helping to maintain stability in equilibrium.

6.12.3 Transparency and Reporting

Transparency is crucial in energy markets. Regulatory frameworks often require participants to disclose data on prices, quantities, and transaction volumes. This transparency fosters trust and facilitates accurate modeling of market equilibrium [23]. Additionally, market monitoring mechanisms are established to detect anti-competitive behaviors, ensuring fair competition.

6.12.4 Market Power and Competition

To prevent monopolistic practices, regulatory frameworks incorporate antitrust regulations that prohibit collusion and other anti-competitive behaviors. Market share limits may also be set for larger players to ensure that competition remains robust, which is vital for maintaining equilibrium in the market.

6.12.5 Consumer Protection

Regulations often establish tariff guidelines to ensure fair pricing for consumers. Consumer protection laws mandate that customers receive clear information about pricing structures and their rights, empowering them to make informed decisions in the market and fostering a more balanced power dynamic between consumers and suppliers.

6.12.6 Environmental and Sustainability Standards

Regulatory frameworks typically include environmental regulations, such as emissions trading schemes and renewable energy standards. These policies encourage sustainable practices among market participants, influencing supply dynamics and shifting market equilibrium in favor of cleaner energy sources.

6.12.7 Grid Reliability and Security

Ensuring grid reliability is a key aspect of regulatory frameworks. Regulations often set reliability standards that participants must adhere to, ensuring supply meets demand and maintaining stability in the market. Capacity markets may also be implemented to ensure that sufficient resources are available during peak demand periods.

6.12.8 Technological Integration

Regulatory frameworks may promote the adoption of innovative technologies, such as smart grids and energy storage systems, which enhance market efficiency. Additionally, data privacy and security regulations are essential as markets become increasingly data-driven, protecting sensitive information and maintaining overall market integrity.

6.13 Benefits of Market Trading in MEHs

- **Price Stability:** Energy trading helps stabilize prices by balancing supply and demand, reducing volatility.
- **Improved Resource Allocation:** Trading directs energy to where it's most needed, maximizing economic efficiency.
- **Enhanced Market Efficiency:** Competitive trading environments lower costs and improve overall market functionality.
- **Incentives for Innovation:** Market competition drives technological advancements and more efficient energy solutions.
- **Risk Management and Hedging:** Trading allows participants to use hedging strategies to mitigate price volatility risks.
- **Facilitation of Renewable Integration:** Energy trading supports the incorporation of variable renewable sources into the grid.
- **Consumer Benefits:** Increased competition often leads to lower prices and better services for consumers.
- **Encouragement of Investment:** A transparent trading market attracts investments by signaling growth opportunities.
- **Flexibility and Responsiveness:** Trading mechanisms enable quick adjustments to changing market conditions and disruptions.
- **Policy Evaluation and Development:** Energy trading aids in assessing and developing policies by simulating various market scenarios.

6.14 Challenges and Limitations of MEHs

- **Market Complexity:** Energy markets involve numerous variables and interactions, making modeling challenging.
- **Data Limitations:** Inaccurate, incomplete, or outdated data can lead to unreliable model outputs.
- **Behavioral Assumptions:** Models often assume rational behavior, which may not reflect real-world participant actions.
- **Regulatory Variability:** Differing regulations across regions can complicate market dynamics and modeling accuracy.
- **Price Volatility:** Rapid price fluctuations can disrupt market equilibrium and affect trading strategies.
- **Market Power Concentration:** Dominant players can manipulate markets, leading to inefficiencies and distorted prices.
- **Time Lags in Adjustment:** Delayed responses to price changes can result in temporary market imbalances.
- **Environmental Externalities:** Traditional models may overlook the impact of environmental costs on market outcomes.
- **Technological Disruptions:** Rapid advancements can change market dynamics faster than models can adapt.
- **Limited Forecasting Accuracy:** Uncertainties in demand and supply forecasting can undermine model reliability and effectiveness.

6.15 Case Studies of Energy Trading in MEHs

6.15.1 Indian Energy Market

Indian energy market, which was previously a regulated market structure, is being transformed to a market-based system in order to create efficiency and transparency. Short term trading of renewable energy is facilitated by the introduction of the Green Term Ahead Market (GTAM) and this will encourage the use of cleaner energy. To make decisions, the price dynamics can be analyzed by equilibrium models. There is also a balance between the renewable and conventional energy sources on the market. Even with the gains, there are still difficulties such as grid integration and policy compatibility.

6.15.2 Nord Pool Power Market

The Nord Pool power market exemplifies the benefits of market coupling in enhancing regional electricity trading. By integrating multiple countries, this model allows for cross-border exchanges that respond dynamically to supply and demand fluctuations [24]. This approach has resulted in lower prices and increased efficiency, making it a valuable case study for understanding how market equilibrium can optimize resource allocation in a multination context.

6.15.3 Electric Reliability Council of Texas (ERCOT) in Texas

In Texas, the ERCOT operates a deregulated energy market where equilibrium models are crucial for managing grid reliability. These models help predict demand spikes and optimize the dispatch of various energy resources, especially during extreme weather conditions. This case highlights the importance of flexibility in energy trading, particularly as the state increases its reliance on renewable sources.

6.15.4 Australian National Electricity Market (NEM)

The Australian NEM provides insights into optimizing grid operations amidst rising renewable energy sources. Market equilibrium models analyze price signals and supply constraints, guiding investments in storage and demand response solutions. This case study demonstrates how equilibrium frameworks can support a reliable and sustainable energy transition, ensuring stability as the market evolves.

6.15.5 California Electricity Market

The California electricity market offers a perspective on volatility and regulatory impacts in energy trading. Following significant policy changes aimed at integrating renewable resources, equilibrium models have been employed to analyze pricing fluctuations and their effects on market stability. This case study emphasizes the need for adaptive strategies in response to regulatory shifts and the increasing share of renewables [25].

6.16 Future Trends in Energy Trading in MEHs

6.16.1 Increased Integration of Renewables

As countries strive for carbon neutrality, the integration of renewable energy sources will become paramount. Market equilibrium models will increasingly focus on forecasting the variability of renewables and their impact on pricing and supply dynamics. Enhanced data analytics and machine learning will aid in improving these predictions.

6.16.2 Decentralized Energy Systems

The rise of decentralized energy systems, such as microgrids and community solar projects, will influence market structures. Equilibrium models will need to adapt to account for local generation and consumption patterns, fostering greater participation from small producers and consumers in trading activities.

6.16.3 Advanced Storage Solutions

The advancement of energy storage technologies will play a critical role in stabilizing supply and demand. Market equilibrium models will incorporate storage dynamics to optimize trading strategies, allowing for more flexible responses to fluctuations in generation and consumption.

6.16.4 Enhanced Regulatory Frameworks

As markets evolve, regulatory frameworks will likely become more sophisticated to address issues like market manipulation, renewable energy credits, and emissions trading. Equilibrium models will need to adapt to these regulatory changes, ensuring that they accurately reflect market conditions and compliance requirements [13].

6.16.5 Digitalization and Smart Grids

The digital transformation of energy systems, including the deployment of smart grids and real-time monitoring technologies, will enhance data availability. This will improve the accuracy of market equilibrium models, enabling more responsive trading strategies and better management of supply and demand.

6.16.6 Global Market Interconnections

Increased globalization of energy markets will lead to more interconnected trading systems. Market equilibrium models will need to account for cross-border exchanges, helping to optimize resource allocation and price stability across regions [25].

6.16.7 Demand Response Programs

The growth of demand response initiatives will influence trading patterns as consumers actively adjust their energy usage in response to price signals. Equilibrium models will integrate these consumer behaviors to improve demand forecasts and trading strategies.

6.16.8 Focus on Sustainability

As environmental concerns rise, energy trading will increasingly incorporate sustainability metrics. Market equilibrium models will evolve to reflect not just economic efficiency but also the environmental impact of trading decisions, aligning with global sustainability goals.

6.17 Conclusion

In energy management and trading systems, the market and energy trading mechanism in medium voltage energy hubs (MV-EHs) is a critical development. MV-EHs enhance economic viability and sustainability by facilitating more efficient and flexible energy trading through the integration of sophisticated technology and decentralized energy supplies. Through these approaches, stakeholders can encourage the switch to renewable energy sources while optimizing energy use, cutting costs, and enhancing system stability. The effective application of these trading mechanisms will be essential to building a robust and adaptable energy ecosystem and promoting cleaner and more efficient energy sources in the future as the energy landscape changes. The efficacy and uptake of MV-EHs will be improved by ongoing innovation and cooperation among industry players, guaranteeing their substantial contribution to the world energy market.

References

1 Shama Naz, I. (2024). A review of peer-to-peer energy trading markets: enabling models and technologies. *Energies* 17: 1702. https://doi.org/10.3390/en17071702.

2 Faia, R., Lezama, F., Soares, J. et al. (2024). Local electricity markets: a review on benefits, barriers, current trends and future perspectives. *Renewable and Sustainable Energy Reviews* 190: 114006.

3 Shah, W.U.H., Hao, G., Yan, H. et al. (2022). The impact of trade, financial development and government integrity on energy efficiency: an analysis from G7-Countries. *Energy* 255: 124507.

4 Faria, A.S., Soares, T., Orlandini, T. et al. (2023). P2P market coordination methodologies with distribution grid management. *Sustain. Energy Grids Netw.* 34: 101075.

5 Zhou, W., Dang, W., Peng, F. et al. (2023). A decentralized peer-to-peer energy trading strategy considering flexible resource involvement and renewable energy uncertainty. *International Journal of Electrical Power & Energy Systems* 152: 109275.

6 Kim, H.J., Chung, Y.S., Kim, S.J. et al. (2023). Pricing mechanisms for peer-to-peer energy trading: towards an integrated understanding of energy and network service pricing mechanisms. *Renewable and Sustainable Energy Reviews* 183: 113435.

7 Liaquat, S., Hussain, T., Celik, B. et al. (2023). Day-ahead continuous double auction-based peer-to-peer energy trading platform incorporating trading losses and network utilisation fee. *IET Smart Grid* 6: 312–329.

8 Wang, J., Li, L., and Zhang, J. (2023). Deep reinforcement learning for energy trading and load scheduling in residential peer-to-peer energy trading market. *International Journal of Electrical Power & Energy Systems* 147: 108885.

9 Qiu, D., Wang, J., Dong, Z. et al. (2023). Mean-field multi-agent reinforcement learning for peer-to-peer multi-energy trading. *IEEE Transactions on Power Apparatus and Systems* 38: 4853–4866.

10 Qiu, D., Xue, J., Zhang, T. et al. (2023). Federated reinforcement learning for smart building joint peer-to-peer energy and carbon allowance trading. *Applied Energy* 333: 120526.

11 Mohseni, S., Pishvaee, M.S., and Dashti, R. (2023). Privacy-preserving energy trading management in networked microgrids via data-driven robust optimization assisted by machine learning. *Sustain. Energy, Grids Netw.* 34: 101011.

12 Bukar, A.L., Hamza, M.F., Ayub, S. et al. (2023). Peer-to-peer electricity trading: a systematic review on current developments and perspectives. *Renew. Energy Focus* 44: 317–333.

13 Kochupurackal, A., Pancholi, K.P., Islam, S.N. et al. (2023). Rolling horizon optimisation based peer-to-peer energy trading under real-time variations in demand and generation. *Energy Systems* 14: 541–565.

14 Liu, J., Zhou, Y., Yang, H., and Wu, H. (2022). Uncertainty energy planning of net-zero energy communities with peer-to-peer energy trading and green vehicle storage considering climate changes by 2050 with machine learning methods. *Applied Energy* 321: 119394.

15 Gunarathna, C.L., Yang, R.J., Jayasuriya, S., and Wang, K. (2022). Reviewing global peer-to-peer distributed renewable energy trading projects. *Energy Research and Social Science* 89: 102655.

16 Muhsen, H., Allahham, A., Al-Halhouli, A. et al. (2022). Business model of peer-to-peer energy trading: a review of literature. *Sustainability* 14: 1616.

17 Azim, M.I., Tushar, W., Saha, T.K. et al. (2022). Peer-to-peer kilowatt and negawatt trading: a review of challenges and recent advances in distribution networks. *Renewable and Sustainable Energy Reviews* 169: 112908.

18 Tsaousoglou, G., Giraldo, J.S., and Paterakis, N.G. (2022). Market mechanisms for local electricity markets: a review of models, solution concepts and algorithmic techniques. *Renewable and Sustainable Energy Reviews* 156: 111890.

19 Zheng, B., Wei, W., Chen, Y. et al. (2022). A peer-to-peer energy trading market embedded with residential shared energy storage units. *Applied Energy* 308: 118400.

20 Dimitriadis, C., Tsimopoulos, E., and Georgiadis, M. (2021). A Review on the complementarity modelling in competitive electricity markets. *Energies* 14: 7133.

21 Zang, H. and Kim, J. (2021). Reinforcement learning based peer-to-peer energy trade management using community energy storage in local energy market. *Energies* 14: 4131.

22 Jegadeeswari, G., Kirubadurai, B., & Lakshmi, D. (2020). Multi carrier based new random pulse width modulation for three phase inverters. at *International Journal of Scientific and Technology Research (IJSTR)*, 9(2), 5425–5432.

23 Bandara, K.Y., Thakur, S., and Breslin, J. (2021). Flocking-based decentralised double auction for P2P energy trading within neighbourhoods. *International Journal of Electrical Power & Energy Systems* 129: 106766.

24 Le Cadre, H., Jacquot, P., Wan, C., and Alasseur, C. (2020). Peer-to-peer electricity market analysis: From variational to generalized Nash equilibrium. *European Journal of Operational Research* 282: 753–771.

25 Guerrero, J., Gebbran, D., Mhanna, S. et al. (2020). Towards a transactive energy system for integration of distributed energy resources: home energy management, distributed optimal power flow, and peer-to-peer energy trading. *Renewable and Sustainable Energy Reviews* 132: 110000.

7

Impact of High or Full Share of RESs and Load-Side Uncertainty on Multi-Vectored Energy Hubs

Stephen Oko Gyan Torto[1], *Rupendra Kumar Pachauri*[1], *and Jai Govind Singh*[2]

[1] *Electrical Cluster, School of Engineering, University of Petroleum and Energy Studies, Dehradun, Uttarakhand, India*
[2] *Department of Energy, Environment and Climate Change, School of Environmental Resource and Development, Asian Institute of Technology, Khlong Nueng, Thailand*

Nomenclature

ΔCO_2	Reduction in carbon emissions (kg)
ΔQ_d	Change in quantity demanded (MWh)
A_t	Actual load at time t (kW)
C_{flat}	Flat rate cost of electricity (USD/MWh)
$C_{off-peak}$	Cost of electricity during off-peak periods (USD/kWh)
C_{peak}	Cost of electricity during peak periods (USD/kWh)
$C_{RTP}(i)$	Real-time price of electricity at time i (USD/MWh)
$C_{storage}$	Required storage capacity (MWh)
E_{actual}	Actual energy consumed (MWh)
$E_{capacity}$	Total storage capacity (MWh)
$E_{curtailed}$	Amount of renewable energy curtailed (MWh)
$E_{discharged}$	Energy discharged from storage (MWh)
E_{loss}	Energy losses in the system (MWh)
$E_{non-RES}$	Energy from non-renewable sources (MWh)
E_{peak}	Peak load (MWh)
E_{RES}	Energy from renewable sources (MWh)
E_{RES}	Emission factor for renewable sources (kWh)
E_{RES}	Energy produced by renewable sources (kWh)
$E_{storage}$	Energy from storage systems (MWh)
E_{total}	Total energy supplied to the grid (MWh)
E_{total}	Total energy produced (kWh)

Energy Management Strategies for Multi-Vectored Energy Hubs to Achieve Low Carbon Societies, First Edition.
Edited by Shubham Tiwari, Jai Govind Singh, Sivaraman Palanisamy, Sharmeela Chenniappan, Rupendra Kumar Pachauri, and Sanjeevikumar Padmanaban.

$E_{totalRES}$	Total renewable energy produced (MWh)
$EF_{non-RES}$	Emission factor for non-renewable sources (kg/kWh)
EF_{RES}	Emission factor for renewable sources (kg/kWh)
F_t	Forecasted load at time t (kW)
$P_{actual}(t)$	Actual power generated (MW)
$P_{flexible}(i)$	Flexible power capacity at time i (kW)
$P_{forecast}(t)$	Forecasted power (MW)
P_{max}	Maximum power output of the RES (MWh)
$P_{shifted}(i)$	Power shifted from peak to off-peak periods at time i (kW)
$P_{total}(i)$	Total power capacity at time i (kW)
Q_d	Original quantity demanded (MWh)
T_{max}	Maximum time the storage system needs to cover the gap in RES output (s)
V_{min}	Minimum voltage level observed (V)
$V_{nominal}$	Nominal voltage level (V)
ε_d	Price elasticity of demand
μ_{load}	Mean load (MW)
μ_{RES}	Mean renewable energy output (MWh)
σ_{load}	Standard deviation of the load (MW)
σ_{RES}	Standard deviation of renewable energy output (MWh)
ΔC_{RTP}	Total cost reduction due to real-time pricing (e.g., USD)
ΔP	Change in price (USD/MWh)
$\in (t)$	Forecasting error at the t (MW)
$E\ (i)$	Energy consumed at time i (MWh)
GRI	Grid reliability index
LVI	Load variability index
$MTBF$	Mean time between failures (h)
$MTTR$	Mean time to repair (h)
n	Number of observations
P	Original price (USD/MWh)
REPI	Renewable energy penetration index
SU	Storage utilization
T	Time period (s)
VSI	Voltage stability index

7.1 Introduction

The global energy landscape is experiencing a significant transition toward renewable energy sources (RES), driven by the urgent need to reduce greenhouse gas (GHG) emissions and mitigate climate change. This transition involves a higher penetration of volatile energy sources, such as solar and wind power, into existing energy systems, including medium voltage electrical heating systems (MV-EHS). While the benefits of such a shift are manifold, including reduced carbon footprints and enhanced sustainability, it also introduces considerable operational challenges primarily due to the variable nature of RES and the unpredictability of load demands.

Energy generation from solar and wind sources is highly dependent on environmental conditions, causing significant uncertainty. Effective forecasting models are crucial for predicting power availability and integrating these sources into the power grid. Developing advanced forecasting technologies is essential for managing unpredictability and efficient operation of MV-EHS. A cooperative energy management approach, as proposed [1], leverages cooperative game theory to enable energy hubs to exchange power effectively. This approach ensures that individual hubs within the network work together to secure maximum profit, which is then distributed fairly to maintain economic stability across the coalition. High-RES penetration can lead to power quality issues and infrastructure resilience. The demand side of energy systems, or load, exhibits its variabilities and uncertainties, which compound the challenges posed by high shares of RES. Load uncertainties, particularly in MV-EHS, can arise from various factors, including consumer behavior, economic activities, and climatic conditions. This uncertainty makes it more challenging to match energy demand with the fluctuating supply from RES, necessitating more flexible and adaptive energy systems. Strategies such as demand response, load forecasting, and dynamic pricing are essential to manage these uncertainties effectively. The transition toward a higher share of RES in MV-EHs brings a complex array of challenges, primarily driven by the uncertain nature of supply and demand. Addressing these challenges requires a holistic approach encompassing advanced forecasting techniques, robust grid management systems, and innovative regulatory frameworks. The successful integration of RES into MV-EHs promises a sustainable energy future. It demands a concerted effort from policymakers, engineers, and the scientific community to navigate the uncertainties associated with this transition.

The imperative to transition toward low-carbon societies is increasingly recognized as crucial in the global fight against climate change. Central to this transition is the role of energy management, which involves the strategic oversight and optimization of energy resources to minimize environmental impact while ensuring energy sufficiency and economic viability. The study [2] framework not only optimizes the cost, emissions, and self-reliability of the coalition but also includes an improved profit distribution scheme that rewards and penalizes microgrids based on their specific attributes, such as their capacity to reduce GHG emissions or their integration with RES. This approach ensures that the microgrids are incentivized to contribute positively to the coalition's goals while maintaining their operational independence. This section elaborates on the importance of effective energy management in achieving low-carbon societies, emphasizing the integration of RES and innovative technologies. Transitioning from fossil fuels to low-carbon solutions will play an essential role, as energy-related carbon dioxide (CO_2) emissions represent two-thirds of all GHG. Equation (7.2) analyzes carbon emission reduction by integrating RES and managing load-side uncertainty in MV-EHs [3]. As noted by the authors in [4], effective energy management strategies reduce GHG emissions, bolster energy security, stabilize energy costs, and support sustainable development. Efficient energy management is crucial as it reduces GHG emissions, enhances energy security, stabilizes energy costs, and supports economic stability and sustainable development. Technological innovation will enable this energy transition, notably in the field of renewable energy.

Demand response management (DRM) is a crucial energy management strategy that adjusts power demand during peak load times, enhancing efficiency and reducing costs. It can flatten peak demand curves by incentivizing consumers to reduce energy usage. As RES like wind and solar increase, robust energy storage solutions like lithium-ion batteries are needed to balance supply and demand. Furthermore, data analytics can facilitate more personalized energy services, enabling utilities to offer tailored energy solutions to consumers based on their specific energy usage patterns and needs. Integrating big data analytics with energy management systems (EMS) allows for predictive maintenance and more efficient energy grid operation, ultimately reducing operational costs and improving service reliability. As noted by [5], applying these innovative approaches in practical settings can dramatically transform the energy landscape, promoting more sustainable and efficient energy usage across various sectors and geographical regions. Figure 7.1 shows the general overview of the impact of load-side uncertainty and RES on multi-vectored energy hubs (MV-EHs) for easy understanding.

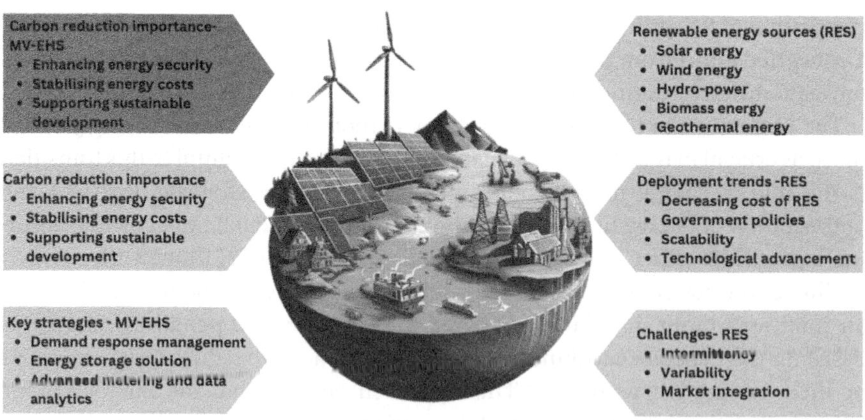

Figure 7.1 General overview of the impact of MV-EHs and RES.

7.1.1 Overview of Multi-Vectored Energy Hubs (MV-EHs)

MV-EHs represent a transformative approach to managing energy flows within an integrated network. These hubs coordinate multiple forms of energy, such as electricity, heat, and gas, optimizing their use based on demand, availability, and environmental impact [31].

Figure 7.2 shows a very simple multi-vectored energy hubs (MV-EHs) diagram designed for easy understanding. It includes the primary components.

7.1.1.1 Core Components of MV-EHS

The architecture of MV-EHs is composed of several key components, as mentioned in Table 7.1, that work together to optimize energy distribution and usage:

7.1.2 Challenges and Opportunities

Aspect	Description
Technical challenges	• The integration of diverse renewable energy sources into the existing grid presents significant challenges due to their intermittent and variable nature. • Advanced grid management technologies, such as energy storage systems and smart grid technologies, are needed for real-time supply and demand balancing, voltage maintenance, and grid stability. • Aging infrastructure needs upgrading to handle new dynamics and ensure cybersecurity and resilience.
Economic considerations	• Transitioning to a low-carbon society necessitates significant financial investment in infrastructure, renewable energy plants, advanced metering systems, and grid management tools. • Economic viability can be influenced by fluctuating market prices for materials, labor, and technologies. • Opportunities for job creation in new industries, reduced technology costs, and economic incentives like tax breaks and subsidies are being explored.
Regulatory frameworks	• Current frameworks may not be equipped to handle high shares of renewable energy. • Policies needed to support renewable adoption, infrastructure modernization, and technology development. • Policymakers can establish stable policies and incentives to encourage green technology investments, set energy efficiency and emissions standards, and guide industries toward sustainable practices.

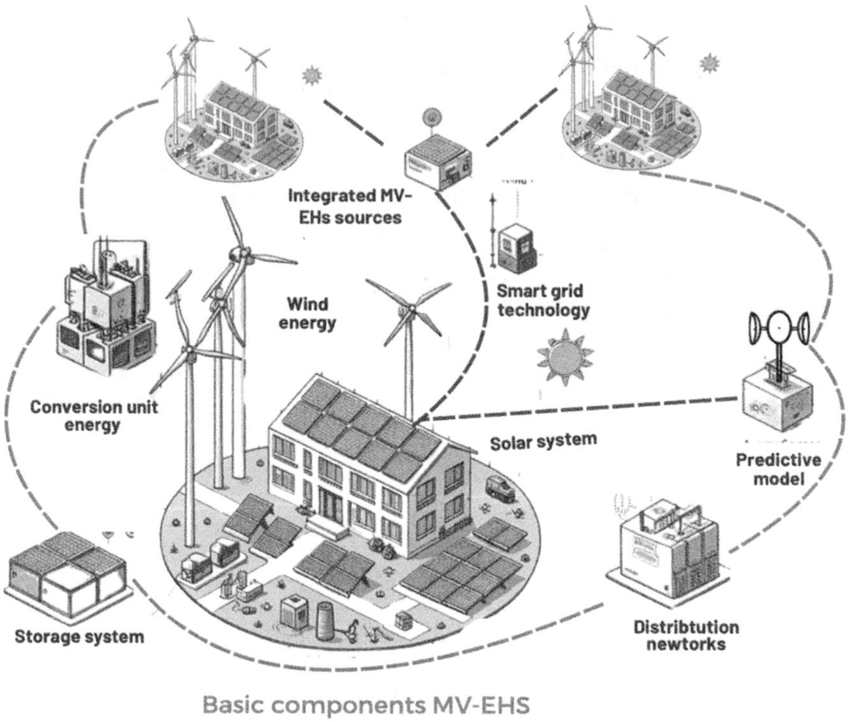

Figure 7.2 Simple diagram of multi-vectored energy hubs (MV-EHs).

Table 7.1 Core components of MV-EHs.

Component	Description
Energy sources	Includes both renewable (solar, wind, and hydro) and non-renewable sources (natural gas and coal) for diversified energy provision.
Energy conversion units	Converts raw energy into usable forms, such as heat pumps, boilers, and electrical inverters.
Energy storage systems	Stores excess energy produced during low-demand periods for peak demand, balancing supply and demand.
Distribution networks	Efficiently manages energy delivery to consumers in various forms.
Interconnected energy systems	Links different energy systems to manage energy flow efficiently across multiple vectors.
Integration of renewable energy sources	Harnesses solar, wind, and bioenergy within the hub, enhancing sustainability and reducing reliance on fossil fuels.
Smart grid technology	Utilizes AI and IoT to predict energy needs and optimize distribution, crucial for dynamic energy management [6].

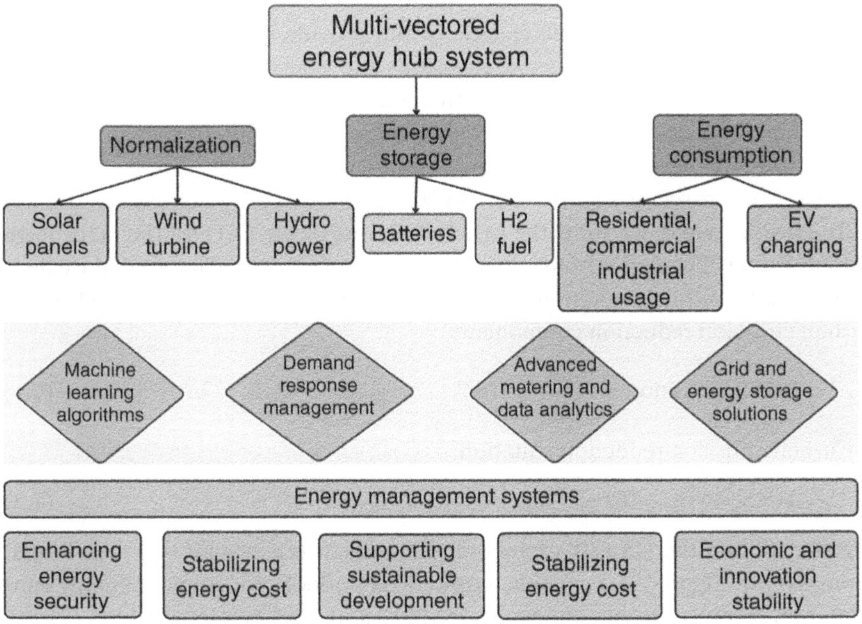

Figure 7.3 MV-EHs overview.

Figure 7.3 shows a very simple MV-EHs diagram designed to show the flow of energy through to the EMS.

7.2 RESs and Their Growing Share

Renewable energy sources (RESs), such as solar, wind, hydro, biomass, and geothermal energy, have become central to the global strategy for reducing reliance on fossil fuels and combating climate change. Solar and wind power, in particular, have witnessed significant growth due to their scalability and decreasing cost of technology. The solar photovoltaics, wind, and battery storage costs have dropped markedly since 2010. However, many recent studies and reports worldwide have not adequately captured such a dramatic decrease.

Solar power, wind energy, and biomass are rapidly expanding RES. Solar panels convert sunlight into electricity, while wind energy is efficient and scalable. Hydroelectric power plants use water movement, while pumped-storage hydroelectricity enhances grid reliability. Biomass, a sustainable energy source, can be produced from wood, crops, or organic waste but requires careful resource management and environmental impact consideration. Additionally, utilizing heat stored beneath the Earth's surface to generate electricity and provide heating through geothermal power plants typically has low emissions and a small land footprint, offering a reliable energy source due to the Earth's constant thermal energy.

The rapid penetration rate of RES is projected to decline further in the near future, bringing new prospects for the widespread penetration of renewables and extensive power-sector decarbonization that previous policy discussions did not fully consider. Here, we show the penetration rate in Eq. (7.1) for the percentage of total energy that comes from renewable sources. It is crucial to assess the extent of renewable energy integration within the overall energy system [7] as well as the carbon emission reduction components.

$$\text{RES Penetration} = \frac{E_{RES}}{E_{total}} * 100\% \tag{7.1}$$

Carbon emission reduction equation:

$$\Delta CO_2 = (E_{RES} \cdot EF_{non{-}RES}) - (E_{RES} \cdot EF_{RES}) \tag{7.2}$$

Note: BP and the US EIA exclude non-marketed renewables (e.g., traditional biomass) [8]. Figure 7.4 compares harmonized outlooks for primary energy consumption from various organizations for 2017 and 2018 baselines. These organizations include

- **BNEF:** New Energy Outlook 2019 (NEO2019), published in June 2019.
- **BP:** Energy Outlook 2019, published in February 2019.

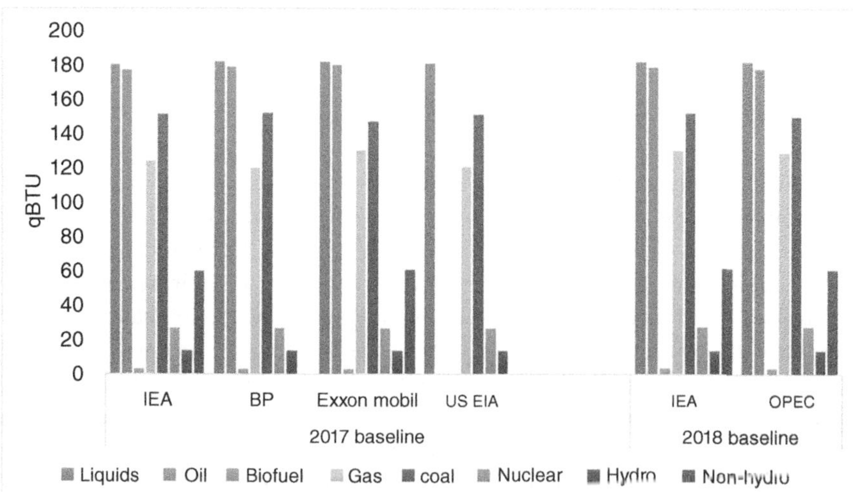

Figure 7.4 Harmonized baseline primary energy consumption (qBtu).

- **Equinor:** Energy Perspectives 2019, published in June 2019.
- **ExxonMobil:** Outlook for Energy 2019, published in August 2019.
- **IEA:** World Energy Outlook 2019 (WEO2019), published in November 2019.
- **IEEJ:** Energy Outlook 2020, published in October 2019.
- **OPEC:** World Oil Outlook 2019 (WOO2019), published in October 2019.
- **Shell:** Sky Scenario, published in April 2018.
- **US EIA:** International Energy Outlook 2019 (IEO2019), published in September 2019.

This chart underscores the shifting landscape of global energy consumption, with a noticeable increase in the share of RES over time. Although traditional energy sources like oil, liquids, and gas continue to dominate, the incremental rise in non-hydro RES points to an evolving energy mix. The data from various leading energy organizations corroborate this trend, highlighting the growing importance of RESs in achieving a sustainable energy future.

7.2.1 Trends in RESs Deployment

The deployment of RES has increased dramatically over the past decade, driven by technological advancements, policy incentives, and growing environmental awareness. Global investments in renewables have consistently risen, leading to an increased share of RESs in the world energy mix.

a) **Decreasing Cost of Solar and Wind:** Solar and wind energy have grown exponentially, with costs falling by over 85% for solar panels and approximately 50% for wind turbines since 2010 [9]. Technological advancements have significantly reduced the costs associated with solar and wind energy, making these technologies competitive with traditional fossil fuels. The decreasing cost of photovoltaic panels and wind turbines has accelerated their adoption and integration into the energy grid.
b) **Technological Advancements:** Technological advancements have significantly reduced the costs associated with solar and wind energy, making these technologies competitive with traditional fossil fuels. The decreasing cost of photovoltaic panels and wind turbines has accelerated their adoption and integration into the energy grid.
c) **Scalability:** Solar and wind energy installations can be scaled from small, individual systems to large, utility-scale power stations. This flexibility allows diverse applications and quicker deployment to meet varying energy demands.
d) **Public and Corporate Commitments:** Increased environmental awareness and commitments from both public and private sectors drive the shift toward renewable energy. Many corporations invest in renewable energy projects to reduce their carbon footprint and improve their sustainability profiles.

7.2.2 Opportunities and Challenges of High RESs Share

The environmental, economic and technological opportunities of high renewable energy shares are highlighted in Table 7.2.

Regulatory and key challenges in merging high renewable energy penetration into existing energy markets is outlined in Table 7.3, necessitating innovative strategies to maintain grid stability.

Table 7.2 Opportunities of high RESs share.

Aspect	Description
Environmental impact	
Reduction in carbon emissions	Renewable energy sources like solar, wind, and hydroelectric power significantly reduce carbon emissions, contributing to international climate goals like the Paris Agreement.
Decrease in air pollution	The operation of RESs involves minimal to no release of air pollutants, reducing the prevalence of respiratory and cardiovascular diseases, leading to a healthier population and potentially lowering healthcare costs.
Conservation of natural resources	RESs generally use less water and land once installed. For instance, solar panels and wind turbines have a relatively small land footprint per unit of electricity generated, especially compared to coal mines or natural gas extraction fields.
Energy security	Renewable energy sources, like solar and wind, are domestically available, reducing imports and providing diversification for countries reliant on imports, thereby reducing supply disruptions and price spikes.
Economic benefits	
Job creation	The renewable energy sector, which involves manufacturing, installing, operating, and maintaining energy systems, is more labor-intensive than fossil fuels, boosting local economies, especially in rural or underdeveloped regions.
Cost reduction	Renewable energy generation costs have significantly decreased in the past decade due to advancements in technology, making them more competitive and lowering energy prices for consumers and businesses.
Innovation and technology development	Renewable energy investments drive technological advancements, promoting economic growth and competitiveness. Efficient solar cells, wind turbines, and energy storage solutions drive the entire technology sector forward.

Table 7.3 Challenges of high-RESs share.

Aspect	Description
Intermittency and variability	Solar and wind energies depend on weather conditions, leading to fluctuating energy outputs requiring sophisticated grid management and storage solutions.
Infrastructure and investment	Developing renewable energy infrastructure requires substantial initial investment, including renewable facilities and transmission network upgrades, and requires coordinated policy support and planning for timely completion.
Market integration	Regulatory and technical challenges arise in integrating high renewable energy shares into existing energy markets, necessitating innovative grid management and market operations to maintain grid stability.

7.3 Handling RES Uncertainty

The uncertainty in RESs is crucial for maintaining grid stability, ensuring reliable energy supply, and achieving energy policy goals. Effective management strategies, including the role of predictive models, are essential for managing this uncertainty. Equation (7.3) measures the variability of the RES, indicating the degree of fluctuation relative to the average output. Figure 7.5 explores the depth of RESs uncertainty in terms of benefits, challenges, and strategies to mitigate. A higher REPI suggests more significant variability, which poses challenges for grid stability

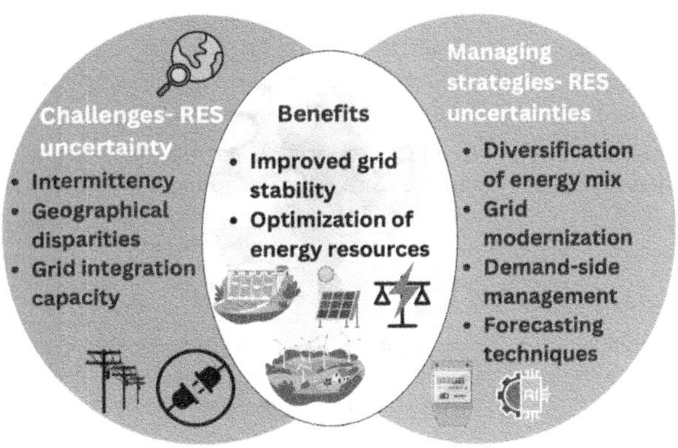

Figure 7.5 Depth of RES uncertainty.

and requires advanced management strategies [10]:

$$REPI = \frac{\sigma_{RES}}{\mu_{RES}}. \tag{7.3}$$

7.3.1 Challenges Posed by RES Variability

The variability inherent to RESs presents significant challenges to the reliability and stability of energy systems. Solar and wind power, mainly, are prone to fluctuations based on weather conditions, time of day, and seasonal changes. These fluctuations make it difficult to predict energy generation levels and can lead to supply–demand imbalances, grid instability, and potential disruptions in energy supply [11].

Table 7.4 Key challenges.

Aspect	Description
Intermittency	
Nature of intermittency	Solar and wind energy's intermittent nature, varying output based on time, cloud cover, and location, and their limited wind speed range pose significant challenges.
Grid management	Managing these fluctuations requires sophisticated forecasting tools and flexible grid management strategies to adjust to changes in energy supply and maintain grid stability. Integrating advanced energy storage systems, such as batteries, is essential for storing excess energy during peak production and releasing it during low production.
Geographical disparities	
Variability of resources	The availability of solar and wind resources varies significantly by region, influenced by local climate conditions and geographic features. Higher latitude regions have less solar exposure, while coastal areas might experience more consistent wind speeds.
Infrastructure requirements	To address energy disparities, interconnected systems, localized energy storage solutions, and renewable sources like hydropower or geothermal are needed, requiring substantial investment in transmission infrastructure.
Grid integration	
Infrastructure upgrades	Traditional grids must undergo significant upgrades to accommodate variable renewable energy sources, requiring a shift toward more flexible systems that can manage frequent power generation and load fluctuations.
Advanced technologies	Advanced control systems and grid technologies, such as smart grids, are crucial for balancing supply and demand, while evolving regulatory frameworks and market structures support these technical changes.

Table 7.5 Strategies for managing RES uncertainty.

Strategy	Description
Demand side management (DSM)	• DSM programs encourage consumers to use renewable energy during peak demand periods, using real-time pricing models to adjust smart appliances and home energy management systems.
Diversification of energy mix	• Combining diverse energy sources like solar, wind, hydropower, biomass, and geothermal, along with geographically diverse renewable facilities, can achieve a more balanced and reliable energy system [7].
Grid modernization	• Incorporating advanced metering infrastructure (AMI) and intelligent control systems for real-time monitoring and management. • Smart grids use digital technology to connect generation, transmission, and consumption parts of the energy system, facilitating better communication and automation.

Table 7.4 details specific technical and geographical challenges including the role of advanced technologies in balancing supply and demand.

7.3.2 Strategies for Managing RES Uncertainty

A combination of technological and policy-based strategies indicated in Table 7.5 can help manage the uncertainty and variability associated with RESs:

7.4 Forecasting Techniques

Improved forecasting techniques, including advanced meteorological models, historical data analytics, machine learning algorithms, and artificial intelligence (AI), are crucial for managing uncertainty in RES. These techniques provide accurate predictions of wind speeds, solar irradiance, and weather conditions, enabling more dynamic and responsive energy management.

7.4.1 Role of Predictive Models in Uncertainty Management

Predictive models based on machine learning and data analytics can help anticipate fluctuations in RES output, providing actionable insights for grid operators. By leveraging weather forecasts and historical energy generation data, predictive models can estimate short-term and long-term energy supply levels with increased accuracy [10].

7.4.1.1 Key Benefits

Improved Grid Stability Predictive models utilize a combination of real-time weather forecasts and historical energy generation data to anticipate changes in energy production from renewable sources such as solar and wind. This predictive capability is crucial for maintaining grid stability. Better predictions allow grid operators to prepare for energy shortages or surpluses, improving grid stability [12].

a) **Proactive Grid Management:** With accurate predictions, grid operators can proactively manage the energy supply, adjusting inputs from various energy sources and activating demand response strategies as needed. For example, suppose a predictive model forecasts a drop in wind energy due to changing weather patterns. In that case, the grid operators can compensate by increasing energy production from other sources or drawing from energy storage.
b) **Preventing Overloads and Blackouts:** By predicting surges in energy production, grid operators can prevent the potential for grid overloads, which can lead to blackouts. Managing these surges might involve temporarily storing excess energy or rerouting it to higher-demand areas.

Optimization of Energy Resources Predictive models also play a critical role in optimizing resource allocation within the energy grid, which can lead to reduced operational costs and lower emissions.

a) **Efficient Resource Scheduling:** By accurately predicting energy availability, utilities can optimize the scheduling of energy resources, ensuring that the most cost-effective and least carbon-intensive options are utilized first. This not only helps in reducing operational costs but also minimizes environmental impact.
b) **Load Balancing:** Advanced predictive models can forecast peak demand periods and adjust the energy system's response accordingly, enhancing the overall efficiency of the grid. This might involve increasing the input from renewables during peak times or leveraging energy storage solutions more effectively.

7.4.2 Energy Storage Systems as a Buffer for RES Variability

Energy storage systems are crucial for integrating RES into the grid, managing variability, and maintaining a stable energy supply. They act as buffers, enhancing power systems' stability and reliability. They could defer the costs needed to improve the transmission and distribution capacity to meet ever-growing power demand. Storage technologies could also play an important role in deregulated markets, like providing arbitrage, increasing the value of renewable power in some of these markets, etc. [13].

$$C_{storage} = P_{\max} \times T_{\max} \tag{7.4}$$

Equation (7.4) determines the storage needed to buffer the variability of RES, ensuring a stable energy supply during periods of low RES generation.

$$SU = \frac{E_{discharged}}{E_{capacity}} \quad (7.5)$$

The utilization of energy storage systems indicates in Eq. (7.5) how effectively the storage energy is used to meet demand. Utilization of these batteries would save capital costs for electric utilities. Due to these reasons, the use of electric drive vehicles (EDV) and other battery energy storage systems (BESS) appears to find applications in power systems. However, a framework needs to be developed to make this technology feasible, which is beneficial for both the utility and the EDV BESS consumer.

7.4.3 Innovative Technologies in RES Uncertainty Handling

The challenge of managing uncertainty in RES calls for innovative technological solutions that enhance grid stability, optimize energy distribution, and facilitate the integration of renewables into the energy system. Advances in digital technologies, materials science, and energy system design have led to the development of several key innovations:

1) **Artificial Intelligence**
   AI is revolutionizing energy systems by analyzing and managing renewable energy variability. It can process vast data from sources like smart grids and weather forecasts, predicting energy generation trends, and optimizing distribution.
2) **Blockchain Technology**
   Blockchain technology offers a decentralized, transparent energy trading system, allowing individual energy producers to sell excess energy directly to consumers. Its transparency and security build trust, facilitating smoother transactions and enabling demand side management (DSM) programs. Real-time price signals enable consumers to adjust their consumption accordingly.
3) **Smart Grids**
   Smart grids use advanced digital technologies to manage electricity more efficiently, enabling real-time supply–demand adjustment and integrating RES like solar and wind. They facilitate distributed energy resources (DERs) integration, stabilizing the grid and maximizing renewable energy utilization. A multifaceted approach involving predictive models, innovative technologies, and flexible storage systems is crucial for a low-carbon society [29–30].

7.5 Impact of High or Full RES Share on MV-EHs

A high or full share of RESs within MV-EHs brings both opportunities and challenges to energy production and distribution. RESs like wind and solar are characterized by intermittent and variable generation, creating unpredictability in energy supply. Figure 7.6 gives a summary of the impact of a high-RES share on MV-EHs.

7.5.1 Effects on Energy Production and Distribution

As power systems have evolved through continuing growth in interconnections, new technologies and controls, and increased operation in highly stressed conditions, different forms of system instability have emerged [14]. For example, voltage stability, frequency stability, and inter-area oscillations have become greater concerns that affect grid reliability than in the past. Equation (7.6) shows the voltage stability index (VSI) used to assess the voltage stability of the grid. Maintaining a VSI close to 1 indicates a stable voltage supply, which is critical when integrating high shares of intermittent RES. Similarly, Eq. (7.7) measures the grid's reliability, factoring in the average operational uptime and repair times. Higher GRI values indicate a more reliable grid [15]. It helps to address necessary changes in the reliability of the grid and the potential role of energy storage,

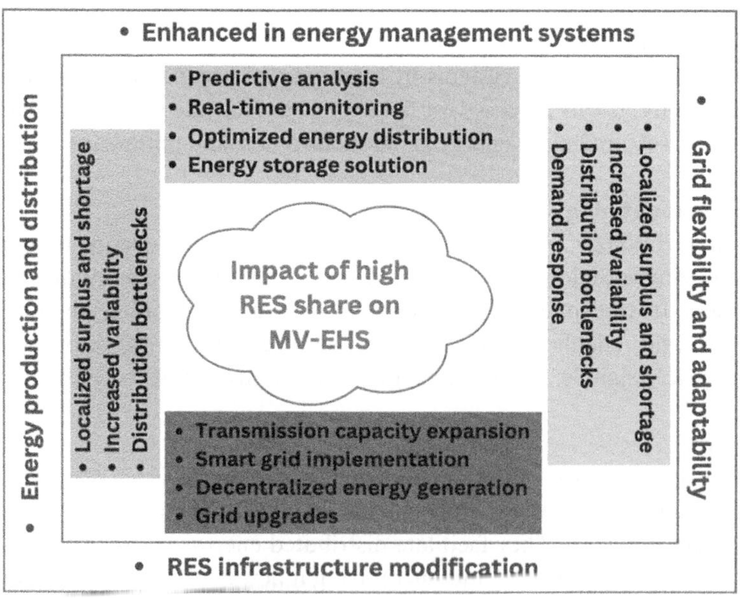

Figure 7.6 Impact of high-RES share on EV-EHS.

with variable renewable energy (VRE) providing 80% of the electricity supply (assuming no transmission-related curtailment).

The inherent intermittency of solar and wind energy sources leads to increased variability in energy production. Unlike conventional power plants that can provide a steady and controllable output, RESs depend heavily on environmental conditions, such as sunlight intensity and wind speed. This variability can lead to significant fluctuations in energy output, posing challenges for maintaining a consistent and reliable energy supply. Geographic disparities in RES availability can lead to uneven energy production across different regions. For example, areas with high wind speeds may generate surplus wind energy, while regions with less wind might experience shortages.

Similarly, solar energy production can vary significantly with latitude and weather conditions, affecting regions differently. Distribution networks may struggle to handle the rapid supply changes, especially if the infrastructure is not equipped for high RES integration [15]. Distribution networks, especially those designed for traditional energy systems, may struggle to handle the rapid and large-scale fluctuations in supply caused by high-RES penetration. If the infrastructure is not adequately upgraded or designed for high levels of RES integration, bottlenecks can occur, leading to inefficiencies and potential disruptions in energy supply.

$$VSI = \frac{V_{\min}}{V_{nominal}} \tag{7.6}$$

$$GRI = \frac{MTBF}{MTBF + MTTR} \tag{7.7}$$

a) **Necessity for Grid Flexibility and Adaptability**
To integrate high RESs into MV-EHs, grid systems need greater flexibility and adaptability. This requires rapid response capabilities and the ability to adapt to changing energy production patterns. A higher flexibility factor indicates more adaptable grids. RES integration costs depend on system flexibility. Equation (7.8) evaluates the grid's ability to adapt to energy supply and demand changes by quantifying the ratio of flexible power capacity to total capacity. As defined by Ref. [15], "system flexibility can be described as the general characteristic of the ability of the aggregated set of generators to respond to the variation and uncertainty in net load."

$$F_{grid} = \frac{\sum_{i=1}^{n} P_{flexible}(i)}{\sum_{i=1}^{n} P_{total}(i)} \tag{7.8}$$

Demand response programs adjust consumer energy consumption to align with variable energy supply, balancing the grid and avoiding costs of power plant start-ups. Energy storage buffers against variability of RES. Upgrades to existing grid infrastructure are necessary for dynamic response to production and demand changes, improving efficiency and reliability. Policies facilitating investment in modern grid technologies can accelerate these upgrades.

b) **Enhancements in EMS**
Advanced EMS are essential for managing high or full shares of RESs in MV-EHs, utilizing data analytics, real-time monitoring, and predictive modeling. This ratio in Eq. (7.9) indicates the percentage of renewable energy generation that is curtailed or wasted due to grid constraints or oversupply. Lowering the curtailment ratio is essential for maximizing the utilization of RES. Below, we explore the key improvements that these enhanced EMS bring to the table:

$$CR = \frac{E_{curtailed}}{E_{totalRES}} \times 100\%. \tag{7.9}$$

Advanced EMS uses predictive analytics to forecast energy generation from intermittent sources like solar and wind, leveraging historical weather data and real-time environmental inputs. This helps mitigate risks associated with variability, enhances system reliability and efficiency, and allows real-time monitoring of energy consumption patterns. Routing energy efficiently between different vectors minimizes losses and maximizes grid stability [16]. Enhanced EMS optimizes energy distribution by efficiently routing it between different vectors such as electricity, heat, and cooling systems. This optimization is key to minimizing energy losses during conversion and transmission and maximizing the energy system's overall efficiency.

c) **Infrastructure Modifications for RES Integration**
Infrastructure modifications are crucial for energy systems to handle high-RES share. Upgrading transmission lines, particularly from wind and solar farms, requires expanding capacity and implementing advanced technologies to reduce transmission losses. This ensures grid stability and efficient transportation of surplus energy. Smart grid technologies, which use digital communications, are also essential for modern energy systems.

7.5.2 Case Studies of MV-EHs With High-RES Share

Table 7.6 presents case studies from different nationals demonstrating diverse strategies for integrating high shares of renewables, improving grid stability, and enhancing resilience.

Table 7.6 Case study of MV-EHs.

Germany's Energiewende 	• Energy transition plan that seeks to achieve a high share of renewables in the energy mix • Diversifying energy sources, implementing innovative grid technologies, and fostering community-based energy projects • Challenges remain Ingrid stability and distribution efficiency due to the intermittent nature of solar and wind energy (Gawel et al., 2013)
India's low RES penetration  Source: Iwayan/ Adobe Stock Photos.	• India relies heavily on coal but is increasingly adopting solar and wind energy. • Innovative projects in decentralized solar microgrids are emerging to provide clean power to remote areas (Subramony et al., 2017) • The implementation of MV-EHS has proven to be an effective strategy for integrating high shares of renewable energy and improving grid resilience • Continued research on scalability, balancing supply variability, enhancing disaster resilience, and improving grid flexibility
Japan energy hub for disaster resilience 	• Following the Fukushima disaster, Japan priotized energy system resilience through its community Energy Management System (CEMs) • Energy system resilience after natural disasters. • Autonomuos microgrids and strategic battery storage improve disaster response and recovery
Denmark's energy transition 	• Denmark is a global leader in intergrating high shares of RESs, particularly wind energy, into its energy system • Over 50% of its electricity produced from wind. The country has invested heavily on renewable energy • Wind energy variability requires careful grid management • The implementation of smart grids and innovative market mechanisms has enable Denmark to stabilize its energy grid despite high wind energy (H. Ã. Lund & Mathiesen, 2009) variability (H. Ã. Lund & Mathiesen, 2009)

7.6 Load-Side Uncertainty and Its Impact on MV-EHs

7.6.1 Understanding Load-Side Uncertainty

Load-side uncertainty refers to the unpredictable variations in energy consumption patterns due to fluctuating user demands. This uncertainty can result from changes in weather, social behaviors, or economic activities, creating challenges for MV-EHs that must balance supply and demand efficiently. Such fluctuations can lead to discrepancies between energy generation and consumption, causing inefficiencies and affecting grid stability [19]. It is plausible that the elasticity of substitution and/or the daily price elasticity differ across customers with different socioeconomic characteristics (e.g., different appliance ownership and different income levels). This affects the elasticity, which also varies between hot and cool days, as stated in Ref. [20]. Equation (7.10) shows how the demand elasticity measures how sensitive the quantity of energy demanded is to change in price as it helps design effective pricing strategies and demand response programs.

Similarly, Eq. (7.11) measures the efficiency of energy usage by comparing the actual energy consumed to the maximum possible consumption over a period. A higher load factor indicates more efficient energy use with fewer peaks and troughs in demand ([21]). Also, Eq. (7.12) measures the variability in consumer energy demand, which is crucial for planning and managing energy systems under load-side uncertainty [19].

$$\varepsilon_d = \frac{\Delta Q_d / Q_d}{\Delta P / P} \tag{7.10}$$

$$Load\ Factor = \frac{E_{actual}}{E_{peak} \times T} \times 100\% \tag{7.11}$$

$$LVI = \frac{\sigma_{load}}{\mu_{load}} \tag{7.12}$$

7.6.2 Key Factors Affecting Load-Side Uncertainty

1) **Weather Variations**
   Temperature fluctuations in residential and commercial buildings impact energy demand for heating and cooling. Unpredictable weather patterns disrupt the balance between forecasted demand and actual energy use. To mitigate these effects, energy systems can integrate weather-responsive technologies like smart thermostats.
2) **Economic Activity**
   Economic activity directly impacts energy consumption, especially in industrial and commercial sectors. Economic growth boosts production, while downturns cause a drop. Energy providers must stay informed and implement flexible policies to adapt to economic changes.
3) **Consumer Behavior**
   Consumer behavior, including high-consumption electronics, home automation, and electric vehicles, impacts load patterns, leading to higher peak

demand times and increased energy consumption. Demand response programs help manage variability and balance the grid, while MV-EHs efficiency is affected.

7.6.3 Demand Side Management Strategies

Demand side management (DSM) represents a critical approach in energy management, focusing on strategies to influence consumer behavior to optimize energy use, reduce peak demand, and ensure a better alignment between consumption and energy supply. DSM strategies help stabilize the grid and contribute to energy conservation and efficiency. Demand response programs that blend customer DSM enable technology investments, and carefully designed MV-EHs time-varying rates can achieve demand impacts that can alleviate the pressure on the power system. Uncertainties in MV–EHS involving fuel prices, RES, and the form of a carbon pricing regime on the horizon emphasize the importance of the demand side resource [20]. Equation (7.13) calculates the cost savings achieved through demand response programs by shifting energy usage from high-cost peak periods to lower-cost off-peak periods. It highlights the economic benefits of DSM strategies in reducing electricity bills and optimizing energy consumption. Here are three pivotal DSM strategies:

$$CS_{DR} = \sum_{i=1}^{n} (P_{shifted}(i) \times (C_{peak} - C_{off-peak})) \tag{7.13}$$

Incentive-based programs motivate consumers to change energy usage patterns, especially during peak demand periods. These programs offer financial rewards like rebates or reduced rates, encouraging consumers to shift energy-intensive activities to off-peak times. Effective communication and education are crucial for high participation rates. Energy efficiency initiatives contribute to environmental sustainability, reduce reliance on non-RES, and lower overall carbon emissions.

7.6.4 Real-Time Pricing and Its Effects on Consumption Patterns

Real-Time Pricing (RTP) is an innovative pricing strategy in the energy market that adjusts electricity prices based on the real-time balance of supply and demand. By reflecting current market conditions in the price per kilowatt-hour, RTP provides a direct economic signal to consumers, influencing their energy consumption patterns to improve grid efficiency and stability. Equation (7.14) quantifies the cost savings achieved by consumers who shift their energy usage in response to real-time pricing signals. It also demonstrates how dynamic pricing can incentivize more efficient energy consumption patterns [20]. Here's how RTP impacts consumer behavior and the broader energy system:

$$\Delta C_{RTP} = \sum_{i=1}^{n} (C_{flat} - C_{RTP}(i)) \times E(i) \tag{7.14}$$

During high demand, typically in late afternoons and early evenings, RTP rates increase, disincentivizing consumers from using energy-intensive appliances. This pricing mechanism discourages excessive electricity use when the grid is most stressed and generation costs are highest. Higher prices during peak demand discourage excessive consumption, reducing grid congestion [20]. RTP encourages consumers to shift their energy-intensive activities to off-peak hours when the prices are lower. This behavior results in cost savings for the consumers and helps flatten the overall demand curve throughout the day.

7.6.5 Technologies for Demand Forecasting and Management

Advanced technologies are essential for accurately forecasting demand and managing load uncertainty. MAPE, as shown in Eq. (7.15), measures the accuracy of load forecasting models, expressing the error as a percentage. Lower MAPE values indicate more accurate forecasts for effective energy management and planning [22].

$$MAPE = \frac{1}{n} \sum_{t=1}^{n} \left| \frac{A_t - F_t}{A_t} \right| * 100\% \tag{7.15}$$

Machine learning algorithms excel at analyzing vast amounts of historical energy consumption data to identify patterns and trends. These algorithms can then use this information to predict future load trends accurately. This predictive capability is particularly valuable in anticipating demand spikes, seasonal fluctuations, or changes in consumption patterns due to economic or social factors.

Unlike traditional statistical methods, machine learning models can continuously improve and adapt as they process new data. This feature makes them incredibly effective in dynamic environments where consumption patterns frequently change. Utilities use machine learning to enhance load forecasting models, which can help in more effective grid management and planning. EMS are sophisticated software platforms that collect data from across the energy grid, including from smart meters, sensors, and other monitoring devices, and use this data to optimize energy distribution.

Figure 7.7 illustrates the overview of load-side uncertainty on MV-EHs.

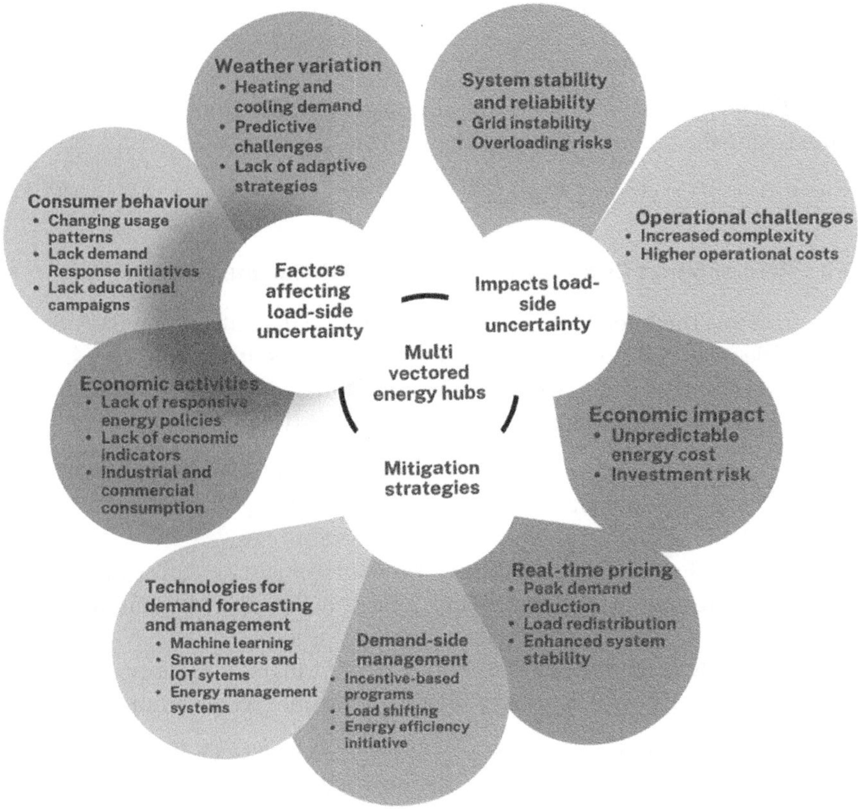

Figure 7.7 Overview of load-side uncertainty on MV-EHs.

7.7 Integrated Energy Management Strategies

7.7.1 Combining Forecasting, RES Handling, and DSM

An integrated energy management strategy combines forecasting, handling RESs, and DSM. Forecasting enables more accurate predictions of both energy demand and supply, which can be aligned through appropriate RES integration and flexible DSM programs. Predictive models and machine learning algorithms can anticipate weather patterns and consumer behavior to accurately forecast renewable energy generation and load demand [22]. Moreover, incorporating energy storage solutions and smart grid technology to manage the intermittency and variability of RESs, as well as implementing dynamic pricing and consumer incentives, can shift demand to align with periods of high renewable energy availability [20]. Equation (7.16) ensures the balance of energy supplied to the grid, accounting for all energy sources, losses, and storage systems:

$$E_{total} = E_{RES} + E_{non-RES} + E_{loss} + E_{storage}. \tag{7.16}$$

Also, Eq. (7.17) calculates the forecasting error at a given time, which is critical for improving the accuracy of energy demand and supply predictions.

$$\in (t) = |P_{actual}(t)| - | P_{forecast}(t) | \tag{7.17}$$

7.7.2 Role of Smart Grids and IoT in Integrated Management

Smart grids and the Internet of Things (IoT) are crucial in integrating EMS. Smart grids enable real-time monitoring and dynamic adjustment of energy flows, while IoT connects devices and sensors that provide granular data for optimizing energy distribution. Key Contributions of Smart Grids and IoT

a) **Enhanced Grid Monitoring:** Smart meters and IoT sensors provide data on energy usage, RES generation, and grid health, enabling quick identification of supply–demand imbalances.
b) **Automated Energy Distribution:** Grid management systems can automatically reroute energy to high-demand areas, stabilize the grid, and maximize efficiency.
c) **Predictive Maintenance:** Continuous monitoring can identify equipment issues early, reducing downtime and preventing unexpected failures.

Figure 7.8 illustrates integrated energy management strategies, which combine forecasting, RES handling, and DSM. It features key elements like predictive models, energy storage, and smart grid technology, all integrated into a centralized system for easy understanding.

7.8 Multi-Vectored Energy Hub

7.8.1 Concept of Energy Hub

The concept of Energy Hub was introduced in Ref. [23]. Energy hub may be considered as a virtual box that can convert a set of input energy carriers into a set of energy demands. This box contains several technologies that can store, transfer, or convert different forms of energy to each other. A general framework of energy hub is shown in Figure 7.1. Different aspects of energy hubs are investigated in the literature, such as economic dispatch of energy hubs [24], demand response and energy hubs [25], the energy hub concept applied to car manufacturing plants [26], and wind power uncertainty modeling in energy hubs [27]. Figure 7.9 illustrates the overview of the energy hub basic components that makeup the framework.

Figure 7.8 Integrated energy management strategies.

The technologies shown in Figure 7.1 are explained as follows:

- It receives power and produces power as well (although at varying voltage levels)—Transformer ($ElectIn_t$)

$$ElectOut_t = \eta_{elec} \cdot ElecIn_t \tag{7.18}$$

- Receives the natural gas ($GasIn_t$) and produces the heat demand ($HeatDemand_t$)—Furnace (F):

$$HeatOut_t = \eta_{furnace} \cdot GasIn_t \tag{7.19}$$

- Combined heat and power (CHP): receives the natural gas ($GasIn_t$) and converts it into heat ($HeatOut_{CHP,t}$) and electricity ($ElecOut_{CHP,t}$) output. The economic dispatch problem for CHP can be presented as follows in Eqs. (7.20a) to (7.20d):

$$ElecOut_{CHP,t} = \eta_{gas-elec} \cdot GasIn_t \tag{7.20a}$$

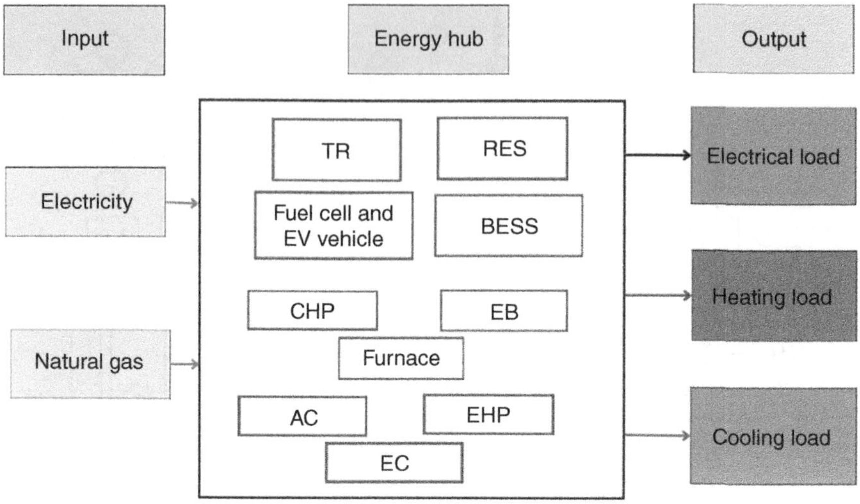

Figure 7.9 The energy hub framework concept.

$$HeatOut_{CHP,t} = \eta_{furnace} \cdot GasIn_t \tag{7.20b}$$

$$ElecOut_{CHP,t} \leq ElecOut_{CHP,\max} \tag{7.20c}$$

$$HeatOut_{CHP,t} \leq HeatOut_{CHP,\max} \tag{7.20d}$$

- Electric power input ($ElecOut_{EHP,t}$) to electric heat pump (EHP) operates on electricity and produces heat demand ($HeatDemand_t$) or cooling demand ($CoolDemand_t$) depending on the selected mode with binary heat ($HeatInd_{EHP,t}$) and cooling ($CoolInd_{EHP,t}$) indicators—EHP. The functioning of EHP is mathematically expressed as follows in Eqs. (7.21a) to (7.21e):

$$CoolDemand_t + HeatDemand_t = COP_{EHP} \cdot ElecOut_{EHP,t} \tag{7.21a}$$

$$HeatMin \cdot HeatInd_{EHP,t} \leq HeatDemand_t \leq HeatMax \cdot HeatInd_{EHP,t} \tag{7.21b}$$

$$CoolMin \cdot CoolInd_{EHP,t} \leq CoolDemand_t \leq CoolMax \cdot CoolInd_{EHP,t} \tag{7.21c}$$

$$HeatInd_{EHP,t} + CoolInd_{EHP,t} \leq 1 \tag{7.21d}$$

$$HeatInd_{EHP,t}, CoolInd_{EHP,t} \subset \{0, 1\} \tag{7.21e}$$

The EHP can operate in either heating or cooling mode. COP_{EHP} denotes the coefficient of performance for cooling and heating of EHPs.

- It absorbs heat and converts it into cooling output ($Q_{CB\_cool(t)}$)—Chiller boiler (CB). The action of the chiller boiler is mathematically represented as follows in Eqs. (7.22a) and (7.22b):

$$CoolDemand_t = \eta_{heat-cool} \cdot HeatDemand_t \tag{7.22a}$$

$$CoolDemand_t \leq CoolDemand_{t,\max} \tag{7.22b}$$

- It can store and discharge electricity—BESS. The BESS operation is articulated mathematically and formulated as follows: $ChargeState_t$ is the state of charge (SoC) indicates the levels of charged ($Charge_t$) and discharged ($Discharge_t$) electricity in BESS mode at time t in Eqs. (7.23a)–(7.23f):

$$ChargeState_t = ChargeState_{t-1} + \left(Charge_t \cdot \eta_{charge} - \frac{Discharge_t}{\eta_{discharge}} \right) \tag{7.23a}$$

$$ChargeMin \leq Charge_t \leq ChargeMax \tag{7.23b}$$

$$DischargeMin \leq Discharge_t \leq DischargeMax \tag{7.23c}$$

$$SOC_{\min} \leq ChargeState_t \leq SOC_{\max} \tag{7.23d}$$

$$ChargeInd_t + DischargeInd_t \leq 1 \tag{7.23e}$$

$$ChargeInd_t, DischargeInd_t \in \{0, 1\} \tag{7.23f}$$

7.8.2 Data

The analyses of the energy hubs exhibit three distinct types of energy demands: electric, heat, and cooling demand. Figure 7.2 presents the energy of the layout of the study. Hourly demand and electricity price data for three distinct energy hub configurations. Table 7.7 provides hourly electricity, heat, and cooling demand data alongside electricity price variations.

- The charging (η_{charge}) and discharging ($\eta_{discharge}$) efficiencies of ESS are assumed to be 0.9. The BESS capacity is $SOC_{\max} = 600$ MWh and $SOC_{\min} = 120$ MWh. The initial stored energy in BESS ($ChargeState_t$) is 120 MWh. The minimum charging and discharging limits are ($ChargeMin$) = 0, and the maximum charging and discharging limits are ($ChargeMax = DischargeMax$) = 120 MWh.
- The transformer efficiency is $\eta_{elect} = 0.98$.

- The CHP efficiency for gas to electricity is $\eta_{gas-elec} = 0.35$ and for gas to heat is $\eta_{furnace} = 0.45$. The CHP system capacity is set to 250 MW.
- For EHP, $HeatMax = CoolMax = 500$ MW. The COP is assumed to be 2.5.
- The furnace efficiency $\eta_{furnace}$ is 0.9. The furnace capacity is 600 MW.
- The chiller boiler has a capacity equal to 500 MW. The efficiency is $\eta_{heat-cool} = 0.95$
- The natural gas price is constant across time period $\lambda_t^{elec} = \$12/\text{MW h}$.
- The fuel cell efficiency ηfc is 0.6 and cost $\beta = \$20$. The fuel cell capacity is 10 MW.

7.8.3 Case Study I

This configuration in case study I, as shown in Figure 7.10, contains a transformer, furnace, and chiller boiler.

- Optimization function:

$$\min OF = \sum \left(\lambda_t^{elec} \cdot ElectOut_t + \lambda_t^{gas} \cdot GasIn_t\right) \tag{7.24}$$

- Electricity demand balance:

$$\eta_{elec} \cdot ElectOut_t = ElecDemand_t \tag{7.25}$$

- Furnace heat output balance:

$$\eta_{furnace} \cdot GasIn_t = HeatOut_{1,t} + HeatOut_{2,t} \tag{7.26}$$

- Chiller boiler cooling demand:

$$\eta_{heat-cool} \cdot HeatOut_{2,t} = CoolDemand_t \tag{7.27}$$

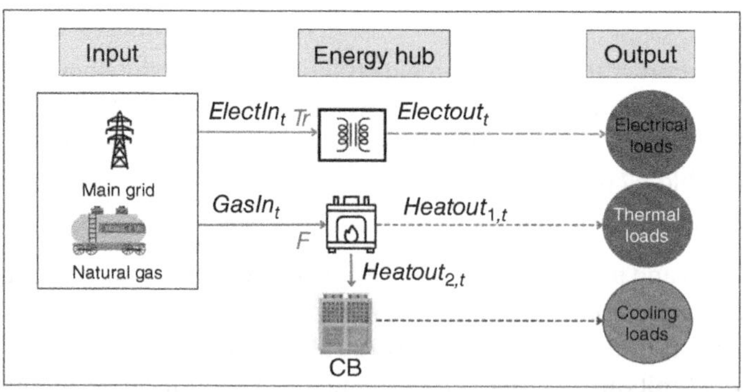

Figure 7.10 The energy hub configuration—transformer, furnace, and CB.
Source: Jemastock / Adobe Stock Photos.

Table 7.7 The hourly demand and electricity price data.

Time	$ElecDemand_t$ (MW)	$HeatDemand_t$ (MW)	$CoolDemand_t$ (MW)	λ_t^{elec} ($/MWh)
T_1	21.4	52.1	11.5	36.7
T_2	23.2	66.7	13.7	40.4
T_3	26.1	72.2	16	38.5
T_4	26.7	78.4	21.4	38
T_5	25.6	120.2	22	40.2
T_6	26.4	83.5	30.8	38.6
T_7	39.5	110.4	38.9	52.3
T_8	47.3	124.3	46.8	67.3
T_9	52.1	143.6	51	70.5
T_{10}	49.1	149.3	48.9	66.2
T_{11}	69.3	154.2	34.8	73.3
T_{12}	62	147.3	32.7	60.8
T_{13}	68	200.7	27.8	63.2
T_{14}	68.6	174.4	32	70.8
T_{15}	56.4	176.5	33.2	63.1
T_{16}	41.3	136.1	34.1	52.5
T_{17}	37.4	108.7	40.8	57
T_{18}	25.4	96.9	43.6	49.2
T_{19}	25.7	89.1	51.5	47.5
T_{20}	21.9	82.5	43.1	49.5
T_{21}	22.4	76.9	36.5	53.1
T_{22}	24.6	66.8	27.7	51.6
T_{23}	22.7	47.2	19.1	50.5
T_{24}	22.6	64.7	11	36.4;

- Heat demand balance:

$$HeatOut_{1,t} = HeatDemand_t \tag{7.28}$$

The integration of transformer, chiller boiler, and furnace shown in Figure 7.10 demonstrates the system's capacity to manage energy demands dynamically and sustainably throughout the 24-hour cycle. The model is formulated in GCode (GAMS) and solved by a linear programming solver. The issue has been resolved, and the aggregate operating expenses amount to $1.73570*10^5$.

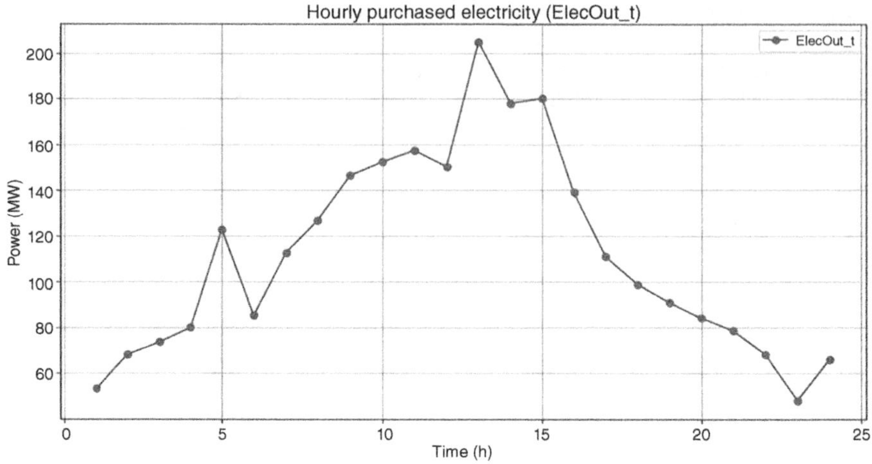

Figure 7.11 The hourly purchased electricity in the energy hub configuration.

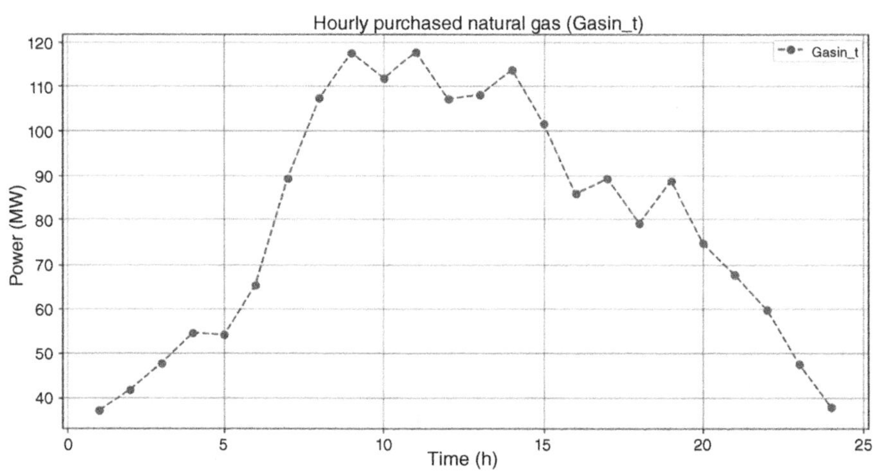

Figure 7.12 The hourly purchase of natural gas in an energy hub configuration.

Figures 7.11 and 7.12 display the hourly purchased electricity and natural gas in energy hub case study I. Two streams will be created from the furnace system's output. The chiller will be supplied by the first, and the heat requirement will be immediately supplied by the second. Figure 7.13 displays the furnace unit's hourly output in energy hub case study I.

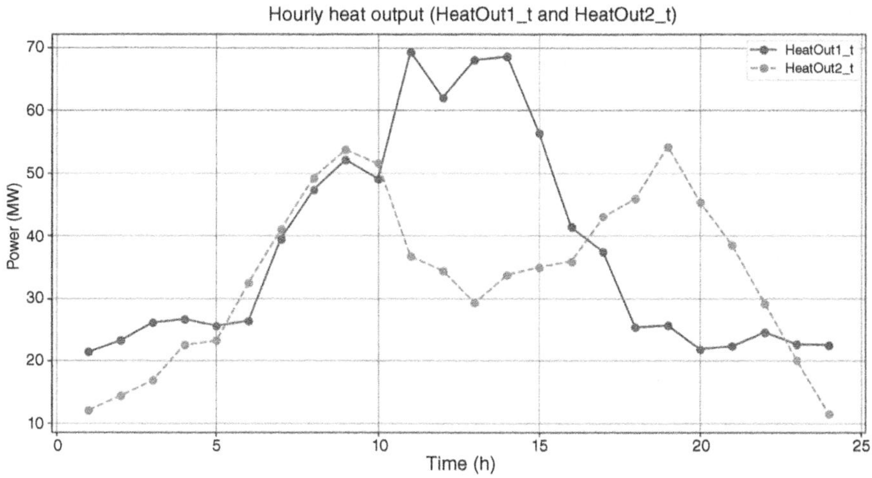

Figure 7.13 The hourly output of the furnace in the energy hub configuration.

7.8.4 Case Study II

This configuration, as shown in Figure 7.14, contains a transformer, furnace, chiller boiler, CHP, BESS, EHP, and BESS.

- Optimization function:

$$\min OF = \sum \left(\lambda_t^{elec} \cdot ElectIn_t + \lambda_t^{gas} \cdot GasIn_t \right) \tag{7.29}$$

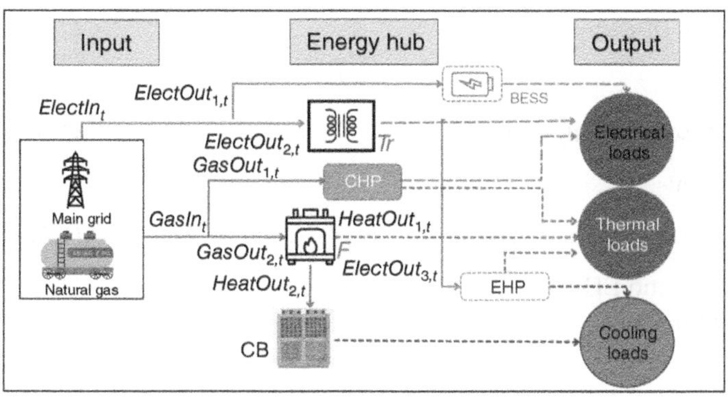

Figure 7.14 The energy hub configuration—transformer, furnace, CB, CHP, BESS, and EHP.

- Electricity demand balance:

$$\eta_{elec} \cdot ElectOut_{2,t} + Discharge_t + \eta_{gas-elec} \cdot GasIn_{1,t} = ElecDemand_t \quad (7.30)$$

- Total electricity output breakdown:

$$ElectIn_t = ElectOut_{1,t} + ElectOut_{2,t} + ElectOut_{3,t} \quad (7.31a)$$

Charging output:

$$ElectOut_{1,t} = Charge_t \quad (7.31b)$$

- SoC update:

$$ChargeState_t = ChargeState_{t-1} + \left(Charge_t \cdot \eta_{charge} - \frac{Discharge_t}{\eta_{discharge}} \right) \quad (7.32a)$$

- Charge and discharge bounds:
Charging bounds:

$$ChargeMin \leq Charge_t \leq ChargeMax \quad (7.32b)$$

Discharging bounds:

$$DischargeMin \leq Discharge_t \leq DischargeMax \quad (7.32c)$$

- SoC limits:

$$SOC_{\min} \leq ChargeState_t \leq SOC_{\max} \quad (7.32d)$$

- Binary constraint for charge and discharge indicators:

$$ChargeInd_t + DischargeInd_t \leq 1 \quad (7.32e)$$

Binary variables:

$$ChargeInd_t, DischargeInd_t \in \{0, 1\} \quad (7.32f)$$

- Gas flow breakdown:

$$GasIn_t = GasIn_{1,t} + GasIn_{2,t} \quad (7.32g)$$

- Heat balance for gas and EHP:
Total heat demand balance:

$$\eta_{furnace} \cdot GasIn_{1,t} + HeatOut_{1,t} + HeatOut_{EHP,t} = HeatDemand_t \quad (7.33)$$

- Furnace heat output:

$$\eta_{furnace} \cdot GasIn_{2,t} = HeatOut_{1,t} + HeatOut_{2,t} \quad (7.34)$$

- Cooling demand for chiller boiler and EHP:

$$\eta_{heat-cool} \cdot HeatOut_{2,t} + CoolOut_{EHP,t} = CoolDemand_t \tag{7.35a}$$

Performance coefficient (COP) application for EHP:

$$CoolOut_{EHP,t} + HeatOut_{EHP,t} = ElecOut_{3,t} \cdot COP \tag{7.35b}$$

- Heat balance for cooling bounds for EHP:
Heat bounds:

$$HeatMin \leq HeatOut_{EHP,t} \leq HeatMax \tag{7.36a}$$

Cooling bounds:

$$CoolMin \leq HeatOut_{EHP,t} \leq CoolMax \tag{7.36b}$$

- Binary mode constraints for EHP (heating and cooling):

$$CoolInd_t + HeatInd_t \leq 1 \tag{7.37a}$$

Binary variable for EHP modes:

$$CoolInd_t + HeatInd_t \in \{0, 1\} \tag{7.37b}$$

The model is formulated in GCode (GAMS) and solved utilizing a mixed integer programming solver. The issue has been resolved, and the aggregate operating expenses amount to $0.84449*$10^5$. Figures 7.15 and 7.16 display the hourly electricity acquired and its distribution between the transformer, EHP, and BESS in energy hub case study II. Two streams will be created from the furnace system's output. The chiller will be supplied by the first, and the heat requirement will be supplied by the second. Figure 7.19 displays the furnace unit's hourly output in energy hub case study II. Figure 7.17 displays the BESS's hourly charge, charging, and discharging status in energy hub case study II. Figure 7.18 displays the EHP's hourly output in energy hub case study II.

7.9 Conclusion and Recommendation

MV-EHs have emerged as a transformative solution for achieving sustainable energy systems. Key findings from previous chapters emphasize:

1) **Integration of Renewable Energy:** MV-EHs can efficiently integrate RESs, such as wind and solar, into existing grids through advanced forecasting, smart grid technology, and energy storage.
2) **Demand Side Management:** DSM strategies and dynamic pricing effectively reduce peak demand and enhance grid stability.

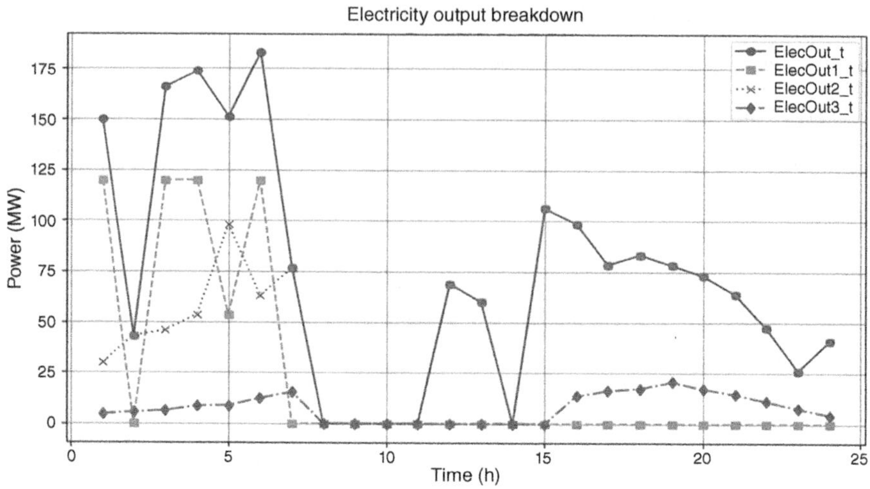

Figure 7.15 The hourly purchased electricity and its division between transformer, EHP, and BESS in energy hub case study II.

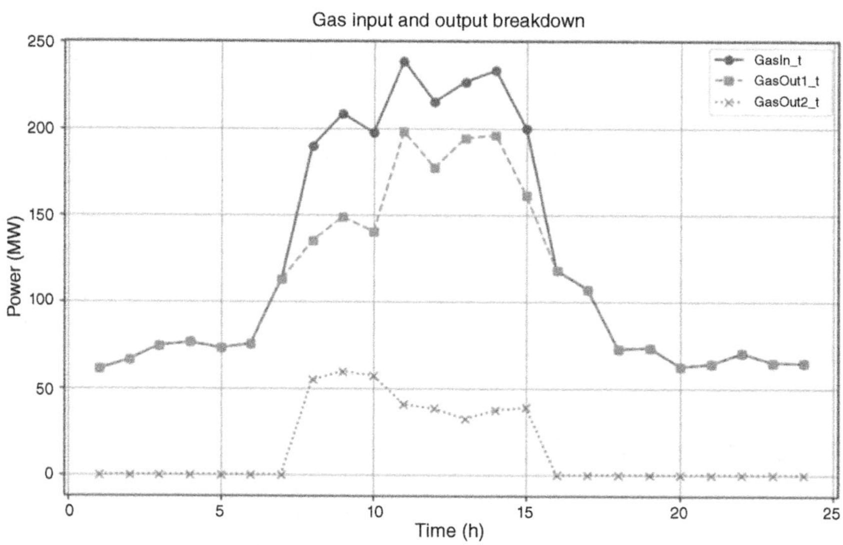

Figure 7.16 The hourly purchased natural gas and its division between furnace and CHP in energy hub case study II.

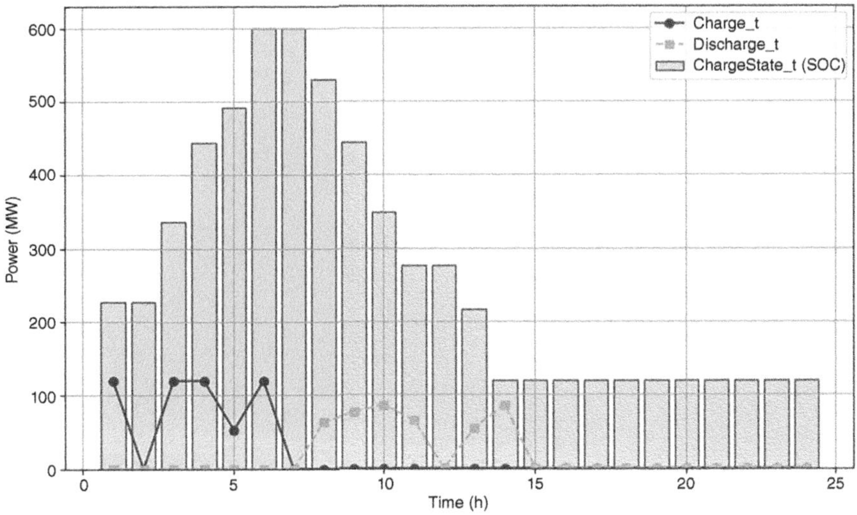

Figure 7.17 The hourly SoC (MWh), charging, and discharging of BESS (MW) in energy hub case study II.

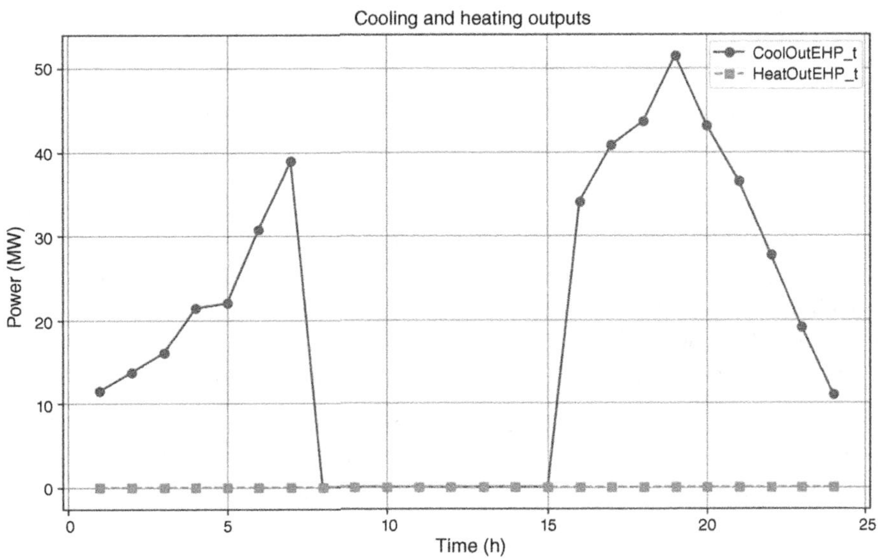

Figure 7.18 The hourly output of the EHP unit in energy hub case study II.

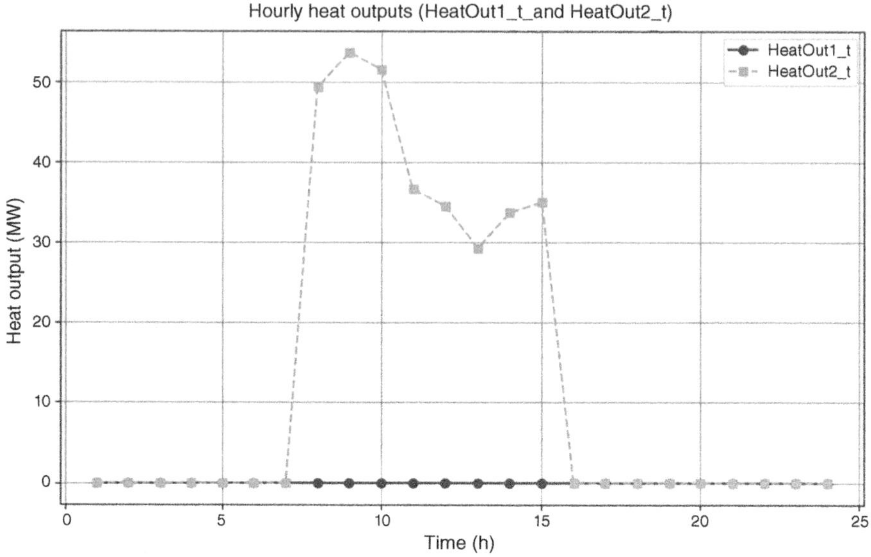

Figure 7.19 The hourly output of the furnace unit in energy hub case study II.

3) **Grid Flexibility and Adaptability:** Modern grids must be flexible to accommodate RES variability and adapt to changing consumption patterns.

Recommendations:

- **Investment in Infrastructure:** Governments should invest in smart grids, advanced metering, and microgrid infrastructure to enhance resilience and efficiency.
- **Supportive Regulatory Frameworks:** Implement market mechanisms that incentivize energy storage, DSM, and renewable energy deployment.
- **Research and Development:** Continued R&D in predictive models, storage technologies, and peer-to-peer energy trading is essential [28].
- **Economic Growth:** The adoption of MV-EHs encourages investments, job creation, and technological advancements in renewable energy sectors [3].

As the world transitions toward sustainable energy futures, MV-EHs will be indispensable for low-carbon societies. Their ability to integrate renewables, balance supply–demand dynamics, and improve grid resilience makes them a cornerstone of energy management strategies. Policymakers, researchers, and industry leaders must work together to develop innovative pathways and implement effective policies that support MV-EHs in meeting global sustainability goals.

References

1 Tiwari, S. and Singh, J.G. (2022). Optimal energy management of multi-carrier networked energy hubs considering efficient integration of demand response and electrical vehicles: a cooperative energy management framework. *Journal of Energy Storage* 51: 104479. https://doi.org/10.1016/j.est.2022.104479.

2 Tiwari, S. and Singh, J.G. (2023). Tri-level stochastic transactive energy management and improved profit distribution scheme for multi-vectored networked microgrids: a multi-objective framework. *Sustainable Cities and Society* 95: 104569. https://doi.org/10.1016/j.scs.2023.104569.

3 Gielen, D., Boshell, F., Saygin, D. et al. (2019). The role of renewable energy in the global energy transformation. *Energy Strategy Reviews* 24: 38–50. https://doi.org/10.1016/j.esr.2019.01.006.

4 Orehounig, K., Mavromatidis, G., Evins, R. et al. (2014). Towards an energy sustainable community: an energy system analysis for a village in Switzerland. *Energy and Buildings* 84: 277–286. https://doi.org/10.1016/j.enbuild.2014.08.012.

5 Mancarella, P. (2014). MES (multi-energy systems): an overview of concepts and evaluation models. *Energy* 65: 1–17. https://doi.org/10.1016/j.energy.2013.10.041.

6 Geidl, M., and Member, S. (2007). Optimal coupling of energy infrastructures. *2007 IEEE Lausanne Power Tech*, Lausanne, Switzerland, pp. 1398–1403. doi: 10.1109/PCT.2007.4538520. 1398–1403.

7 Lin, J. (2010). Rapid cost decrease of renewables and storage accelerates the decarbonization of China's power system. *Nature Communications* 2020: 1–9. https://doi.org/10.1038/s41467-020-16184-x.

8 Newell, R. G., & Raimi, D. (2022). Global Energy Outlook Comparison Methods: 2022 Update. *Resources for the Future*, July, 1–30.

9 Couture, T., Busch, H., Hansen, T., and Leidreiter, A. (2019). REN21 – Renewables in Cities 2019 Global Status Report – Preliminary Findings.

10 Lopes, F.M., Silva, H.G., Salgado, R. et al. (2018). Short-term forecasts of GHI and DNI for solar energy systems operation: assessment of the ECMWF integrated forecasting system in southern Portugal. *Solar Energy* 170: 14–30. https://doi.org/10.1016/j.solener.2018.05.039.

11 Lund, H., Alberg, P., Chang, M. et al. (2018). The status of 4th generation district heating: research and results. *Energy* 164: https://doi.org/10.1016/j.energy.2018.08.206.

12 Lara Fanego, V. (2017). Assessment and forecasting of solar resource: applications to the solar energy industry.

13 Divya, K.C. and Østergaard, J. (2009). Battery energy storage technology for power systems-an overview. *Electric Power Systems Research* 79 (4): 511–520. https://doi.org/10.1016/j.epsr.2008.09.017.

14 Kundur, P., Paserba, J., Ajjarapu, V. et al. (2004). Definition and classification of power system stability. *IEEE Transactions on Power Systems* 19 (3): 1387–1401. https://doi.org/10.1109/TPWRS.2004.825981.

15 Kondziella, H. and Bruckner, T. (2016). Flexibility requirements of renewable energy based electricity systems—a review of research results and methodologies. *Renewable and Sustainable Energy Reviews* 53: 10–22. https://doi.org/10.1016/j.rser.2015.07.199.

16 Gawel, E., Strunz, S., and Lehmann, P. (2013). Germany's energy transition under attack: is there an inscrutable German Sonderweg? *Nature and Culture* 8: 121–133. https://doi.org/10.3167/nc.2013.080201.

17 Subramony, B.V.A., Doolla, S., and Chandorkar, M. (2017). Microgrids in India. *IEEE Electrification Magazine* 5: 47–55.

18 Lund, H.Ã. and Mathiesen, B.V. (2009). Energy system analysis of 100% renewable energy systems—the case of Denmark in years 2030 and 2050. *Energy* 34: 524–531. https://doi.org/10.1016/j.energy.2008.04.003.

19 Albadi, M.H. (2008). A summary of demand response in electricity markets. *Electric Power Systems Research* 78: 1989–1996. https://doi.org/10.1016/j.epsr.2008.04.002.

20 Faruqui, A. and Sergici, S. (2017). Household response to dynamic pricing of electricity—a survey of the experimental evidence.

21 Lund, H. (2007). Renewable energy strategies for sustainable development. *Energy* 32 (6): 912–919. https://doi.org/10.1016/j.energy.2006.10.017.

22 Kavousi-Fard, A., Samet, H., and Marzbani, F. (2014). A new hybrid modified firefly algorithm and support vector regression model for accurate short term load forecasting. *Expert Systems with Applications* 41 (13): 6047–6056. https://doi.org/10.1016/j.eswa.2014.03.053.

23 Geidl, M., Koeppel, G., and Favre-Perrod, P. (2007). The energy hub–a powerful concept for future energy systems. *Third Annual Carnegie Mellon Conference on the Electricity Industry* (13–14 March 2007). ETH Zurich, Switzerland: Klaus Fröhlich Power Systems and High Voltage Laboratories.

24 Beigvand, S.D., Abdi, H., and La Scala, M. (2017). A general model for energy hub economic dispatch. *Applied Energy* 190: 1090–1111. https://doi.org/10.1016/j.apenergy.2016.12.126.

25 Batić, M., Tomašević, N., Beccuti, G. et al. (2016). Combined energy hub optimisation and demand side management for buildings. *Energy and Buildings* 127: 229–241. https://doi.org/10.1016/j.enbuild.2016.05.087.

26 Kampouropoulos, K., and Andrade, F. (2016). Energy hub optimization applied on car manufacturing plants. *2016 IEEE ANDESCON*, Arequipa, Peru, pp. 1–4. 10.1109/ANDESCON.2016.7836236.

27 Dolatabadi, A., Mohammadi-Ivatloo, B., Abapour, M., and Tohidi, S. (2017). Optimal stochastic design of wind integrated energy hub. *IEEE Transactions* *on Industrial Informatics* 13 (5): 2379–2388. https://doi.org/10.1109/TII.2017.2664101.

28 IRENA (2019). Global energy transformation: The REmap transition pathway (Background report to 2019 edition), International Renewable Energy Agency, Abu Dhabi.

29 Duan, C.C., S. and Liu, T.C.B. (2011). Smart energy management system for optimal microgrid economic operation. *IET Renewable Power Generation* 5: 258–267. https://doi.org/10.1049/iet-rpg.2010.0052.

30 Torto, S.O.G., Pachauri, R.K., and Singh, J.G. (2025). Enhancement of grid stability and energy innovation by harnessing the power of the load curtailment. *Electrical Engineering*. https://doi.org/10.1007/s00202-025-03070-2.

31 Torto, S.O.G., Pachauri, R.K., Singh, J.G. et al. (2024). Networked Micro-grid Topologies for Transactive Energy Management System: An overview and Future Perspectives. *Scientific African* e02488: https://doi.org/10.1016/j.sciaf.2024.e02488.

8

Green Energy, Greener Future: Bioenergy's Role in Carbon Reduction

Nilay Kumar Sarker and Prasad Kaparaju

School of Engineering and Built Environment, Griffith University (Nathan Campus), Nathan, Queensland, Australia

8.1 Introduction

Approximately 80% of the world's population, amounting to six billion individuals, resides in nations that predominantly import fossil fuels [1]. This dependence subjects these populations to the vagaries of international political dynamics and crises. In stark contrast, every country possesses the inherent potential to exploit renewable energy sources, a potential that remains largely untapped. According to projections by the International Renewable Energy Agency (IRENA), renewable energy could provide up to 90% of global electricity by 2050 [2]. The adoption of renewable energy technologies not only mitigates the risks associated with fossil fuel dependency but also promotes economic diversification, fosters job creation, and contributes to poverty reduction.

Currently, renewable energy sources represent the most economical option for power generation in many regions globally [3]. The last decade has seen a significant reduction in the costs associated with these technologies. From 2010 to 2020, the cost of solar electricity decreased by 85%, while onshore and offshore wind technologies experienced cost reductions of 56% and 48%, respectively [3]. These declining costs enhance the appeal of renewable energy, particularly in low- and middle-income countries that are expected to drive the bulk of new electricity demand. Consequently, renewable energy is poised to dominate new electricity supply in the coming years, potentially accounting for 65% of the global electricity market by 2030 and facilitating the decarbonization of up to 90% of the power sector by 2050 [3]. This shift could significantly reduce carbon emissions and play a crucial role in combating climate change. Despite a temporary increase in solar and wind power costs due to elevated commodity and freight prices in

Energy Management Strategies for Multi-Vectored Energy Hubs to Achieve Low Carbon Societies, First Edition.
Edited by Shubham Tiwari, Jai Govind Singh, Sivaraman Palanisamy, Sharmeela Chenniappan, Rupendra Kumar Pachauri, and Sanjeevikumar Padmanaban.

2022 and 2023, these renewable sources are likely to become more competitive as fossil fuel prices are experiencing even sharper increases, according to the International Energy Agency (IEA) [4].

Bioenergy, derived from biomass, is pivotal in the transition to a low-carbon future. Biomass encompasses a diverse array of organic materials, including agricultural residues, energy crops, forest residues, animal wastes, organic municipal solid waste (MSW), and algae. The potential of bioenergy lies in its capacity to supplant a substantial portion of fossil fuels, thereby mitigating greenhouse gas (GHG) emissions and promoting sustainable development [5]. Specifically, biomass includes any organic material originating from plants and animals, such as agricultural residues (e.g., corn stover, wheat straw, and rice husks), energy crops (e.g., switchgrass, miscanthus, and hybrid poplar), woody biomass and forest residues (e.g., logging residues, unused mill residues, and whole-tree biomass), animal wastes (e.g., livestock manure processed for biogas production), and the biogenic portion of organic MSW (e.g., food scraps and yard trimmings). Additionally, microalgae and cyanobacteria are highly efficient in converting sunlight and nutrients into biomass [5].

Several technologies are employed to convert biomass into bioenergy. Thermochemical conversion processes, such as combustion, gasification, and pyrolysis, transform biomass into energy, syngas, or bio-oil. Biochemical conversion involves the use of enzymes and microorganisms to produce biofuels like ethanol and biodiesel. Chemical conversion, particularly transesterification, is used to produce biodiesel from fats and oils. Moreover, anaerobic digestion of organic materials like animal manure and MSW generates biogas, which can be utilized for heat and power [5].

Bioenergy offers numerous environmental and economic benefits. In terms of GHG emission reductions, the current bioeconomy displaces approximately 2.4% of fossil energy consumption, avoiding 116 million tons of CO_2-equivalent emissions annually. By 2030, it is projected to displace 9.5% of fossil energy consumption and avoid up to 446 million tons of CO_2-equivalent emissions annually [5]. Economically, bio-based activities generated over $48 billion in revenue and supported 285,000 jobs in 2014. Projections for 2030 suggest that the bioeconomy could expand direct revenue to nearly $259 billion and create 1.1 million jobs [5]. In this chapter, we will explore the role of bioenergy in achieving a low-carbon society.

8.2 The Significance of Bioenergy

8.2.1 Limitations of Wind and Solar Energy

Over the next five years, renewable power capacity is expected to escalate, with photovoltaic (PV) solar and wind energy comprising a staggering 96% of new installations. This trend is driven by the lower generation costs of these renewables compared to both fossil and nonfossil alternatives in most nations, alongside sustained policy support. Projections suggest that by 2028, solar PV and wind capacity additions will more than double those of 2022, achieving a record-breaking total of nearly 710 GW. Both policy and commercial interests have favored solar and wind over other renewable options globally. For instance, in 2022, renewable sources accounted for 32% of Australia's total electricity output, with solar and wind contributing significantly. This marked the highest proportion of renewables in the nation's electricity mix on record, surpassing the previous year and even the historical highs of the mid-1960s brought about by the Snowy Mountains hydroelectric project. Solar and wind have significantly fueled the growth in renewable energy production over the past decade. In 2022 alone, small-scale solar generation increased by 18%, with an annual average growth of 22% since 2015. Similarly, wind energy grew by 11% in 2022 and has averaged a 14% annual increase since 2015.

However, transitioning to an energy system heavily reliant on solar and wind presents numerous challenges. A primary issue is the mismatch between the intermittent nature of these energy sources and the constant demand for electricity, which can lead to reliability concerns. Solar and wind power exhibit considerable daily and seasonal variability, often leading to periods when demand surpasses supply. Integrating these sources effectively requires substantial energy storage solutions and sophisticated grid management to accommodate long-duration supply shortages [6]. Moreover, the variability and regional specificity of these resources demand extensive infrastructure development. Economic and geopolitical factors also significantly influence the scalability and implementation of large-scale renewable systems. To counteract these challenges, it is often necessary to incorporate complementary technologies like demand management systems or alternative energy sources, highlighting the complexities of moving toward a sustainable, renewable-dominated energy paradigm [6].

In Australia, the shift toward solar and wind power, crucial for the nation's renewable energy future, is fraught with challenges due to the inherent unpredictability of these resources. Instances of simultaneous low solar and wind outputs, especially during winter, pose serious reliability issues. This variability is further complicated by meteorological patterns and large-scale climate phenomena such as the El Niño-Southern Oscillation, which impact the predictability and intensity of energy shortages. Such complexities necessitate integrated planning and robust management strategies to mitigate the climatic impacts on renewable energy production, ensuring reliable energy supply continuity in Australia [7].

8.2.2 Benefits of Bioenergy

In the evolving landscape of renewable energy, bioenergy presents a distinct contrast to the more prevalent solar and wind options. Despite facing stiff competition, biomass holds a pivotal role in climate action strategies, particularly under the targets set by the Paris Agreement to limit global temperature increases to well below 2 °C. Bioenergy's unique advantage lies in its ability to provide continuous baseload power, a stark difference from the intermittent nature of solar and wind sources. This attribute is crucial as societies pivot from fossil fuels toward more sustainable energy solutions. Moreover, bioenergy is particularly suited to applications in high-energy-demand sectors such as aviation and shipping, offering a renewable alternative to conventional fuels [8].

Bioenergy also leverages underutilized or degraded lands that are unsuitable for traditional agriculture, thus minimizing competition with food production. Currently, bioenergy contributes significantly to both the global primary energy supply and the renewable energy sector, underscoring its importance. In regions with existing bioenergy infrastructures, it remains a cost-effective solution for reliable power generation, despite the challenges and higher initial costs associated with emerging renewable technologies [8].

However, the sustainability of bioenergy is not without challenges. Issues such as land use efficiency and the carbon emissions associated with bioenergy production necessitate careful management, and these factors might lead to a reduced role in the global energy mix by mid-century as other renewable technologies advance [8].

Taking the specific case of Vietnam, the utilization of bioenergy from rice straw illustrates numerous benefits. Rice straw, a byproduct of rice cultivation, not only offers a sustainable energy source through CO_2 sequestration via photosynthesis but also provides a solution to the environmental degradation caused by its traditional open burning practices. This alternative promotes local energy sources, enhancing socioeconomic development within rural communities and reducing reliance on imported fossil fuels. The strategic use of rice straw for energy can also contribute significantly to national energy security and align with global sustainability objectives [9]. However, the integration of rice straw into the energy sector requires addressing several technological, environmental, and institutional barriers to ensure that bioenergy remains a viable and environmentally sound option in the renewable energy portfolio [9].

8.2.3 Strategies for Low-Carbon Transition Via Bioenergy

Bioenergy, particularly when integrated with carbon capture and storage (BECCS), offers a promising route toward achieving negative carbon dioxide emissions and facilitating a transition to a low-carbon economy [10]. Central to this approach is the sustainable production of biomass, which necessitates policies that promote the cultivation of energy crops on marginal or degraded lands. This strategy helps avoid competition with food production and reduces impacts on biodiversity. For bioenergy to be effectively integrated into the wider energy landscape, a holistic policy framework is imperative. Such a framework should include economic incentives like subsidies for bioenergy producers, tax benefits for investments in clean technologies, and a comprehensive carbon pricing mechanism to enhance the economic competitiveness of bioenergy relative to fossil fuels.

Additionally, policy support is crucial for advancing research and development in bioenergy technologies and for supporting the infrastructural developments required for its large-scale implementation. Furthermore, the integration of bioenergy within the electricity and transportation sectors could be accelerated through specific regulatory mandates, such as renewable portfolio standards and biofuel blending requirements. These measures would significantly boost the adoption of bioenergy, diversifying energy sources and diminishing reliance on fossil fuels [10].

8.3 Integrating Bioenergy into National and Regional Emission Reduction Plans: A Case Study Approach

8.3.1 Bioenergy in Europe: A Regional Overview

Europe's progression in incorporating biofuels into its energy framework demonstrates a dynamic synthesis of innovation, regulatory frameworks, and sustainability considerations [11]. This transition from first-generation biofuels, which primarily utilize food-based sources like vegetable oils and animal fats, to more advanced biofuels underscores the region's commitment to reducing GHG emissions and bolstering energy security. The adoption of second-generation biofuels, derived from nonedible lignocellulosic biomass, and third-generation biofuels, based on algal cultivation, marks a critical shift toward minimizing the competition between biofuel production and food supply. Furthermore, the exploration of fourth-generation biofuels, involving genetically modified organisms and the valorization of waste products, reflects an ambitious stride toward enhanced efficiency and sustainability [11].

The European Union's legislative cornerstone, the Renewable Energy Directive, facilitates these technological advancements by mandating significant cuts in GHG emissions and promoting renewable energy utilization [11]. This regulatory scaffolding not only spurs technological innovation but also addresses the complex sustainability challenges—encompassing ecological, economic, and social facets—associated with the expansion of biofuels. Consequently, Europe's biofuel initiative is not merely about diversifying energy sources and achieving environmental mandates but also about driving the broader transition toward a sustainable, low-carbon future.

8.3.2 Bioenergy Initiatives in the Nordic Countries

Bioenergy is central to the ambitious plans of Nordic countries to shift toward sustainable and low-carbon energy systems. Denmark, Finland, Iceland, Norway, and Sweden have made strategic moves to elevate the role of bioenergy, extensively using it to replace conventional fossil fuels in electricity generation, heating, and transportation. In Finland and Sweden, bioenergy, derived primarily from local biomass and waste, not only powers renewable electricity generation but also fuels expansive district heating networks [12]. These networks significantly improve energy efficiency and reduce urban carbon footprints, showcasing the Nordic commitment to innovative energy solutions that capitalize on local resources for sustainable development.

In the transportation sector, the transition from oil-based fuels to biofuels is seen as crucial, with expectations for a significant increase in biofuel usage by 2050 to curb emissions from this notably carbon-intensive sector [12]. The Nordic strategy highlights bioenergy as a flexible and adaptable energy source, essential for meeting diverse energy demands across various sectors. Nonetheless, the expansion of bioenergy faces hurdles such as the sustainability of biomass production and the need for technological advancements in bioenergy conversion processes. Concerns about land use, biodiversity conservation, and the carbon balance of bioenergy production are critical, requiring meticulous environmental management and innovation in conversion technologies. This approach aims to maximize energy output while minimizing ecological impacts, demonstrating the Nordic countries' proactive role in modelling sustainable energy transitions on the global stage [12].

8.3.3 Bioenergy Contributions to Emission Reductions in Nigeria

In Nigeria, the strategic implementation of bioenergy is crucial for advancing toward a sustainable and low-carbon future. Analyses conducted with the Nigerian Energy Calculator 2050 (NECAL2050) suggest that bioenergy, sourced from agricultural residues, forest byproducts, and waste materials, is integral to reducing the nation's dependency on fossil fuels [13]. Faced with escalating energy demands and the imperative to decrease GHG emissions, bioenergy emerges as a renewable solution that capitalizes on Nigeria's rich biomass resources. These resources, including residues from extensive agricultural activities and forest biomass, are presently underexploited.

Transitioning to bioenergy could markedly reduce carbon emissions, especially in energy-intensive sectors such as transportation, where biodiesel and bioethanol have the potential to replace conventional petrol and diesel for vehicle fuel. This shift is poised to not only mitigate environmental impacts but also enhance energy security by diversifying Nigeria's energy portfolio and diminishing oil reliance.

However, unlocking the full potential of bioenergy hinges on establishing a supportive policy framework that encourages sustainable biomass production without compromising food security or contributing to deforestation. It also requires investment in bioenergy technologies, infrastructure development for biomass collection and processing, and efforts toward community engagement and acceptance. By addressing these challenges through thorough planning and collaborative stakeholder engagement, bioenergy can become a pivotal element of Nigeria's low-carbon development strategy, fostering economic growth while aligning with environmental sustainability goals [13].

8.3.4 Uganda's Approach to Bioenergy

In Uganda, the strategic deployment of bioenergy systems presents a viable path toward a sustainable, low-carbon future. The effectiveness and sustainability of bioenergy are largely dependent on the scale of implementation. Smaller, community-centric bioenergy projects are particularly valued for their compatibility with the socioeconomic dynamics of local communities. These initiatives utilize local biomass resources such as agricultural residues and wood waste, converting them into energy without the need for extensive infrastructure changes. By adopting decentralized systems, Uganda could reduce energy transmission losses and mitigate the ecological impacts often associated with large-scale bioenergy operations, which typically require significant land modification and can disturb local ecosystems [14].

However, several challenges impede the proliferation of bioenergy in Uganda. A primary concern is the lack of capital investment for advancing bioenergy technologies, which could yield higher efficiencies and lesser environmental impacts. Additionally, the country faces the need for a supportive regulatory framework that not only facilitates bioenergy development but also safeguards against negative consequences on food security and land use practices. Competing land use demands, particularly the competition for scarce arable land between bioenergy and agriculture, pose a significant challenge [14].

Environmental management is also a critical aspect of bioenergy production. It is essential to implement sustainable practices that prevent biodiversity loss, wisely manage water resources, and maintain soil health, especially in large-scale projects. Moreover, potential social disruptions and displacement must be addressed through inclusive planning and active community engagement, ensuring that bioenergy projects enhance local development without depleting social or environmental resources [14].

To address these complexities, Uganda could benefit from policies that promote small-scale projects and drive innovation in bioenergy technologies. Such policies would support the gradual expansion of bioenergy capacity, allowing for the careful assessment and mitigation of impacts in a controlled and sustainable fashion. Integrating bioenergy into broader national energy and rural development strategies could enable Uganda to harness this renewable resource to fulfill its energy requirements while pursuing its low-carbon and development objectives [14].

8.3.5 The Role and Potential of Bioenergy in the United States

In 2019, renewable energy sources accounted for nearly 8% of the total energy supply in the United States, with bioenergy comprising approximately 60% of this renewable share. Bioenergy's role has evolved over the decades, particularly with the growth of liquid biofuels such as bioethanol and biodiesel, which are now integral to the transport sector [15].

Liquid biofuels, primarily bioethanol and biodiesel, have seen substantial growth. Bioethanol, used as an oxygenate in gasoline since the 1980s, expanded significantly from 230 petajoules (PJ) in 2003 to over 1200 PJ by 2012. By 2019, bioethanol represented around 8% of gasoline consumption in terms of energy. Biodiesel, while lower in volume compared to bioethanol, also grew markedly from 40 PJ in 2010 to approximately 200 PJ by 2013, stabilizing at around 3–4% of diesel consumption [15].

Solid biofuels remain a dominant bioenergy form, with stable use around 2.4 exajoules (EJ), primarily in industrial applications. Biogas, predominantly sourced from landfill gas, and renewable MSW each contribute around 0.15 EJ. Despite modest growth, these sources hold considerable potential for expansion based on domestic biomass availability [15].

The United States' extensive biomass resources present significant opportunities for expanding bioenergy. Solid biomass, biogas, and renewable MSW are underutilized, and there is potential for substantial growth in these areas. Enhancing the utilization of these resources can further reduce reliance on fossil fuels and contribute to emission reduction goals [15].

The United States aims to achieve net-zero carbon emissions by 2050, with specific targets for clean electricity, transportation, and heating sectors. By 2035, the country targets 100% clean electricity, significantly reducing carbon pollution. In the transport sector, bioenergy plays a crucial role, with policies like the renewable fuel standard (RFS) mandating the use of renewable fuels, including biofuels, to replace or reduce petroleum-based fuels [15].

8.3.6 Bioenergy's Role and Potential in India

India, the world's third largest energy consumer, has been aggressively pursuing renewable energy strategies to mitigate its heavy reliance on fossil fuels, which not only strains the national economy due to import dependence but also contributes significantly to carbon emissions and air pollution. In 2019, renewable energy sources accounted for 22% of India's total energy supply, with bioenergy representing 88% of this share. This high contribution is primarily due to the extensive use of biomass in residential heating and cooking. However, India's bioenergy landscape is evolving, with significant potential for expansion in industrial applications, power generation, and transportation [16].

Solid biofuels, including agricultural residues and woody biomass, constitute the majority of bioenergy use in India. Traditional biomass use, such as in residential heating, is decreasing, being replaced by more modern and efficient forms of biomass energy. For instance, the use of solid biomass for power production has doubled since 2010, indicating a shift toward more sustainable bioenergy practices [16].

The use of liquid biofuels, although currently modest, is growing. Bioethanol, primarily produced from sugarcane molasses, has seen increased utilization in transportation, representing 3% of gasoline consumption by energy in 2019. The National Policy on Biofuels-2018 aims to achieve 20% ethanol blending in petrol by 2025–2026, leveraging various feedstocks, including surplus rice and maize [16].

Biogas production, primarily from MSW and agricultural residues, is an area with substantial growth potential. The government's "Sustainable Alternative Towards Affordable Transportation" (SATAT) initiative encourages the production and use of compressed biogas (CBG), with plans to establish 5000 CBG plants producing 15 million metric tons per annum by 2023–2024 [16].

India's vast agricultural sector provides a significant source of biomass, with opportunities to expand bioenergy production from agricultural residues, forestry waste, and urban waste. The government's policy framework supports this through various incentives and regulatory measures aimed at enhancing bioenergy's role in the energy mix.

8.3.7 The Role and Potential of Bioenergy in the United States

In 2019, renewable energy sources accounted for 9.7% of China's total energy supply, with bioenergy representing a significant portion of this share. Bioenergy's contribution can be categorized into solid biofuels, biogas, and liquid biofuels, each playing a distinct role in different sectors of the economy [17].

Solid biofuels, including agricultural and forestry residues, make up over 90% of bioenergy use in China. Traditional biomass use for residential heating and cooking has been declining, replaced by oil and gas. However, the use of solid biofuels for electricity and industrial heating has been increasing, indicating a shift toward modern bioenergy applications. This includes the use of pellets for industrial heating and electricity production, which has doubled since 2010 [17].

Biogas production in China has stabilized around 300 PJ, accounting for 16.6% of the total bioenergy share. Biogas is primarily used for residential cooking and electricity generation. The potential for biogas to replace natural gas in residential and industrial applications is significant, especially as waste management systems improve and more feedstock becomes available [17].

Transport biofuels, mainly bioethanol, represent about 7.2% of the bioenergy in China. Although the current share of biofuels in the transport sector is less than 1%, there is substantial room for growth. The government has implemented several policies to promote bioethanol and biodiesel production, aiming to increase their use in the coming years [17].

China's extensive agricultural sector provides a vast resource base for bioenergy production. The potential for expanding bioenergy is substantial, particularly through the utilization of agricultural residues, forestry waste, and MSW. With appropriate policy support and technological advancements, bioenergy can significantly contribute to China's renewable energy targets [17].

8.3.8 Bioenergy's Role and Potential in Brazil

Bioenergy plays a pivotal role in Brazil's renewable energy landscape. In 2019, bioenergy contributed significantly to the country's total energy supply, which was heavily dominated by renewable sources. The primary bioenergy resources include sugarcane ethanol, biodiesel, and solid biofuels like sugarcane bagasse [18].

Brazil's ethanol industry is one of the most advanced globally, with ethanol production reaching 36 billion liters in 2019. Ethanol is primarily used in the transport sector, where it is blended with gasoline at a mandatory 27% ratio (E27) and used in pure form (E100) in flex-fuel vehicles. The biodiesel sector, although smaller, is growing, with a mandatory blend of 12% (B12) in diesel. In 2019, biodiesel production was 5.9 billion liters, with soybean oil being the predominant feedstock [18].

Solid biofuels, mainly sugarcane bagasse, are integral to Brazil's bioenergy landscape. Bagasse is used for electricity and heat production in the sugar and ethanol industries, contributing to both onsite energy needs and the national grid. The use of bagasse for cogeneration has made significant inroads, enhancing energy efficiency and reducing GHG emissions [18].

Brazil's potential for bioenergy expansion is substantial, driven by its vast agricultural sector and supportive policy framework. The country's Nationally Determined Contribution (NDC) under the Paris Agreement sets ambitious targets for bioenergy [18]:

- Increase the share of sustainable bioenergy in the energy matrix to approximately 18% by 2030.

- Expand ethanol and biodiesel production, with a greater proportion of advanced biofuels, including cellulosic ethanol.
- Enhance the role of bioenergy in electricity generation, aiming for a significant increase in the use of biomass, wind, and solar energy.

8.4 Developing Policy Frameworks to Support Bioenergy

8.4.1 Policy Strategies in the Nordic Countries

The Nordic countries—Sweden, Denmark, Finland, and Norway—have tailored their biofuels policies to align with their distinct energy needs, environmental priorities, and national capabilities. Each nation has established strategic documents and policy instruments that not only adhere to European Union directives, where applicable, but also address their unique national concerns and resources.

Sweden: A Comprehensive Biofuel Strategy [19]

Sweden's biofuel policy is integrated within its broader environmental and energy strategies, targeting a future powered by sustainable and renewable energy. The Swedish government champions the development of third-generation biofuels using nonfood feedstocks, demonstrating a robust commitment to minimizing fossil fuel reliance. Substantial government investment supports research, development, and deployment (RD&D), facilitating significant advancements in biofuel technologies. Sweden's strategic documents underscore the convergence of energy and environmental goals, as evidenced by public funding mechanisms and tax exemptions designed to promote biofuel usage.

Denmark: Strategic Integration and Export Orientation [19]

Denmark's biofuel policies are closely tied to its commitments to combat climate change and its role as a leading exporter of renewable energy technologies. Danish biofuel strategies merge EU compliance with national objectives, aiming for a full transition to renewable energy by 2050. The emphasis is on the production and export of biofuels rather than domestic consumption, maintaining Denmark's position at the forefront of global renewable energy technology markets. Government-funded RD&D is supported by strategic alliances between public institutions and the private sector.

Finland: Leveraging Industrial Synergies [19]

In Finland, biofuel policy is significantly influenced by the country's strong forestry and paper industries, capitalizing on existing expertise and infrastructure. The Finnish government encourages the development of second-generation biofuels, particularly focusing on integrating these technologies into existing industrial processes. RD&D funding in Finland is geared toward technological breakthroughs in gasification and pyrolysis of wood residues, enhancing both energy security and environmental sustainability within its industrial sectors.

Norway: Balancing Non-EU Compliance and Environmental Standards [19]

Norway, while not an EU member, ensures its biofuel policies are closely aligned with EU directives to maintain market compatibility and uphold environmental standards. Norwegian policy concentrates on reducing emissions in the transportation sector, employing measures like tax exemptions and incentives for cleaner fuel technologies. The strategic documents demonstrate a careful balance, aiming to integrate renewable energy sources while supporting the nation's substantial oil and gas industries.

Impact of RD&D Funding [19]

RD&D funding in the Nordic countries is crucial in molding the biofuel landscape. In Sweden and Finland, where innovation is prioritized, funds are directed toward developing novel biofuel technologies and integrating them with existing industries. Denmark focuses on technology development and exporting its renewable energy solutions, while Norway allocates its RD&D funding to harmonize its traditional energy sectors with renewable initiatives.

The diversity of biofuels policy across the Nordic countries illustrates the complex interplay between national priorities, natural resource allocations, and global environmental commitments. While all four nations strive to augment the use of renewable energy sources, their strategies reflect distinct national priorities and capabilities. Understanding these differences is essential for stakeholders in the energy sector, policymakers, and researchers committed to crafting and implementing sustainable energy policies.

8.4.2 United Kingdom: Policies Driving Bioenergy

The trajectory of the United Kingdom toward a reduced carbon footprint is intricately connected with the evolving landscape of bioenergy technology and pertinent policies. The challenges in realizing a low-carbon future through bioenergy are multifaceted, spanning the synchronization of policy frameworks with technological and market developments, the creation of supportive infrastructures, and the mitigation of societal apprehensions regarding novel technologies. Nonetheless, these challenges also offer opportunities for systemic transformation and the advancement of sustainable energy protocols, contingent on a concerted enhancement of policy backing and societal endorsement of bioenergy alternatives [20].

Emphasized in the findings are lessons for policymakers, underscoring the imperative of cultivating an enabling environment for bioenergy through dynamic policies tailored to the distinctive requisites and capabilities of diverse regions, such as Wales. These policies should be geared toward transitioning bioenergy from a peripheral innovation to a cornerstone of the energy framework, advocating a comprehensive perspective that integrates not only technological elements but also social, economic, and environmental considerations [20].

8.4.3 Bioenergy Policy Landscape in Brazil

Brazil's biofuel policy capitalizes on its agricultural capabilities, particularly the extensive cultivation of sugarcane, which is a cornerstone of its ethanol production industry. Originating as a strategic response to the 1970s oil crisis, the ProAlcohol initiative laid the foundation for what has evolved into a sophisticated ethanol fuel sector. This approach is highlighted by strict ethanol blending mandates, which have been established as high as 27% ethanol in gasoline [21]. Complementing these mandates are economic incentives aimed at promoting the uptake of flex-fuel vehicles, which are designed to operate on high ethanol blends, thereby enhancing the integration of biofuels into the national energy matrix.

In addition to regulatory mandates, Brazil is committed to the sustainability of its bioenergy sector through the RenovaBio program [21]. Launched specifically to support the blending mandates, RenovaBio focuses on reducing carbon emissions from biofuels. Furthermore, the Brazilian government actively supports the biofuel industry through significant investments in research and development, particularly targeting next-generation biofuels. This sustained investment underscores Brazil's dedication to maintaining its leadership in biofuel technologies, integrating environmental concerns with technological advancement.

8.4.4 India's Framework for Bioenergy Development

India's bioenergy policy framework includes several initiatives to promote the use of bioenergy across different sectors [16]:

- **Electricity Act of 2003 and Amended National Tariff Policy-2016:** These policies mandate state electricity regulators to promote electricity generation from nonconventional sources, including bioenergy.
- **Policy on Biomass Utilization for Power Generation (2017):** This policy encourages the cofiring of biomass with coal in thermal power plants, thus integrating bioenergy into the mainstream power sector.
- **National Policy on Biofuels-2018:** This policy expands the scope of raw materials for biofuel production and sets ambitious blending targets for ethanol and biodiesel.

India's biofuel policy, though a relatively recent development, has experienced swift expansion, motivated by the dual goals of securing energy resources and reducing environmental degradation. The cornerstone of this strategy is the National Policy on Biofuels, which sets ambitious targets for biofuel integration, including a 20% ethanol blending mandate by 2025 and initiatives to enhance biodiesel consumption. To circumvent competition with the food supply, the policy advocates for the utilization of nonfood feedstocks, such as agricultural residues. Additionally, the framework is supported by government mechanisms, including fixed ethanol pricing based on feedstock type and substantial subsidies aimed at fostering the establishment of new biofuel production facilities. Together, these measures reflect India's robust commitment to cultivating a sustainable and expansive biofuel industry [21].

8.4.5 Bioenergy Policy Framework in the United States

As the world's premier biofuel producer, the United States adopts a comprehensive and expansive strategy to biofuel production that utilizes a diverse array of feedstocks and technological approaches, leveraging its extensive agricultural capacity and advanced technological infrastructure. Integral to this strategy is the RFS, which compels the inclusion of specific volumes of biofuels within the national fuel supply, effectively stimulating demand. Additionally, the biofuel sector is bolstered by enduring tax credits and financial incentives that support both traditional and innovative biofuel technologies. A significant emphasis is also placed on fostering innovation through robust government funding for research into biofuel technologies, with a particular focus on the development of advanced biofuels, such as cellulosic ethanol and renewable diesel, further solidifying the U.S. position at the forefront of the global biofuel industry [21].

8.4.6 Bioenergy Policies in China

China has established a robust policy framework to support the development of bioenergy [17]:

- **Renewable Energy Law of the People's Republic of China (2006):** This law provides the foundation for renewable energy development, including bioenergy, outlining the general conditions for integrating renewable sources into the national energy mix.
- **Blue-Sky Action Plan (2018):** Aims to reduce emissions of major air pollutants and GHGs through various measures, including promoting bioenergy.
- **Action Plan for Reaching Carbon Dioxide Peak before 2030 (2021):** Sets targets for increasing the share of nonfossil energy to 25% by 2030 and reducing carbon intensity by 65% compared to 2005 levels.

China is investing heavily in research and development to enhance bioenergy technologies. This includes improving the efficiency of biomass conversion, developing new biofuel production processes, and exploring innovative uses of bioenergy. Notable advancements include breakthroughs in cellulosic ethanol technology and the development of high-yield bioethanol production processes [17].

The future of bioenergy in China is promising, with significant potential for growth in various sectors. Key challenges include enhancing the efficiency of biomass conversion technologies, improving supply chain logistics, and ensuring sustained policy support. Addressing these challenges will be crucial for maximizing the contribution of bioenergy to China's energy transition.

8.4.7 Comparative Analysis of Global Bioenergy Policies

Bioenergy policies worldwide reflect diverse national priorities and resources. The Nordic countries—Sweden, Denmark, Finland, and Norway—tailor their policies to their unique energy needs and environmental goals, integrating EU directives with national strategies. Sweden focuses on third-generation biofuels and significant RD&D investments, while Denmark emphasizes renewable energy exports and strategic public-private alliances. Finland leverages its forestry industry for second-generation biofuels, and Norway aligns with EU directives to reduce transportation emissions. The UK fosters an enabling environment for bioenergy through dynamic policies and infrastructure support. Brazil capitalizes on its sugarcane industry with strict ethanol blending mandates and the RenovaBio program to reduce carbon emissions. India promotes bioenergy through policies encouraging nonfood feedstocks and substantial subsidies, aiming for a 20% ethanol blend by 2025. The United States, the leading biofuel producer, employs the RFS and robust financial incentives to support diverse biofuel technologies, focusing on advanced biofuels. China's policies, including the Renewable Energy Law and the Blue-Sky Action Plan, set ambitious targets for nonfossil energy, supported by heavy investment in bioenergy R&D. RD&D funding in these countries drives technological advancements and integration of biofuels into existing industries, illustrating the complex interplay between national priorities and global environmental commitments. Understanding these diverse strategies is crucial for stakeholders in the energy sector to develop and implement sustainable energy policies.

8.5 Overcoming Barriers to a Low-Carbon Society Through Bioenergy

Although bioenergy holds considerable promise for a low-carbon future, its integration into society is impeded by a myriad of technical, economic, and regulatory challenges. One of the critical hurdles is the technical and economic viability of carbon capture and storage (CCS), essential for bioenergy with carbon capture and storage (BECCS) technologies [10]. The high costs and unproven scalability of CCS technologies pose substantial barriers that necessitate targeted policy interventions. These policies must encourage technological innovation and facilitate cost reductions, potentially through financial support for research and the implementation of pilot projects.

Additionally, the environmental and social repercussions associated with bioenergy, such as land-use changes, water resource management, and community impacts, demand stringent environmental regulations and responsible project execution. The regulatory environment surrounding bioenergy is often fraught with uncertainties that can discourage investment and impede the sector's growth. To overcome these barriers, governmental policies need to be clear, consistent, and supportive, ensuring long-term stability and fostering a conducive growth environment within the bioenergy sector [10].

Effective policies should establish definitive guidelines for sustainable land use and provide incentives for adopting low-carbon technologies, alongside mechanisms that engage and provide fair compensation to affected communities. Furthermore, regulatory frameworks must be flexible and adaptive to the rapid advancements in bioenergy technology and changing market conditions. Such adaptability is crucial for the sector's sustainable evolution and its integration into broader national energy strategies [10].

8.6 Explorative Case Studies on Bioenergy Implementation

8.6.1 Innovations in Lignocellulosic Biomass Utilization

The utilization of lignocellulosic biomass for bioenergy products holds significant potential yet encounters formidable challenges in fostering a low-carbon society. This pathway, an alternative to fossil fuels, offers the advantages of renewability and the capability to substantially reduce GHG emissions [22]. Notably, bioenergy derived from lignocellulosic materials such as pellets and biomass-based electricity significantly lowers GHG emissions in comparison to conventional energy sources, thereby facilitating a transition toward a low-carbon economy. Nevertheless, the deployment of these technologies faces considerable hurdles, particularly in economic terms. For instance, the production of pyrolysis oil often results in negative net present values, as revealed by the article's technoeconomic assessments. Additionally, while the environmental impact of these technologies is generally lower than that of fossil fuels, there is considerable variability in GHG emissions among different bioenergy products. Some, like pyrolysis oil production, still generate relatively high levels of emissions. Addressing these multifaceted challenges necessitates robust policy support, continued technological advancements, and substantial financial investments to improve the efficiency and sustainability of bioenergy production processes. A nuanced, integrated approach is essential to fully realize the economic and environmental benefits of incorporating lignocellulosic biomass into our energy systems while mitigating associated risks [22].

8.6.2 Bioenergy Development in Taiwan

To transition to a low-carbon society, Taiwan is emphasizing waste-to-energy (WTE) technologies to diminish its dependence on imported energy and to reduce GHG emissions. Significant progress has been made in enhancing the efficiency of MSW incineration plants, with increased power generation achieved through technological upgrades and operational improvements, thereby transforming waste into a viable energy source [23]. Legislative and policy frameworks, including the Renewable Energy Development Act, play a pivotal role by providing regulatory incentives that integrate WTE technologies into Taiwan's national energy strategy. Furthermore, Taiwan's WTE initiatives are aligned with broader sustainable development goals, aiming for net-zero emissions by 2050 and adopting practices that bolster the circular economy, where waste materials are reused and recycled for maximum value [23].

However, the advancement of WTE technologies in Taiwan faces several challenges. Financial constraints are significant, with the costs of setting up and operating WTE facilities often exceeding those of conventional energy sources. Technological barriers also persist, necessitating further innovation to enhance the energy efficiency of WTE processes and manage the variability in waste composition. Additionally, WTE plants can encounter public opposition due to concerns over emissions and other environmental impacts, making it essential for these facilities to meet stringent environmental standards to gain public trust. To overcome these obstacles, increased governmental support through subsidies, grants, or tax incentives is crucial. Public-private partnerships could facilitate the development and deployment of advanced WTE technologies, while community engagement and transparency are necessary to address public concerns and highlight the benefits of WTE. Lastly, investing in research and development is essential for uncovering new technologies that can optimize the efficiency and environmental footprint of WTE facilities. By addressing these challenges and implementing necessary actions, Taiwan can effectively advance toward becoming a low-carbon society through strategic engagement with WTE technologies [23].

8.6.3 Bioenergy in Eastern Africa: Challenges and Opportunities

An article highlights the substantial potential for Eastern African countries in the international liquid biofuel market [24]. These regions, favored by climatic conditions ideal for bioenergy crop cultivation, are well-positioned to meet the increasing global demand for transportation fuels. Despite this potential, the growth of the biofuel sector is impeded by the absence of robust market dynamics and supportive government policies. The economic feasibility of biofuel production significantly depends on external influences, including policies and market opportunities from industrialized countries, while internal policy gaps and disjointed strategies restrict effective sector development.

The article points out the environmental benefits of biofuel production, such as emission reductions and decreased dependency on fossil fuels. It also addresses socioeconomic challenges, including land-use conflicts and technical difficulties due to inadequate domestic frameworks. It calls for improved governance and clear policies that facilitate local land use and integrate biofuel production into traditional agricultural practices.

Additionally, the article underlines the importance of both small- and large-scale biofuel projects. It suggests that learning from global best practices and enhancing international cooperation could significantly propel sustainable biofuel adoption in Eastern Africa. Embracing successful global models can accelerate the region's move toward sustainable energy solutions, thus contributing to broader environmental objectives and boosting local economic growth.

8.6.4 Bioenergy in Australia: Current Status and Future Potential

Australia has substantial biomass resources, including forestry and cropping residues and bagasse from sugarcane processing. The study by Middelhoff et al. (2022) identified 1.0 million tonnes of bagasse, 16.6 million tonnes of forestry residues, and 28.7 million tonnes of stubble residues available for bioenergy production across over 223 prospective sites [25]. High-resolution biomass resource maps were developed to facilitate the optimized siting of bioenergy plants, considering spatial constraints and transport distances. As of 2020, bioenergy contributes approximately 1.4% of Australia's electricity output, lagging behind other OECD countries with around 2.4%. The underdevelopment of the bioenergy sector is attributed to a lack of detailed information on biomass availability and potential [25]. To address this, the Australian Government commissioned a bioenergy roadmap to promote the sector's growth. The roadmap highlights bioenergy's role in providing dispatchable electricity, especially crucial for grid stability and integration of variable RE sources.

The integration of bioenergy into Australia's energy mix relies on accurate resource mapping and optimized siting. Using a geographic information system (GIS)-based algorithm, a study created high-resolution biomass resource maps (5 × 5 km) for bagasse, forestry, and stubble residues [25]. The siting algorithm identified potential bioenergy sites by considering proximity to transmission infrastructure and transport distances. This approach ensures that bioenergy plants are strategically located to maximize resource utilization while minimizing transport emissions and costs. The study identified several prospective sites for new bioenergy plants, leveraging unutilized biomass resources. For example, in Queensland, three new sites were identified with a combined potential of over 1.1 million tonnes of bagasse. In New South Wales, one new site was identified with a potential of 31,853 t of bagasse. The siting algorithm prioritized locations within close proximity to existing transmission lines to reduce infrastructure costs.

8.7 Bioenergy Coupled with Carbon Capture and Storage (BECCS)

8.7.1 Advancing Negative Emissions in Pursuit of a Low-Carbon Society

Bioenergy with Carbon Capture and Storage (BECCS) emerges as a critical technology for climate change mitigation, offering the potential for negative carbon emissions. This innovative approach combines biomass utilization for energy production with carbon capture and storage techniques, effectively removing CO_2 from the atmosphere. As global efforts intensify to meet stringent climate targets, the significance of BECCS is underscored by initiatives like the Paris Agreement, which aims for a net-zero GHG emissions scenario in the second half of this century.

A recent comprehensive review outlines essential research priorities for advancing BECCS, spanning technical, economic, and social dimensions. These priorities include developing pilot and demonstration projects to establish evidence-based practices and gather data on the scalability and effectiveness of BECCS. Additionally, conducting thorough economic assessments and life-cycle analyses is crucial to understanding the financial and environmental sustainability of BECCS. Moreover, increasing public awareness and acceptance are vital, as societal support is necessary for the broader application of this technology. Expanding research into biogas cofiring and hydrogen production from biomass could also enhance the versatility and economic viability of BECCS [26].

Effective stakeholder engagement is pivotal for bridging gaps between market needs and industrial capabilities, involving collaboration among industry, academia, and policymakers. This integrative approach is essential for addressing the complex challenges of BECCS and fostering an environment conducive to innovative solutions. To ensure BECCS can contribute meaningfully to global climate mitigation, it must be supported by robust policies that promote research and development, subsidize pilot projects, and facilitate its integration into climate strategies. These policies should not only incentivize BECCS deployment but also consider its social and environmental impacts to ensure it contributes positively to community development and biodiversity conservation.

In conclusion, BECCS represents a forefront technology in the array of negative emission strategies, playing a vital role in achieving global climate objectives. By focusing on the identified research priorities and fostering strong collaboration among all stakeholders, BECCS can transition from a theoretical mitigation strategy to a practical and indispensable element of our energy future. Its development represents a comprehensive endeavor that spans technological, economic, social, and environmental aspects, each critical to its success and sustainability [26].

8.7.2 Exploring Local-Scale Opportunities for Carbon Reduction

Bioenergy with carbon capture and storage (BECCS) is a critical technology in the quest for a low-carbon future, merging bioenergy production with CO_2 capture and storage to potentially generate negative emissions. This process not only helps remove CO_2 from the atmosphere but also produces energy, aligning with international climate objectives and fostering sustainable community practices. At the community scale, BECCS introduces both challenges and opportunities. The acceptance of BECCS within communities is shaped by factors such as the origins and applications of the captured CO_2, local economic repercussions, and the surrounding policy framework. Despite the promise of BECCS, public understanding is generally low; where it exists, perceptions are frequently influenced by local media and the immediate perceived benefits, like job creation and regional economic development [27].

Governance and policy play essential roles in the successful deployment of BECCS, as they must safeguard community interests and ensure local benefits from these projects while still aligning with national and global climate agendas. The literature emphasizes the necessity of a dual approach that both protects communities from associated risks and enhances their ability to benefit from BECCS initiatives. This involves fair land use practices, equitable economic returns, and transparent project planning and execution. Furthermore, there is a critical need for empirical research into the impacts of BECCS at the community level, with a particular focus on understanding the community's desire for and potential involvement in the BECCS value chain, aiming to integrate BECCS sustainably into local economies and ecosystems in a way that supports environmental justice and sustainability [27].

In conclusion, integrating BECCS within community-led environmental and economic frameworks presents a significant opportunity to achieve both local and global sustainability objectives. Effective implementation of BECCS at the community level demands cooperative efforts from researchers, policymakers, and community stakeholders to meticulously address the socioeconomic and environmental facets of sustainability, thereby making BECCS a foundational element for a resilient, low-carbon future [27].

8.7.3 BECCS Implementation Examples from Tanzania

In sub-Saharan Africa, Tanzania offers a distinctive case study due to its extensive biomass resources and feasible underground CO_2 storage capacities. Tanzania's capacity for BECCS is largely driven by its agricultural sector and vast expanses of both arable and nonarable land suitable for energy crop cultivation, such as sugarcane for bioethanol production. Theoretical studies support the presence of considerable CO_2 storage potential in regions that are both less populated and less industrialized, facilitating a combination of energy production and carbon sequestration [28]. Despite these advantages, the implementation of BECCS in Tanzania faces significant obstacles. Institutional and policy barriers, such as weak governance and unpredictable land management systems, complicate the deployment of this technology. Additionally, socioeconomic and ethical issues, including potential land conflicts and concerns over using local resources for the benefit of more developed nations, could lead to societal resistance and ethical dilemmas [28].

Addressing these concerns requires a multi-faceted approach. The technical and market challenges—like the lack of necessary infrastructure for carbon capture and the need for market development for biomass-based products—must be overcome through robust governmental support and international cooperation. Moreover, ethical considerations, such as the equitable distribution of resources and ensuring that local communities benefit from BECCS projects, are paramount. For BECCS to be successfully integrated into Tanzania's environmental strategy, a comprehensive framework involving clear policy directives, community engagement, and transparency is essential. Only through such a concerted effort can BECCS contribute effectively to Tanzania's pursuit of a sustainable, low-carbon future [28].

8.8 Conclusion

The integration of bioenergy into a low-carbon economy presents a promising yet complex challenge. This chapter has outlined the multifaceted role of bioenergy in mitigating climate change through sustainable practices and technological innovations. As countries strive for energy independence and environmental sustainability, bioenergy stands out as a versatile solution capable of providing a continuous energy supply and supporting the circular economy. However, the successful deployment of bioenergy technologies requires robust policy support, public acceptance, and continued research and development to overcome the economic and technical barriers currently faced. Ultimately, the strategic implementation of bioenergy, along with other renewable resources, is crucial for achieving global targets of reducing carbon emissions and fostering a sustainable, low-carbon society.

References

1. IRENA. A New World: the Geopolitics of the Energy Transformation. 2019.
2. Fast-Track Energy Transitions to Win the Race to Zero [press release]. 2021.
3. IRENA. WORLD ENERGY TRANSITIONS OUTLOOK 2022: 1.5° C PATHWAY. 2022.
4. IEA. Renewable Energy Market Update: outlook for 2022 and 2023. 2022.
5. Rogers, J.N., Stokes, B., Dunn, J. et al. (2017). An assessment of the potential products and economic and environmental impacts resulting from a billion ton bioeconomy. *Biofuels, Bioproducts and Biorefining*. 11 (1): 110–128.
6. Tong, D., Farnham, D.J., Duan, L. et al. (2021). Geophysical constraints on the reliability of solar and wind power worldwide. *Nature Communications*. 12 (1): 6146.
7. Richardson D, Pitman AJ, Ridder NN. Climate influence on compound solar and wind droughts in Australia. npj *Climate and Atmospheric Science*. 2023;6(1):184.
8. Reid, W.V., Ali, M.K., and Field, C.B. (2020). The future of bioenergy. *Global Change Biology*. 26 (1): 274–286.
9. Cuong, T.T., Le, H.A., Khai, N.M. et al. (2021). Renewable energy from biomass surplus resource: potential of power generation from rice straw in Vietnam. *Scientific Reports*. 11 (1): 792.
10. Bibas, R. and Méjean, A. (2014). Potential and limitations of bioenergy for low carbon transitions. *Climatic Change*. 123 (3): 731–761.
11. Darda, S., Papalas, T., and Zabaniotou, A. (2019). Biofuels journey in Europe: currently the way to low carbon economy sustainability is still a challenge. *Journal of Cleaner Production*. 208: 575–588.
12. Sovacool, B.K. (2017). Contestation, contingency, and justice in the Nordic low-carbon energy transition. *Energy Policy*. 102: 569–582.
13. Dioha, M.O., Emodi, N.V., and Dioha, E.C. (2019). Pathways for low carbon Nigeria in 2050 by using NECAL2050. *Renewable Energy Focus*. 29: 63–77.
14. Buchholz, T. and Volk, T.A. (2012). Considerations of project scale and sustainability of modern bioenergy systems in Uganda. *Journal of Sustainable Forestry*. 31 (1–2): 154–173.

15 Luc Pelkmans LB, Jim Spaeth. Country Report 2021: USA. IEA Bioenergy; 2021.

16 Pelkmans, L. (2021). *Country Report 2021: India*. IEA Bioenergy.

17 Dou Kejun WZ, Ren Dongming. Country Report 2021: China. IEA Bioenergy; 2021.

18 Ekbom T. Assessment of Successes and Lessons Learned for Biofuels Deployment: status of Biofuels Policies and Market Deployment in Brazil, Canada, Germany, Sweden and the United States. IEA Bioenergy TCP; 2023.

19 Olsen, D.S., Klitkou, A., and Eerola, A. (2013). *Analysis of biofuels policy in the Nordic countries*. Oslo: TOP-NEST.

20 De Laurentis, C. (2015). Innovation and policy for bioenergy in the UK: a co-evolutionary perspective. *Regional Studies.* 49 (7): 1111–1125.

21 IEA. Biofuel Policy in Brazil, India and the United States. 2023.

22 Wang, Y., Wang, J., Zhang, X., and Grushecky, S. (2020). Environmental and economic assessments and uncertainties of multiple lignocellulosic biomass utilization for bioenergy products: Case studies. *Energies.* 13 (23): 6277.

23 Tsai, W.-T., Tsai, C.-H., and Shen, Y.-H. (2023). Waste-to-energy for circular low-carbon society in Taiwan: a case study. *Journal of Material Cycles and Waste Management.* 25 (4): 1876–1886.

24 Miftah, F.K. and Mutta, D. (2024). Potential markets and policies for sustainable liquid biofuel production with emphasis to Eastern Africa countries: a review. *Energy, Sustainability and Society.* 14 (1): 1.

25 Middelhoff, E., Madden, B., Li, M. et al. (2022). Bioenergy siting for low-carbon electricity supply in Australia. *Biomass and Bioenergy.* 163: 106496.

26 Stavrakas, V., Spyridaki, N.-A., and Flamos, A. (2018). Striving towards the deployment of bio-energy with carbon capture and storage (BECCS): a review of research priorities and assessment needs. *Sustainability.* 10 (7): 2206.

27 Buck, H.J. (2019). Challenges and opportunities of bioenergy with carbon capture and storage (BECCS) for communities. *Current Sustainable/Renewable Energy Reports.* 6 (4): 124–130.

28 Hansson, A., Fridahl, M., Haikola, S. et al. (2020). Preconditions for bioenergy with carbon capture and storage (BECCS) in sub-Saharan Africa: the case of Tanzania. *Environment, Development and Sustainability.* 22 (7): 6851–6875.

9

Perspectives of Integrating Bioenergy into the Energy Mix in Developing Nations: A SWOT Analysis

Toyese Oyegoke

Chemical Engineering Department, Ahmadu Bello University, Zaria, Nigeria

9.1 Introduction

The energy sector in developing nations is plagued by various crises and challenges that have become a part of daily life for residents [1]. These challenges encompass issues such as inconsistent power supply, frequent collapses of national power grids [2–5], dwindling water levels in hydropower plant dams, and fluctuating government energy policies. Some of the countries affected and encountering such challenges include Bangladesh [6], Papua New Guinea, Yemen, Nepal, Burkina Faso, the Dominican Republic, Nigeria, Sri Lanka [7], Venezuela [8, 9], Puerto Rico [10], the Philippines, Niger, Pakistan [11], and many others [12–14]. Consequently, the populace in these communities faces difficulties in affording local power generation plants, leading them to resort to traditional methods like burning wool or firewood for illumination and cooking and, more recently, the use of charcoal for heating [15], which poses a risk of promoting deforestation if not managed properly amidst rising fossil fuel prices.

The significance of energy supply for achieving sustainable living and industrial development cannot be overstated. However, the inadequate attention from governments in these countries has hindered development, increased unemployment rates, and resulted in the shutdown of numerous production companies in these communities. Industrialization heavily relies on a consistent energy supply, and the absence of it, coupled with reliance on local power generators, severely impacts the economic viability of production processes.

In many developing nations, reliance on fossil-based fuels like gasoline, diesel, and kerosene, along with their generating plants, is the primary energy option. This reliance has gradually increased their share in the energy mix, contributing

Energy Management Strategies for Multi-Vectored Energy Hubs to Achieve Low Carbon Societies, First Edition.
Edited by Shubham Tiwari, Jai Govind Singh, Sivaraman Palanisamy, Sharmeela Chenniappan, Rupendra Kumar Pachauri, and Sanjeevikumar Padmanaban.

to the growing global greenhouse emissions that expose communities to drastic climate changes, increased flooding, and wildfires.

Hence, there is an urgent need to explore long-term, technically and economically viable solutions to address the challenges faced by residents in these countries. Some researchers have proposed alternative energy sources in the literature, including solar power, wind power, geothermal power, and other renewables like bioenergy, which hold promise in alleviating energy poverty prevalent in developing countries. These renewable sources offer infinite availability for continuous usage, making them suitable options for energy production. For instance, communities in desert areas with high sunshine levels can harness solar and wind power, albeit with the requirement of expensive energy storage devices like batteries, which may pose affordability challenges to rural or developing community residents. Conversely, promoting bioenergy sources offers numerous advantages, such as meeting the energy needs of developing countries and enhancing living standards, particularly in rural communities.

Bioenergy, derived from biomass through biochemical, thermochemical, or combined processes, has a long history of use, especially in rural areas of African and Asian countries, where it is used for heating, cooking, and other purposes [16, 17]. Modern bioenergy technologies are designed to minimize carbon emissions, contributing to efforts to reduce global greenhouse gas emissions. The literature highlights various technologies [17–27] that convert biomass into energy sources, such as bioethanol, biomethane, biodiesel, biomethanol, biohydrogen, and other biochemicals, demonstrating the versatility of bioenergy in energy generation. Additionally, bioenergy has been shown in numerous studies [28–31] to offer various benefits, including better sewage and solid waste management, environmental cleanliness, improved community health, waste value addition, sustainable energy provision, air quality improvement, and potential mitigation of climate change [32].

Given the heavy reliance of residents in developing countries on agriculture, particularly in rural communities, and the persisting energy crises due to governments' inability to address energy supply challenges, including the fact that the fossil-based sources or traditional approaches commonly explored by residents of such communities as energy sources have environmental implications, this report employs SWOT analysis to evaluate the strengths (S), weaknesses (W), opportunities (O), and threats (T) associated with integrating bioenergy into the national energy mix as a possible alternative renewable energy source. This initiative aims to strengthen sustainable energy supply, environmental sustainability, and national development in these countries' communities. By identifying S, W, O, and T, the report presents an effective managerial strategy to facilitate the successful deployment of the bioenergy integration initiative, accompanied by recommendations. SWOT analysis serves as a valuable approach in decision-making and strategic planning, making it apt for assessing the integration of bioenergy into the national energy mix.

9.2 Literature Review

9.2.1 Overview of Bioenergy

Bioenergy is a form of renewable, alternative, low-carbon, and green energy system that involves the conversion of biomass (plant- or animal-based materials) to energy as well as sequestering atmospheric carbon and many other environmental and socioeconomic advantages that it offers to the world. According to EERA [33] and Rogers et al. [34], bioenergy can be said to be one of the different resources available for meeting our energy demand and, as a renewable form of energy, can be derived from organic materials better known as biomass, which can be used to produce transportation fuels, heat, electricity, and products.

This form of energy system or source strongly promotes global climate change targets, reduces greenhouse gases, and promotes a sustainable environment. A survey of the literature classified bioenergy technologies into three classes: thermochemical, biochemical, and combined process technologies. Figure 9.1 diagrammatically summarizes the different pathways, or established technological routes, for sourcing bioenergy using biomass resources.

9.2.1.1 Thermochemical Process Technology

This type of technology is energy-intensive and entails energy-requiring processes like gasification, pyrolysis, and direct combustion. It is diagrammatically represented in Figure 9.2. The process technology majorly involves the decomposition of biomasses in the presence or absence of air to produce light gases. It also involves breaking down bonds between carbon, hydrogen, and oxygen in the biomass molecular structure to release the chemical energy stored during

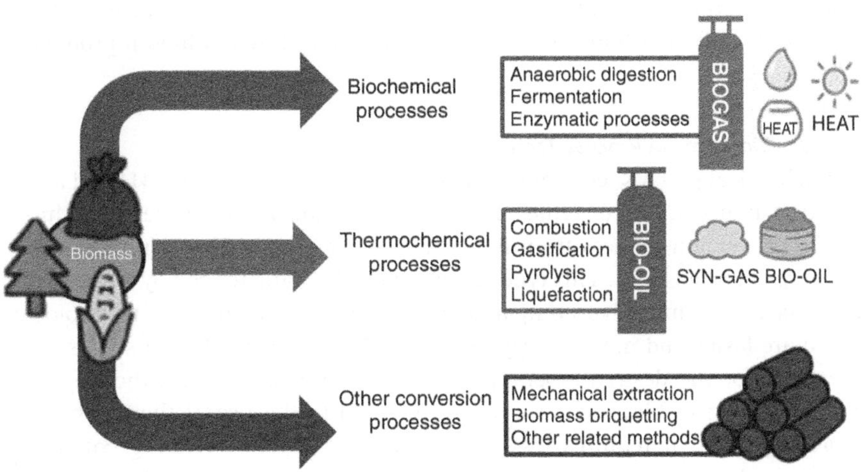

Figure 9.1 An overview of different possible bioenergy production routes.

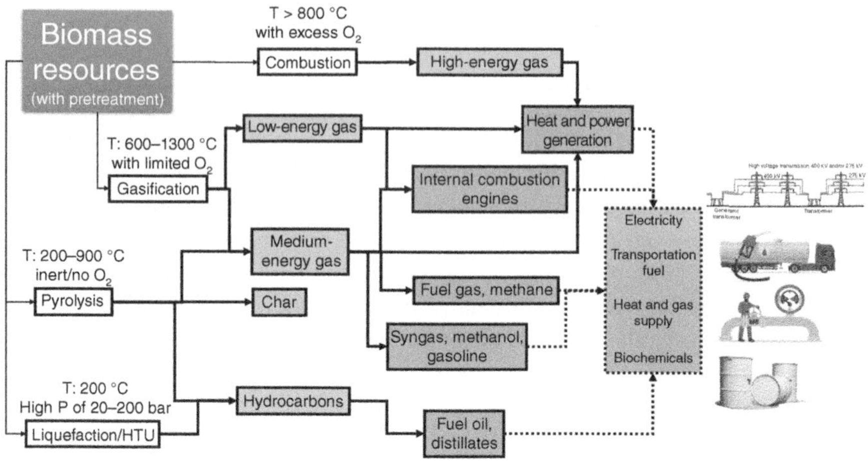

Figure 9.2 Overview of thermochemical process technology. *Source:* kontur-vid / Adobe Stock Photos; Feodora / Adobe Stock Photos; missisya / Adobe Stock Photos.

photosynthesis [35]. It has also been seen as an efficient means of converting biomass into biofuel [36].

The thermochemical process operating in the presence of an excessive supply of air and > 800 °C is known as direct combustion; one operating in the presence of a limited air supply and 700–1100 °C is known as gasification; one operating in the absence of air and 350–220 °C is referred to as pyrolysis, and liquefaction operates at a lower temperature of 250–400 °C and high pressure of 4–22 MPa [35]. Following the varying operating conditions of the different technologies, combustion leads to heat production, gasification leads to gas production like syngas and other light hydrocarbons, pyrolysis leads to char and bio-oil production [37, 38], and liquefaction (or hydrothermal) leads to bio-oil production [35, 36].

9.2.1.2 Biochemical Process Technology

In biochemical process technologies like fermentation [18, 23, 39–41] and anaerobic digestion [29, 42], enzymes and microorganisms are often deployed in the energy conversion process from lignocellulosic materials to specific biofuels or bioenergy products [43]. This type of technology is characterized by being slower and less energy-intensive compared to thermochemical process technologies, which are faster and more energy-intensive. Despite being slow, it shows better selectivity for the desired bioenergy product, like bioethanol synthesized from the breakdown of biomass resources, compared to the use of thermochemical approaches, and it requires less energy due to its mild operating conditions. A typical scheme presenting a biomass conversion process during the production

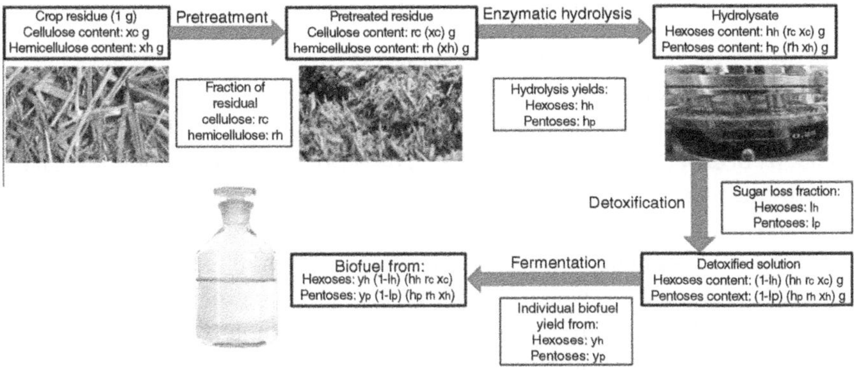

Figure 9.3 Overview of a biochemical-based conversion process for biomass to bioethanol, a bioenergy product. *Source:* [43] / with permission of Elsevier.

of bioenergy is shown in Figure 9.3, including the fermentation process for obtaining bioethanol fuels.

Other biochemical approaches [43, 46, 47] include anaerobic digestion (shown in Figure 9.4), which is a widely adopted technology for producing bioenergy through an enzymatic process that breaks down biomass or organic material in a reactor, often referred to as a biodigester. The enzymatic process involves a series of biochemical reactions that lead to the production of biogas [47]. Biogas, a bioenergy product, can be used to fuel gas cookers, heaters, and power plants, addressing energy challenges and poverty in developing countries, especially rural communities. During this conversion process, biomethane, biohydrogen, and carbon dioxide are produced, along with some impurities such as hydrogen sulfide, which are eliminated during gas upgrading processes.

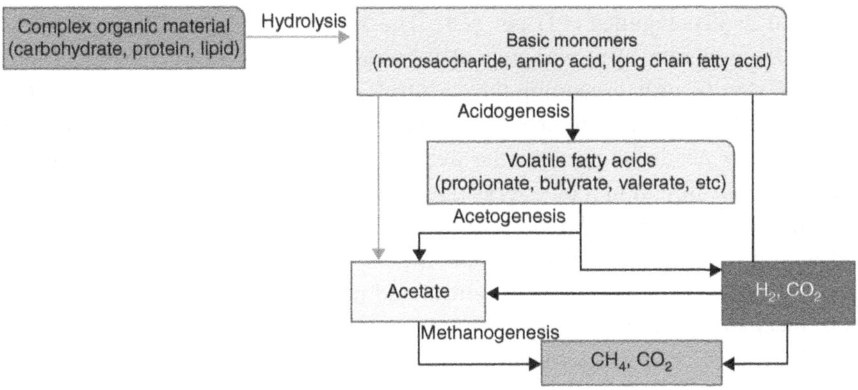

Figure 9.4 A schematic diagram for four-step anaerobic digestion of biomass into biogas. *Source:* [44, 45].

9.2.2 State of Art in Bioenergy Integration in Developing Nations

A survey of research in the literature shows that the integration of bioenergy in developing nations has continued to be a subject of attention, where some works refer to it as the fuel of tomorrow. Singh and Singh's [48] research has reiterated the possibility of experiencing heightened food insecurity and exploitation of local residents by the bioenergy producers as possible adverse effects. Arndt et al. [49] identified land access, neighboring countries' collaboration, and openness for regional importation of production utilities and feedstock as possible obstacles that countries like South Africa need to overcome as they strive to promote the integration of bioenergy in their country. Arndt et al. [49] further emphasized the necessity of ensuring equity, fairness, and socially acceptable procedures for land investment to enable the ease and smooth running of biofuel production processes.

As a way of diverting attention away from the use of food crops, Oyegoke et al. [50] emphasize how different wastes, particularly agro-waste, could be potentially used for biofuels and energy generation as a measure for enhancing waste management. In the same vein, Qyyum et al. [51] analyzed the potential of converting biowaste into bioenergy as a means to drive sustainable economic growth in developing countries. It uses the Product Space Model to assess the income potential of biowaste and suggests industrial policies based on the model's indicators. The study identifies Argentina, Brazil, India, Thailand, and Ukraine as key countries for economic assessment, highlighting their potential for bioenergy production and revenue generation. The findings suggest that these countries can benefit from focusing on bioenergy production to improve their economic prospects.

To gain insight into the worth of energy realizable via bioenergy integration, Tolessa [52] estimates Ethiopia's bioenergy potential from various biomass sources for 2020–2021, with forest residue contributing 56.01%, crop residue 28.29%, livestock waste 15.36%, and municipal solid waste 0.33% of the total estimated 2955 petajoules (PJ) per year. The report shows that Ethiopia could potentially generate approximately 819.7 terawatt hours (TWh) of electricity annually, significantly exceeding the country's current energy consumption. The findings underscore the importance of biomass resources in meeting Ethiopia's future energy needs, suggesting the need for integrated development research and the establishment of a bioenergy database to fully exploit these resources. In a similar trend, Dirisu et al. [53] assessed bioenergy usage in developing nations, highlighting its potential for sustainable economic growth. The study emphasizes the need to optimize bioenergy resources and proposes using the clean development mechanism for emission reduction. It underscores bioenergy's widespread availability and its role in achieving Sustainable Development Goals 7 (it focuses on ensuring universal access to reliable, affordable, and sustainable energy) [54, 55] and 11 (it focuses on making cities and human settlements inclusive, resilient, safe, and sustainable) [56]. It offers valuable insights for policymakers on formulating effective policies for sustainable development through bioenergy utilization.

Other works, like Alsaleh et al.'s [57] study, analyzed the technical efficiency of the bioenergy industry in the EU28 region from 1990 to 2013 using data envelopment analysis. The focus was on the efficiency of bioenergy production as a key factor in its selection as a renewable and sustainable energy source. Findings from their study show that there is higher technical efficiency in developing countries compared to developed countries, while scale efficiency tends to be higher in developed countries. The higher efficiency reported for developed nations could possibly be due to the advanced technologies and stable regulatory environments, which can facilitate better efficiency. Falcone [58] examines the challenges and opportunities faced by developing countries like Albania, Brazil, India, and Kenya in adopting sustainable energy policies, where the importance of shifting towards renewable energy sources to meet climate targets and reduce the harmful effects of fossil fuels was emphasized. The study further highlights the progress these countries have made in adopting renewable energy and suggests solutions to overcome financial, institutional, and technological obstacles. It also stresses the need for cooperation among nations and sharing knowledge to speed up the transition to sustainable energy. The work suggests that policymakers prioritize social, environmental, and economic aspects for a fair transition to sustainable energy.

9.2.3 Previous Studies on SWOT Analysis of Bioenergy Integration

Zhen and Man [59] carried out a SWOT analysis to assess the development of biomass energy in Heilongjiang Province, China. Their study identifies internal strengths and weaknesses, as well as external opportunities and threats, related to biomass energy development. The analysis highlights the province's abundant biomass resources and the potential for economic growth through biomass energy utilization. The study emphasizes the need for strategic planning to capitalize on opportunities and address challenges, suggesting that a comprehensive strategic system is crucial for promoting economic development and adjusting the energy structure in the region.

In Uganda, Okello et al. [60] employed the SWOT-Analytical Hierarchy Process (AHP) to appraise bioenergy technologies, which was effective in evaluating stakeholder preferences and identifying factors influencing the selection of bioenergy systems. Findings from the study suggested that Jatropha could be accepted for household energy, while stakeholders viewed charcoal as unsustainable due to environmental threats. The study further highlights the importance of suitable policies to increase the adoption of bioenergy technologies in Uganda, along with improving the availability of skilled personnel for their dissemination.

The SWOT analysis of bioenergy production in Moldova was also carried out by Soimu [61], which shows that a rushed shift to bioenergy, specifically bioethanol, could negatively impact the country's economic and social systems. Despite Moldova's biomass potential, the country lacks the technological and legislative capacity to implement a large-scale bioethanol production strategy. Importantly, the analysis suggests that Moldova should start with small-scale bioethanol production for self-consumption in agriculture, focusing on restructuring farms and ensuring a balance between biofuel production and food crop prices. The study emphasizes the need for government control to prevent a food crisis and excessive prices for grain, food, and bioenergy resources.

SWOT analysis has also found applications in other approaches to energy generation, like thermal power and others. Using the approach, Niyibizi [62] recognizes Africa as a continent blessed with abundant renewable energy resources that can largely meet its energy needs, with renewables becoming increasingly cost-competitive compared to fossil fuels. Despite this, access to electricity remains low, with only a small percentage of the population having direct access. The analysis highlights the progress being made in many African countries in deploying renewable energy technologies and further emphasizes the need to address challenges such as limited access to financing, inadequate infrastructure, and policy barriers to fully harness the continent's renewable energy potential. Other work is by Shaikh et al. [63], who employed the method to analyze the existing energy projects in Sindh, Pakistan, focusing on their current conditions and future roadmap for sustainable electric power generation. The study explores the strengths, weaknesses, opportunities, and threats associated with these projects, particularly in the context of renewable energy sources like wind, solar, and conventional energy. Key findings reveal that while there is significant potential in these energy sources, there are challenges hindering their development, such as technological constraints and environmental impacts. The study emphasizes the importance of strategic energy policies that prioritize financial profitability, energy security, and environmental protection to promote the valuable utilization of renewable energy sources in Sindh.

To understand the potential of solar thermal energy in Colombia, Belancur et al. [64] employed SWOT analysis to reveal its prospects, which ascertained that Colombia has abundant renewable resources, while only a small fraction of its energy supply comes from nonconventional renewable energies. The study further established the necessity to promote solar thermal systems for heat generation alongside photovoltaic energy. The analysis indicates that Colombia could benefit significantly from solar heat, with over 1.3 million households in cold-climate areas gaining access to hot water or reducing firewood use. Implementing solar stoves could help rural households shift away from firewood for food preparation. However, to realize these benefits, the report stresses the importance of strengthening legislation, providing financial support, and addressing political uncertainty and corruption, which could hinder the adoption of renewable energy projects in the country.

A survey of the literature [51, 57–64] shows that the use of SWOT analysis for the analysis of energy production or generation processes is not a new idea. The survey has shown different forms of its applications in the literature, especially in bioenergy and other renewable energy applications. However, most of the reports have consistently deployed it only for the identification of the strengths, weaknesses, opportunities, and threats involved in the case or scenarios being analyzed. Most of the literature gives insignificant attention to the innovation of strategies for effectively implementing the idea or scenarios evaluated using a SWOT matrix [65] with the identified S, W, O, and T to design strengths-opportunities (SO), strengths-threats (ST), weaknesses-opportunities (WO), and weaknesses-threats (WT) strategies, which has the potential of taking advantage of opportunities, avoiding threats, introducing new opportunities by reducing weaknesses, and avoiding threats by minimizing weaknesses. This report seeks to implement it to innovate a better strategy for managing and sustaining a bioenergy generation process in developing nations blessed with vast biomass resources.

9.3 Study Strategy

9.3.1 Data Sources

Quantitative and qualitative data employed in the analysis are sourced mainly from journal articles, conference proceedings, internet web reports, government-published documents, and other forms of reports in the literature.

9.3.2 Method of Analysis

In deploying the SWOT analysis methodology in this study, we begin by selecting a group of developing nations to focus on. To achieve this, we initially aim to shortlist at least one country from each continent based on a list of countries with the corresponding frequency of power failures or national grid collapses. We shortlist developing countries with the highest frequency of national grid collapse (or power outage index) across continents (Africa, Asia, Central America and the Caribbean, Europe, the Middle East, North America, Oceania, and South America) using the World Bank Enterprise Survey Data provided in the literature [12]. Subsequently, this set of shortlisted countries is examined as typical cases of developing countries in the analysis to identify the strengths, weaknesses, opportunities, and threats involved in integrating bioenergy into the nation's energy mix, along with proposing possible strategies for effectively implementing these initiatives.

9.4 SWOT Analysis

To begin, the study examines the rankings of countries based on the average frequency of power outages experienced by firms or establishments in a typical month worldwide. The top five countries with the highest power outage values were then shortlisted from each continent, as detailed in Table 9.1. The study also identified the top five countries with the highest power outages globally, which include Pakistan (75.20), Bangladesh (64.50), Papua New Guinea (41.90), Iraq (40.90), and Yemen (38.80).

Analyzing the level of power outages reported across continents, it can be observed that the trend of power outages across continents follows this order: Asia (Pakistan) > Oceania (Papua New Guinea) > Africa (Nigeria) > Central America and Caribbean (Dominican Republic) > North America (Mexico) > Europe (Albania). This analysis confirms that Asian countries are at the forefront of countries with frequent power outages worldwide, with some Oceania and African countries following closely. North American and European countries experience the least power outage frequencies, while other continents like Asia, Oceania, and Africa face significant power outages. The frequency of power outages experienced by these countries underscores the importance of exploring alternative energy sources like bioenergy to complement existing ones in their energy mix.

An effective understanding of the identified strengths (S), weaknesses (W), opportunities (O), and threats (T) involved in integrating bioenergy in such countries' energy mix is essential for innovating implementation strategies to address threats and capitalize on opportunities presented by these initiatives, as presented in subsections 9.4.1–9.4.4, respectively.

9.4.1 The Strengths of Bioenergy Integration in Developing Nations

In our analysis of effective measures for successfully integrating bioenergy into the energy mix of low-income nations or developing countries like Pakistan, Bangladesh, Nigeria, the Dominican Republic, and so on, as presented in Table 9.1, some potential advantages or strengths the initiatives tend to offer are as follows:

9.4.1.1 Abundance of Biomass Resources

Many low-income nations or developing countries are richly blessed with vast resources for agricultural and forestry activities, providing a large and sustainable source of biomass for bioenergy production. In such countries, a significant portion of the population depends on agriculture, particularly those in rural areas. Food production and its byproducts can be utilized for bioenergy production.

Table 9.1 Countries ranking for power outages averagely experienced by firms in a typical month.

Rank	Oceania	Value	South America	Value		Value	Europe	Value
1	Papua New Guinea	41.90	Guyana	8.50	Mexico	1.60	Albania	1.50
2	Solomon Islands	4.90	Suriname	2.80			Bosnia and Herzegov-ina	1.30
3	Samoa	4.60	Venezuela	2.60			Romania	1.20
4	Tonga	2.00	Paraguay	1.70			North Macedonia	1.10
5	Vanuatu	1.60	Brazil	1.60			Greece	0.90

Rank	Central America and the Caribbean	Value	Asia	Value	Africa	Value	World	Value
1	Dominican Republic	7.40	Pakistan	75.20	Nigeria	32.80	Pakistan	75.20
2	St. Kitts and Nevis	4.20	Bangladesh	64.50	Central African Republic	29.00	Bangladesh	64.50
3	St. Lucia	3.00	Iraq	40.90	Benin	28.00	Papua New Guinea	41.90
4	Dominica	2.80	Yemen	38.80	Niger	22.00	Iraq	40.90
5	Antigua and Barbuda	2.80	India	13.80	Congo	21.50	Yemen	38.80

Source: Adapted from [12].

However, there is a growing effort to discourage the use of food-based crops and instead focus on non-edible, non-food crops and food waste, which are often available as waste in open landfills or burned during farm or bush clearing processes, posing environmental risks. Embracing the initiative of establishing bioenergy production plants in these communities would help eliminate food and other biomass wastes often deposited in open landfills and discourage the burning of biomass wastes, especially agro-waste. Hence, bioenergy production would add value to these countries' biomass resources.

In Pakistan, Shahani et al. [66] also estimated that waste biomass could only contribute up to 9% of Pakistan's energy demand by 2030, which would in turn reduce the country's reliance on costly fossil fuels. Mahmood et al. [67] further state that Pakistan produces 230 billion tons of biomass annually, including crop residues, animal waste, municipal solid waste, and forest residues. Which has the capacity of sustaining the bioenergy derived from waste biomass that could potentially replace imported energy sources, such as coal, oil, natural gas, electricity, and LPG, and also reduce the financial burden on the country of the provision of subsidies for expensive imported energy products. The potential of Pakistan's biomass resources to help the government to meet their nation's energy demand was similarly ascertained by another report by Khan et al. [68], estimating that these resources could produce 20,709 MW of bioelectricity and 12,615 million m^3 of biogas annually [68], and Mahmood et al. [67] report that deploying the currently nonproductive land currently for energy crop production could potentially produce 2500–3000 MW of energy. However, there is a call for research and development of the conversion process technologies, and government incentives would facilitate the integration process. Similar findings were also established by Guzman-Bello et al. [69] for the Dominican Republic, showing that the country is blessed with sufficient non-edible bioresources like sugarcane, rice, coconut, and other agro-wastes, which can aid in meeting their energy demand via bioenergy integration. Parra et al. [70] indicate that sugarcane and its wastes remain the major energy resource currently considered extensively for bioenergy production in Ecuador, and it was also reported that 222,060.71 ha of land have been identified as available for growing these dedicated bioenergy crops, which has been estimated to show a potential 8603 GWh annual electricity production.

9.4.1.2 Environmental Benefits and Improved Waste Management

Bioenergy production offers a transformative solution for efficient community waste management while addressing environmental challenges. In many developing countries where agriculture and forestry are prevalent, abundant biomass resources serve as a sustainable source of bioenergy. By converting agricultural residues, food wastes, and other biomass materials into bioenergy, communities can significantly reduce waste sent to open landfills [50, 71]. This practice not only reduces the release of harmful greenhouse gases and pollutants from decomposing waste but also minimizes the risk of disease vectors breeding in these sites.

The integration of bioenergy into developing nations' energy mixes can help reduce greenhouse gas emissions and mitigate climate change [72]. Bioenergy production involves using organic materials, which, when burned, release carbon dioxide that was recently absorbed during plant growth, creating a closed carbon cycle [53]. This contrasts with fossil fuels, which release carbon sequestered for millions of years, contributing to the greenhouse effect and climate change [72, 73].

Moreover, transitioning from traditional cooking methods, like burning firewood, to modern bioenergy stoves, such as bioethanol or biomethane gas cookers, brings significant environmental benefits [74, 75]. These stoves produce fewer emissions and soot than traditional stoves, improving indoor air quality and reducing health risks for residents. Additionally, using bioethanol, biomethane, or biodiesel power generators instead of fossil fuel-powered ones helps reduce carbon emissions and mitigate climate change impacts [18, 72].

In essence, the adoption of bioenergy not only addresses energy poverty but also promotes sustainable waste management practices, enhances air quality, and fosters community health and well-being.

9.4.1.3 Energy Security

The introduction of bioenergy into the energy mix of developing countries would significantly diversify their energy sectors, reducing the risk associated with heavy reliance on a single energy source. For example, countries heavily reliant on hydroelectric power face substantial disruptions in the event of plant failures or decreased water levels in dams. Similarly, European countries heavily reliant on Russian gas for their power plants encountered significant challenges during the Ukraine crisis [76, 77], highlighting the importance of seeking alternative gas supplies.

Diversifying the energy mix of developing countries through the integration of bioenergy is essential for complementing existing energy sources. The Ukrainian crisis [78] underscored the significance of energy diversification for ensuring a country's energy security.

Moreover, bioenergy production can enhance energy access and security in developing countries [53, 79, 80]. By diversifying the energy mix and reducing reliance on imported fossil fuels, countries can become more self-sufficient in meeting their energy needs. This is particularly important in regions prone to geopolitical tensions or fluctuations in global energy markets, as it reduces vulnerability to external supply disruptions.

Therefore, diversifying the energy mix of developing countries has the potential to reduce the monopoly of a single energy source, promote national energy security, and decrease dependency on imported fossil fuels or fossil-based energy sources.

9.4.1.4 Local Energy Production, Rural Development, and Social Acceptance

Integrating bioenergy into the energy mix of developing nations can significantly promote local energy production. Unlike other plants that require proximity to water sources, such as hydroelectric plants, most developing nations are blessed with vast land and biomass resources, making it feasible to establish bioenergy production plants in rural or urban communities. Local energy production in these communities has the potential to reduce power supply costs, minimize losses during power transmission, and enhance energy access in rural areas where traditional power infrastructure is lacking. Moreover, the experience of power outages in Niger due to the Nigerian government's suspension of energy supply following a military coup highlights the vulnerability of countries dependent on external sources for energy [81, 82]. Promoting the integration of bioenergy into the energy mix of such countries could help reduce dependence on neighboring countries for power supply, thereby enhancing their energy security by promoting country independence for energy supply.

Furthermore, bioenergy production can create local job opportunities and stimulate economic growth in rural areas. Establishing bioenergy production plants requires a significant workforce for operations, maintenance, and feedstock cultivation, providing employment opportunities for local communities. Additionally, revenue generated from bioenergy production can be reinvested in local infrastructure, education, and healthcare, further benefiting the community. Overall, the integration of bioenergy into the energy mix of developing countries offers a multifaceted solution to various energy and economic challenges, contributing to sustainable development and improving the quality of life for communities. Bioenergy projects often enjoy high levels of social acceptance due to their perceived environmental and economic benefits, which can facilitate project implementation. For example, community members, NGOs, or the World Bank [83, 84] may support bioenergy projects because they reduce reliance on fossil fuels, create local jobs, and contribute to environmental sustainability by reducing greenhouse gas emissions.

9.4.2 The Weaknesses or Limitations of Bioenergy Integration Initiatives

In addition to understanding the advantages (or strengths) that the initiative of integrating bioenergy into our energy mix could offer, it is vital to understand the weaknesses (or limitations) of the initiative. This section presents some potential weaknesses that could be experienced in the implementation.

9.4.2.1 Lack of Infrastructure and High Initial Investment Costs

Due to the poor level of technological advancement in many of these low-income and developing countries like Bangladesh [85–88], the design and fabrication of the necessary infrastructure required for bioenergy production, which could include processing facilities, transportation networks, and storage capabilities, might suffer some challenges and could introduce some setbacks to the bioenergy integration processes.

Following the facts that most of these developing countries heavily depend on imported products and technology [89, 90] and the worth of these countries' exchange rates for a dollar is extremely high [91–96], these practices could set the pace of the bioenergy integration backward due to the high level of investment that the capital cost would require for purchasing all forms of processing facilities for the start-up and running of the plant.

If not carefully managed, it could possibly contribute to financial losses in the project and scare people from investing in it, despite being an environmentally efficient approach for combating rising global warming (caused by fossil fuels and related energy sources) and diversifying developing countries' energy mixes.

9.4.2.2 Technological Challenges

The successful integration of bioenergy into developing nations' energy mixes hinges on effectively managing and adopting imported technologies designed for biomass conversion and bioenergy production. However, this poses a dilemma: whether the residents or citizens of these countries can effectively operate and maintain the equipment, given the complexity of biomass conversion processes, which typically require specialized knowledge and dedicated equipment. Unfortunately, such experts are often in short supply in many developing countries. The lack of local expertise may lead to increased labor costs due to the need to recruit foreign experts. To address this challenge, training programs and capacity-building efforts are essential.

Another technological challenge lies in the compatibility of specific imported technologies with locally sourced feedstock. The availability of diverse feedstock types can affect the efficiency and effectiveness of bioenergy production processes, particularly in developing countries where infrastructure for fabricating process equipment is lacking.

Furthermore, the cost of acquiring and implementing foreign bioenergy production technologies can be prohibitively high for developing nations with limited financial resources. This financial barrier could hinder the widespread adoption of bioenergy technologies in these countries.

9.4.2.3 Limited Public Awareness and Acceptance

Many residents and governments of developing nations have shallow or no knowledge of bioenergy and its prospects. This lack of knowledge and public awareness has contributed to the initiative's low adoption rate across these nations.

The masses are largely unaware that promoting this initiative would help add value to our waste, enhance the waste management approach, promote community health and safety, and contribute to the achievement of environmental sustainability. This suggests that research institutes, academic institutes, and government agencies, like the national orientation agencies in developing countries, are not adequately educating the public about all these.

9.4.2.4 Competition for Land Usage, Policy, and Regulatory Challenges

The use of biomass for bioenergy production often competes with food production or other land uses, creating potential conflicts in land allocation. The high demand for bioenergy feedstocks, which offer significant returns for farmers, can lead to dilemmas in land use decisions [97–99]. This competition raises concerns about food supply security, as agricultural land may be diverted from food production to bioenergy feedstock cultivation. In some cases, this competition can lead to higher food prices or scarcity, affecting local communities and economies.

Regulatory challenges further exacerbate these issues. Some developing nations' regulatory agencies struggle to effectively regulate land allocation, leading to conflicting land use practices [100–102]. This can result in inefficient use of land resources and potential environmental degradation. Additionally, inconsistent or inadequate policies and regulations related to bioenergy production create uncertainty for investors and project developers, hindering the growth of the bioenergy sector in these countries.

9.4.2.5 Resource Availability and Market Access

The availability of biomass resources poses a significant challenge to the integration of bioenergy into the energy mix of developing nations. Unlike developed nations, where year-round agriculture is common, many developing nations rely on seasonal agriculture or cultivate feedstocks intermittently. This seasonal nature of agriculture can limit the availability of biomass for bioenergy production [103, 104]. Furthermore, the lack of infrastructure for irrigation farming in developing nations further complicates the production of both food and energy crops, affecting overall bioenergy production.

Accessing international bioenergy markets can also be challenging for developing countries, limiting their potential for export revenues. Limited access to foreign currency for purchasing production equipment is another barrier [105] to the development of the bioenergy sector. Unequal public access to dollars and other foreign currencies can hinder investment in bioenergy projects, delaying the growth of the sector and limiting its potential impact on energy security and economic development.

9.4.3 The Opportunities for Bioenergy Integration

Some of the opportunities available for the promotion of bioenergy integration in developing nations' energy mixes are presented herein.

9.4.3.1 Government Incentives and Policies

Government incentives and policies play a crucial role in encouraging investment in bioenergy production [106–108]. These measures include reduced taxation on production, waivers of customs duties on imported production utilities and equipment, and subsidies for bioenergy products. In many developing countries, incentive packages have been implemented to attract both local and state government investment in projects that support environmental sustainability and reduce dependency on imported fossil fuels.

These incentives and policies have significantly reduced the cost of bioenergy products, making them more affordable for the public [107]. This has led to increased adoption of local technology and production methods, resulting in a gradual reduction in the use of fossil fuels and the monopolization of the energy sector in many countries. Moreover, governments often provide support and incentives specifically for renewable energy projects, creating a favorable policy environment for the further development of bioenergy.

9.4.3.2 Growing Demand for Eco-friendly Sustainable Energy Sources and Their Export Potential

The global push for greener, cleaner, and sustainable energy sources [109–111] is gaining momentum in response to escalating environmental challenges. Both developed and developing nations are grappling with the consequences of heightened greenhouse gas emissions, including severe climatic shifts, flooding, wildfires, and other environmental disasters [112–114]. This mounting concern for environmental preservation is steering a global transition towards cleaner energy alternatives, such as bioenergy. Consequently, there is a surging demand for bioenergy products as countries seek to diminish their dependence on fossil fuels and embrace more sustainable energy alternatives [115]. This shift is not solely driven by environmental consciousness but also by the economic advantages and energy security that bioenergy promises.

This escalating demand for bioenergy products offers substantial prospects for investors and project developers. With the bioenergy market rapidly expanding, there exists a promising potential for high returns on investments. This prospect is particularly enticing in developing nations where the adoption of bioenergy technologies is still nascent, presenting significant opportunities for growth [53, 116]. Moreover, as the global focus on sustainability and environmental stewardship intensifies, the demand for bioenergy products is expected to surge in the foreseeable future. As governments and industries worldwide prioritize clean energy solutions, the bioenergy market is primed for substantial expansion, making it an appealing investment avenue for those seeking to capitalize on the transition to a more sustainable energy landscape. Furthermore, low-income or developing countries endowed with abundant biomass resources can leverage these assets by exporting bioenergy products to international markets, thereby generating substantial export revenues.

9.4.3.3 Potential for Collaboration with International Organizations and Donor Agencies

Projects related to bioenergy production have the potential to attract funding and technical assistance from international organizations and donor agencies. There is a growing global effort to promote the deployment of clean and sustainable energy sources, and bioenergy projects align well with these objectives. Organizations such as the World Bank [83, 84], United Nations agencies, and various federal, state, and local governments are increasingly interested in partnering in the execution of bioenergy production projects. Such partnerships can help reduce reliance on fossil fuels, promote cleaner air, and create safer and more conducive living environments. Additionally, collaboration with international organizations and donor agencies can provide access to expertise, technology, and financial resources that may otherwise be unavailable, making bioenergy projects more feasible and impactful.

9.4.3.4 Improved Waste Management Technologies and Environmental Benefits

Integrating bioenergy into a nation's energy mix, particularly in developing nations, offers a unique opportunity to manage waste effectively. By utilizing waste-to-energy or bioenergy technologies, communities can add value to their generated waste [29, 30, 43, 50, 117]. These technologies not only improve environmental sanitation practices but also enhance the cleanliness of our surroundings, contributing to better public and community health. This holistic approach can lead to a reduction in residents' illness and mortality rates, improving overall well-being in the country's communities. Moreover, bioenergy production offers significant environmental benefits. It can help reduce greenhouse gas emissions, thereby mitigating climate change effects. Additionally, by converting waste into energy, bioenergy projects can improve soil fertility and reduce the pressure on forests, promoting ecosystem health and biodiversity conservation.

9.4.3.5 Economic Diversification, Job Creation, and Community Development

Establishing a bioenergy sector can significantly diversify the economy, reducing reliance on traditional sectors like agriculture or mining. This diversification can lead to a more resilient and sustainable economy that is less vulnerable to fluctuations in commodity prices or market demands. One key benefit of bioenergy projects is their potential to create employment, especially in rural areas where biomass resources are abundant. By providing job opportunities in operations, maintenance, and feedstock cultivation, bioenergy projects can stimulate economic growth and improve livelihoods in these communities [49, 53, 67]. Furthermore, bioenergy projects can contribute to community development by investing in local infrastructure. This investment not only supports the bioenergy project itself but also enhances the overall quality of life in the community. Improved infrastructure, such as roads, schools, and healthcare facilities, can have long-lasting benefits for residents, fostering a more prosperous and sustainable community.

9.4.3.6 Energy Access, Technology Innovation, and Local Content Development

Bioenergy projects play a crucial role in enhancing energy access, particularly in rural and remote areas with limited or no grid connectivity. By providing a reliable source of energy, these projects improve the quality of life and enable economic activities in these communities, driving local development. Moreover, bioenergy projects stimulate technology innovation and knowledge transfer. They create opportunities for local industries and research institutions to develop and adapt bioenergy technologies, fostering a culture of innovation. This not only benefits the bioenergy sector but also contributes to the overall growth of the local economy.

9.4.4 The Potential Threats to Bioenergy Integration

In the context of bioenergy integration in low-income nations or developing countries, some potential threats that could be encountered during the implementation of bioenergy integration programs in the developing nations' energy mixes are presented herein.

9.4.4.1 Competition from Low-Cost Fossil Fuels

The competition with low-cost fossil fuels presents a significant challenge for bioenergy projects, particularly in developing countries where fossil fuels are often subsidized. Governments in these countries may not have implemented sufficient incentives and support policies to promote bioenergy, further complicating the competition. This competitive landscape could hinder the economic viability of bioenergy projects, making it challenging for them to thrive. The ability to effectively compete with low-cost fossil fuels is crucial for the success of bioenergy projects in these regions. Winning this competition consistently is essential for ensuring the positive economic viability of such projects. However, failure to do so could result in financial losses for investors and jeopardize the future of bioenergy initiatives in these countries.

9.4.4.2 Policy and Regulatory Uncertainty and Social Resistance

Developing nations often experience fluctuating policies and regulations, especially with changing political administrations. These shifts in government stance can significantly impact the economic viability and sustainability of bioenergy integration programs in these countries. Such uncertainties may deter private investors from committing capital to these projects as the feasibility and profitability of bioenergy production become less predictable. In addition to policy uncertainties, bioenergy projects may face resistance from local communities or environmental groups. Concerns related to land use, biodiversity conservation, and other environmental or social impacts can lead to opposition against bioenergy initiatives. Addressing these concerns and ensuring transparent communication with stakeholders is crucial for the successful implementation of bioenergy projects in developing nations.

9.4.4.3 Vulnerability to Climate Change and Extreme Weather Events

Developing nations often rely on seasonal agriculture, which is increasingly threatened by the changing climatic conditions attributed to global warming. The unpredictability of weather patterns poses a significant challenge to the viability of bioenergy projects in these regions, as it can lead to insufficient bioenergy feedstock for regular production cycles. Climate change can have a direct impact on the availability and quality of biomass resources, further complicating the reliability of bioenergy production. Extreme weather events, such as droughts, floods, and storms, can disrupt the supply chain of biomass feedstock, making it difficult for bioenergy projects to maintain consistent production levels. Adaptation strategies, such as diversifying feedstock sources or implementing more resilient agricultural practices, may be necessary to mitigate the risks posed by climate change to bioenergy projects in developing nations.

9.4.5 Key Deductions from the SWOT Analysis Results

The SWOT analysis of integrating bioenergy into the energy mix of developing nations reveals significant opportunities and challenges, as shown in Table 9.2. The strengths include abundant biomass resources, environmental benefits like improved waste management, and the potential for energy security through diversification. Moreover, bioenergy projects can stimulate local economies, create jobs, and promote social acceptance. However, weaknesses such as high initial investment costs, technological challenges, and limited public awareness need to be addressed.

On the other hand, opportunities lie in government incentives, growing global demand for sustainable energy, and collaboration with international organizations. Improved waste management technologies and economic diversification

Table 9.2 Summary of the findings made from the SWOT analysis.

Strengths (S)	Weaknesses (W)
• Abundance of biomass resources • Environmental benefits and improved waste management • Energy security • Local energy production, rural development, and social acceptance	• Lack of infrastructure and high initial investment costs • Technological challenges • Limited public awareness and acceptance • Competition for land usage, policy, and regulatory challenges • Resource availability and market access
Opportunities (O)	**Threats (T)**
• Government incentives and policies • Growing demand for eco-friendly, sustainable energy sources and their export potential • Potential for collaboration with international organizations and donor agencies • Improved waste management technologies and environmental benefits • Economic diversification, job creation, and community development • Energy access, technology innovation, and local content development	• Competition from low-cost fossil fuels • Policy and regulatory uncertainty and social resistance • Vulnerability to climate change and extreme weather events

further enhance the prospects of bioenergy integration. Nonetheless, threats like competition from low-cost fossil fuels, policy uncertainties, and climate change impacts pose significant challenges. Addressing these issues requires a holistic approach involving supportive policies, technological innovation, and public awareness campaigns to realize the full potential of bioenergy in developing nations.

9.4.6 Exploration of Effective Implementation Strategy for Bioenergy Integration in Developing Nations

To effectively implement the bioenergy integration program in developing nations, it is essential to capitalize on strengths, address weaknesses, leverage opportunities, and mitigate threats. One strategy is to focus on SO by utilizing the abundance of biomass resources and promoting government incentives and policies that support bioenergy production. This approach can help maximize the environmental and economic benefits of bioenergy projects. Another strategy is the ST approach, which involves leveraging the environmental benefits of bioenergy to mitigate competition from low-cost fossil fuels. By emphasizing the positive environmental impact of bioenergy, countries can enhance public acceptance and overcome regulatory uncertainties.

Additionally, the WO strategy can be applied by investing in infrastructure development and technological innovation to overcome the lack of expertise and high initial investment costs. This approach can help developing nations build capacity and create a conducive environment for bioenergy projects. Finally, the WT strategy involves addressing limited public awareness and acceptance by implementing public education campaigns and engaging stakeholders. By raising awareness about the benefits of bioenergy and addressing concerns, countries can foster greater acceptance and support for bioenergy integration initiatives.

9.4.7 SO Strategy

Bioenergy projects offer significant benefits, including abundant biomass resources, energy security, local energy production, improved waste management, environmental sustainability, and rural development. These strengths can be leveraged to explore opportunities such as export potential, energy access in rural communities, economic diversification, local content development, and government incentives.

Governments must implement supportive policies and provide incentives to promote bioenergy production, such as tax breaks, subsidies, and customs duty waivers. It is essential to implement policies that encourage the use of biomass resources for bioenergy production and provide financial incentives for bioenergy projects.

9.4.8 ST Strategy

The strength of bioenergy production can be utilized to avoid or mitigate threats. For example, using biomass waste instead of food crops for bioenergy production can help manage waste and reduce competition with low-cost fossil fuels. Other threats like vulnerability to climate change, social resistance, and regulatory uncertainties can be addressed by promoting irrigation farming, raising public awareness, and ensuring energy security and poverty alleviation, which can attract and sustain supportive government policies and regulations for bioenergy production in developing countries.

Governments must further diversify their energy sources by integrating bioenergy to reduce reliance on imported fossil fuels. It is also vital to develop policies that encourage the use of bioenergy alongside other renewable energy sources to ensure energy security and reduce dependency on fossil fuels.

9.4.9 WO Strategy

Addressing weaknesses identified in developing nations, such as limited public awareness, can be achieved through public education campaigns using both private and public media outlets. Government policies can also regulate land allocation to prevent competition between food and energy crops. Furthermore, promoting local technology development can reduce the high initial investment costs and discourage equipment importation, thus favoring the exportation of bioenergy products and technologies, improving environmental sustainability, creating jobs, and fostering community development.

Moreover, to effectively implement the bioenergy initiative for harnessing the opportunities offered by the initiative, governments and industry stakeholders must invest in infrastructure development and technology transfer to overcome technological challenges and improve efficiency, which is one of the major challenges (or weaknesses) in the program. It is also vital to establish training programs and capacity-building efforts to enhance local expertise and ensure the compatibility of imported technologies with locally sourced feedstock. Moreover, it is required to fund public education campaigns to raise awareness about the benefits of bioenergy and address concerns by engaging with communities, NGOs, and the media to promote the environmental and economic benefits of bioenergy projects and ensure transparency in communication. In addition, the promotion of local content development in bioenergy projects has the potential to stimulate economic growth and create job opportunities and has the capability of encouraging the use of locally sourced feedstock and labor in bioenergy production to enhance local economic development and reduce reliance on imported goods and services.

9.4.10 WT Strategy

Minimizing weaknesses, such as limited infrastructure and technological challenges, can help prevent threats like competition from fossil fuels and uncertain government policies. Ensuring accessible infrastructure reduces the initial capital investment cost, making bioenergy products cheaper than fossil energy sources. Public education by relevant government agencies can also enhance public acceptance, preventing social resistance and attracting supportive government policies for bioenergy integration.

Moreover, to mitigate the impact of climate change on bioenergy production, governments should implement measures such as promoting resilient agricultural practices and developing adaptation strategies that diversify feedstock sources and ensure the sustainability of biomass resources in the face of changing climatic conditions. Lastly, to ensure regulatory certainty and policy stability, governments can provide stability to attract investment and ensure the long-term viability of bioenergy projects. They are also essential to establishing clear and consistent policies that support bioenergy integration and provide a favorable investment climate for investors and project developers.

9.4.11 Implementation Strategies Summary

In conclusion, effective implementation of bioenergy integration in developing nations requires strategies that capitalize on strengths, address weaknesses, leverage opportunities, and mitigate threats. Implementing these strategies can maximize the benefits of bioenergy projects and create a sustainable energy future for developing nations.

9.5 Integration of Bioenergy into Energy Hubs

The transition from fossil fuels, which are less clean, to cleaner and more sustainable energy systems in developing nations necessitates the efficient integration of diverse energy sources. The role of energy hubs in promoting the efficient integration of these energy sources is vital [118–120]. This is because energy hubs can serve as nodes for managing and distributing various forms of energy, particularly through renewable sources, offering a strategic solution for optimizing energy usage.

To briefly highlight the concept of energy hubs, they are advanced systems designed to integrate and manage multiple energy sources, also known as energy vectors, such as electricity, heat, gas, and liquid fuels, within a single platform. The system coordinates the supply, conversion, and storage of energy from various sources, enhancing overall system efficiency and energy security [121]. In developing nations, where there is a need to integrate bioenergy into the country's energy mix to meet diverse energy needs across different sectors of the economy, the introduction of energy hubs would offer a flexible approach. This would allow for the exploration of different energy mixes to meet the needs of both urban and rural areas while promoting the use of local and renewable resources. Other benefits of setting up energy hubs in societies include the promotion of proper integration of a range of energy sources, efficient optimization, and flexibility. These features would give energy hubs an edge in meeting the energy demands of various communities within a nation.

9.5.1 Role of Bioenergy in Energy Hubs

Understanding bioenergy as a class of energy derived from the exploration of a range of biomass resources through various chemical processes—such as gasification, combustion, pyrolysis, and liquefaction—can lead to the production of diverse energy outputs that can be seamlessly integrated into an energy hub. Bioenergy can be integrated into energy hubs in three basic ways: electricity generation, fuel production, and combined heat and power (CHP) generation, as shown in Figures 9.2–9.4.

Electricity Generation: Bioenergy systems, such as biomass combustion and gasification, generate electricity that can be fed directly into the energy hub [122–125]. Many advanced countries, such as China, Germany, Japan, the United States, and Brazil, have functional electricity generation plants using biomass energy sources. Among the few developing countries that have established such plants is Bangladesh [126]. Literature surveys have shown that electricity generated from biomass sources can be used for immediate consumption, stored in batteries, or distributed by energy hubs across the grid to meet fluctuating demands.

Heat and Power Generation: CHP systems utilizing bioenergy can provide both electricity and heat to the energy hub, as shown earlier in Figure 9.2. The heat generated can be used for district heating, industrial processes, or stored for later use. Literature surveys have shown that some countries have already integrated the advantageous functions of bioenergy into their energy hubs for meeting heating and cooling needs within their districts [127, 128]. Additionally, some studies have revealed the potential benefits of integrating bioenergy into energy hubs for CHP generation using biomass resources from developing countries like Sri Lanka [66] and Anticosti Island [129], which show potential for meeting or helping to meet the countries' energy demands.

Fuel Production: Fuels derived from biomass are often referred to as biofuels. These biofuels can be produced via transesterification [24, 130–132], a combination of hydrolysis and fermentation [18, 21, 133–137], or pyrolysis and liquefaction processes [37, 50, 138, 139], as shown earlier in Figures 9.3 and 9.4. These fuels are useful for transportation and backup power generation within the energy hub. They can be stored and used when other energy sources are scarce, ensuring a continuous energy supply.

9.5.2 Benefits of Integrating Bioenergy in Developing Nations' Energy Hubs

Integrating bioenergy into energy hubs would present developing nations with several advantages. These advantages include the opportunity for the countries' energy hubs to adapt to varying energy demands by switching between different energy vectors, such as electricity, heat, and biofuels. This flexibility is particularly valuable in developing nations, where energy demand patterns may vary widely between urban and rural areas. Another opportunity presented by integrating bioenergy into the countries' energy hubs is the reduction in dependence on imported fossil fuels and the contribution to local economic development. The use of bioenergy also helps manage waste streams by converting them into valuable energy outputs and aids in reducing greenhouse gas emissions by replacing fossil fuel-based energy with renewable bioenergy. This contributes to achieving national and international climate goals.

9.5.3 Challenges and Considerations

Integrating bioenergy into energy hubs offers significant benefits but also introduces a range of challenges that must be addressed. A key challenge is technology integration, which involves aligning various technologies such as biomass processing units, CHP systems, and energy storage solutions. This process requires substantial investment in research, development, and capacity building to ensure seamless operation. Additionally, infrastructure development poses a critical challenge, particularly in developing nations where establishing the necessary systems for biomass collection, transportation, and processing can be complex. Addressing these infrastructure needs is essential for unlocking the full potential of bioenergy in energy hubs. Furthermore, effective policy and regulatory support are crucial for successful integration. Policies must provide incentives for renewable energy use and establish clear guidelines for developing bioenergy infrastructure.

Despite these challenges, the integration of bioenergy into energy hubs represents a promising strategy for developing nations aiming to transition to sustainable energy systems. By leveraging local biomass resources and advanced energy management technologies, energy hubs can enhance energy security, reduce emissions, and foster economic development. Future research and policy efforts should focus on overcoming the challenges associated with technology integration, infrastructure development, and regulatory support to fully realize the potential of bioenergy within energy hubs.

9.6 Recommendations, Summary, and Suggestions for Future Studies

9.6.1 Recommendations

Based on the SWOT analysis and findings of this study, several key recommendations can be made for policymakers, industry stakeholders, and investors in low-income nations or developing countries to maximize the benefits of bioenergy integration and minimize the risks:

9.6.1.1 Policy and Regulatory Frameworks

Governments should develop clear and consistent policy frameworks that support the development of bioenergy projects. This includes providing incentives such as tax breaks, subsidies, and feed-in tariffs to attract investors and stimulate growth in the bioenergy sector. Additionally, policies should prioritize sustainable land use practices to prevent competition between food and energy crops.

9.6.1.2 Capacity Building and Technology Transfer

To address the lack of expertise and infrastructure in developing countries, capacity-building programs should be implemented to train the local workforce in bioenergy technologies and operations. International cooperation and technology transfer agreements can help facilitate the adoption of advanced bioenergy technologies in these regions.

9.6.1.3 Financial Support and Investment

Financial institutions and donors should provide funding and investment opportunities for bioenergy projects in developing countries. This can be in the form of grants, loans, or venture capital to help mitigate the high initial investment costs associated with bioenergy projects and encourage private sector involvement.

9.6.1.4 Public Awareness and Stakeholder Engagement

Efforts should be made to raise public awareness about the benefits of bioenergy and engage stakeholders, including local communities, NGOs, and government agencies, in the decision-making process. This can help build support for bioenergy projects and ensure their long-term success.

9.6.1.5 Research and Development

Continued research and development efforts are needed to improve the efficiency and sustainability of bioenergy technologies. This includes developing new feedstock sources, improving conversion processes, and exploring innovative uses for bioenergy by-products.

By implementing these recommendations, policymakers, industry stakeholders, and investors can help create a conducive environment for bioenergy integration in developing countries, leading to sustainable energy systems, economic growth, and environmental conservation.

9.6.2 Summary of Key Findings

The study on bioenergy integration in developing nations has highlighted several key findings. First, developing countries possess abundant biomass resources that can be harnessed for bioenergy production, offering a sustainable alternative to fossil fuels. Second, the environmental benefits of bioenergy, including reduced greenhouse gas emissions and improved waste management, make it a valuable tool for mitigating climate change and promoting sustainable development. Third, while bioenergy projects face challenges such as high initial investment costs and competition from low-cost fossil fuels, strategic policies and incentives can help overcome these barriers and unlock the full potential of bioenergy integration. Overall, the findings suggest that bioenergy has the potential to play a significant role in the energy transition of developing nations, offering a path towards cleaner, more sustainable energy systems.

The implications of this study for energy policy and sustainable development are significant. Policymakers in developing nations can use these findings to develop and implement effective strategies for promoting bioenergy integration. This includes creating supportive policy frameworks, providing incentives for investment in bioenergy projects, and enhancing public awareness and acceptance of bioenergy technologies. Additionally, the study underscores the importance of international collaboration and knowledge sharing to accelerate the adoption of bioenergy in developing countries. By prioritizing bioenergy in their energy policies, governments can not only improve energy access and security but also drive economic growth, create jobs, and mitigate environmental impacts.

9.6.3 Suggestions for Future Research

Future research focusing on bioenergy in similar contexts should consider several key areas. First, there is a need for further analysis of the socioeconomic impacts of bioenergy integration, including its effects on local communities, livelihoods, and land use patterns. Second, research should explore innovative technological solutions and business models that can enhance the viability and scalability of bioenergy projects in developing nations. Third, studies should investigate the potential synergies between bioenergy and other renewable energy sources, such as solar and wind power, to create integrated energy systems that are resilient and sustainable. Finally, there is a need for longitudinal studies to assess the long-term sustainability and effectiveness of bioenergy integration in achieving energy security and environmental goals in developing countries.

References

1 Guan, Y., Yan, J., Shan, Y. et al. (2023). Burden of the global energy price crisis on households. *Nature Energy* 8: 304–316. https://doi.org/10.1038/s41560-023-01209-8.

2 Dare, O. (2024). National grid collapses again. *The Punch*. https://punchng.com/national-power-grid-collapses-again/.

3 Okechukwu, N. (2024). National grid collapses again, throws Nigerians into blackout. *The Punch*. https://punchng.com/national-grid-collapses-again-throws-nigerians-into-blackout/.

4 Izuaka, M. (2024). Again, Nigeria's electricity grid collapses. *Premium Times*. https://www.premiumtimesng.com/news/top-news/746687-again-nigerias-electricity-grid-collapses-4.html?tztc=1.

5 Channels Television (2024). Blackout as DisCos Report Fresh National Grid Collapse Channels Television. Channels TV News. https://www.channelstv.com/2024/04/15/blackout-as-discos-report-fresh-national-grid-collapse/ (accessed 18, April 2024).

6 Al-Mahmood SZ (2014). Bangladesh power restored after nationwide blackout. *The Wall Street Journal*.

7 NDTV (2016). Sri Lanka Faces Nationwide Power Blackout; Water Supply Affected. NDTV News World. https://www.ndtv.com/world-news/sri-lanka-faces-nationwide-power-blackout-water-supply-affected-1286773#:~:text=Sri%20Lanka%20Faces%20Nationwide%20Power%20Blackout%3B%20Water%20Supply%20Affected,-Advertisement&text=Colombo%3A%20Sri%20Lanka%20today%20witnessed,parts%20of%20the%20island%20nation (accessed 18, April 2024).

8 Smith S, Sanchez F, Torchia C (2019). Venezuela buckles under massive power, communications outage. *AP News*.

9 Sequera V, Buitrago D (2019). Venezuela, blaming U.S. for six-day blackout, orders diplomats to leave. *Reuters*.

10 Criss D. (2018). Puerto Rico's power outage is now the second-largest blackout in history. CNN. https://edition.cnn.com/2018/04/16/us/puerto-rico-blackout-second-largest-globally-trnd#:~:text=Well%20now%2C%20here's%20another%20one,electricity%20have%20been%20lost%20there (accessed 18, April 2024).

11 Masood S. (2021). Much of Pakistan Loses Power in Massive Blackout. *The New York Times*.

12 Index Mundi (2024). Countries ranked by Power outages in firms in a typical month (number). Index Mundi Report. https://www.indexmundi.com/facts/indicators/IC.ELC.OUTG/rankings (accessed 18, April 2024).

13 Utility Bidder (2024). Power Cut Countries. Utility Bidder Report. https://www.utilitybidder.co.uk/compare-business-energy/power-cut-countries/ (accessed 18, April 2024).

14 Wikipedia (2024). List of major power outages. Wikipedia. https://en.wikipedia.org/wiki/List_of_major_power_outages (accessed 18, April 2024).

15 NAN (2024). Nigerians Return to Charcoal Stoves, Firewood and Matters Arising. *The Leadership*.

16 Liu, T., Miao, P., Shi, Y. et al. (2022). Recent advances, current issues and future prospects of bioenergy production: a review. *Science of the Total Environment* 810: 152181. https://doi.org/10.1016/J.SCITOTENV.2021.152181.

17 Demirbas, A.H. and Demirbas, I. (2007). Importance of rural bioenergy for developing countries. *Energy Conversion and Management* 48: 2386–2398. https://doi.org/10.1016/J.ENCONMAN.2007.03.005.

18 Oyegoke, T., Obadiah, E., Adah, F. et al. (2022). Trends of progress in setting up biorefineries in developing countries: a review of bioethanol exploration in nigeria. *Journal of Renewable Energy and Environment* 9: 37–52. https://doi.org/10.30501/JREE.2021.278037.1197.

19 Oyegoke, T. (2022). Green fuels: concepts, benefits, and studies in Nigeria. *Bulgarian Chemical Communications* 54: 29–39. https://doi.org/10.34049/bcc.54.1.5458.

20 Oyegoke, T., Sardauna, M.Y., Abubakar, H.A., and Obadiah, E. (2022). Exploration of biomass for the production of bioethanol: "a process modeling and simulation study." *Renewable Energy Research and Application* 3 (1): 1–19. https://doi.org/10.22044/RERA.2020.10287.1042.

21 Abemi A, Oyegoke T, Dabai FN, Jibril BY (2018). Technical and economic feasibility of transforming molasses into bioethanol in Nigeria. *Faculty of Engineering National Engineering Conference*, Zaria: Ahmadu Bello University, p. 145.

22 Sardauna MY (2019). Comparative study of bioethanol production from different from biomass via the use of process simulation approach. Thesis. Ahmadu Bello University Zaria.

23 Ajayi, O.O., Onifade, K.R., Onadeji, A., and Oyegoke, T. (2020). Techno-economic assessment of transforming sorghum bagasse into bioethanol fuel in Nigeria: 1 - process modeling, simulation, and cost estimation. *Journal of Engineering Studies and Research* 26: 154–164.

24 Ibraheem, K.A., Oyegoke, T., Geoffrey, T.T. et al. (2020). Biodiesel production from neem seed oil: catalyst synthesis, the effect of time, and temperature on biodiesel yield. In: *19th Congress NIMACON: "Innovative Development Of Materials For Economic Diversification" held at OAU* (ed. J.O. Olawale, O.O. Ige, D.A. Isadare, and B. Babatope). Nigeria: Materials Science and Technology Society of Nigeria (MSN).

25 Toledo Arana, J., Torres, J.J., Acevedo, D.F. et al. (2019). One-step synthesis of CaO-ZnO efficient catalyst for biodiesel production. *International Journal of Chemical Engineering* 2019: https://doi.org/10.1155/2019/1806017.

26 Kumar Tiwari, A., Kumar, A., and Raheman, H. (2007). Biodiesel production from jatropha oil (Jatropha curcas) with high free fatty acids: an optimized process. *Biomass Bioenergy* 31: https://doi.org/10.1016/j.biombioe.2007.03.003.

27 Ajala, E.O., Ajala, M.A., Odetoye, T.E., and Okunlola, A.T. (2019). Synthesis of solid catalyst from dolomite for biodiesel production using palm kernel oil in an optimization process by definitive screening design. *Brazilian*

Journal of Chemical Engineering 36: https://doi.org/10.1590/0104-6632.20190362s20180516.

28 Singh, E., Kumar, A., Mishra, R., and Kumar, S. (2022). Solid waste management during COVID-19 pandemic: recovery techniques and responses. *Chemosphere* 288: https://doi.org/10.1016/J.CHEMOSPHERE.2021.132451.

29 Moustakas, K., Sotiropoulos, D., and Vakalis, S. (2021). Evaluation of the biogas potential of agricultural biomass waste for energy applications in Greece: a case study of the western Greece region. *Waste Management and Research* 39: 438–447. https://doi.org/10.1177/0734242X20970607.

30 Igwebuike, C.M. and Oyegoke, T. (2024). Decarbonizing our environment via the promotion of biomass methanation in developing nations: a waste management tool. *Pure and Applied Chemistry* 96 (5): 651–670. https://doi.org/10.1515/pac-2023-1109.

31 Mishra, K., Siwal, S.S., Nayaka, S.C. et al. (2023). Waste-to-chemicals: green solutions for bioeconomy markets. *Science of the Total Environment* 887: 164006. https://doi.org/10.1016/J.SCITOTENV.2023.164006.

32 Cavallaro, E., Sessa, M.R., and Malandrino, O. (2023). Renewable energy communities in the energy transition context. *International Journal of Energy Economics and Policy* 13: 408–417. 10.32479/IJEEP.14230.

33 EERA (2024). Bioenergy basics. Department of Energy. https://www.energy.gov/eere/bioenergy/bioenergy-basics (accessed 19, April 2024).

34 Rogers, J.N., Stokes, B., Dunn, J. et al. (2017). An assessment of the potential products and economic and environmental impacts resulting from a billion ton bioeconomy. *Biofuels, Bioproducts and Biorefining* 11: 110–128. https://doi.org/10.1002/BBB.1728.

35 Ram, M. and Mondal, M.K. (2022). Biomass gasification: a step toward cleaner fuel and chemicals. In: *Biofuels and Bioenergy: Opportunities and Challenges* (ed. B. Gurunathan, R. Sahadevan, and Z.A. Zakaria), 253–276. Elsevier https://doi.org/10.1016/B978-0-323-85269-2.00008-3.

36 Zhang, J. and Zhang, X. (2019). The thermochemical conversion of biomass into biofuels. In: *Biomass, Biopolymer-Based Materials, and Bioenergy: Construction, Biomedical, and Other Industrial Applications* (ed. D. Verma, E. Fortunati, S. Jain, and X. Zhang), 327–368. Woodhead Publishing https://doi.org/10.1016/B978-0-08-102426-3.00015-1.

37 Babu, B.V. (2008). Biomass pyrolysis: a state-of-the-art review. *Biofuels, Bioproducts and Biorefining* 2: 393–414. https://doi.org/10.1002/BBB.92.

38 Tanger, P., Field, J.L., Jahn, C.E. et al. (2013). Biomass for thermochemical conversion: targets and challenges. *Frontiers in Plant Science* 4: 43282. https://doi.org/10.3389/FPLS.2013.00218/BIBTEX.

39 Orji, J., Aleke, K., Chika, E. et al. (2016). Bioethanol Production from Corncob Hydrolysed by Cellulase of Aspergillus niger using Zymomonas mobilis and Saccharomyces cerevisiae isolated from Palm Wine. *International Journal of Current Research in Biosciences and Plant Biology* 3: 39–45. https://doi.org/10.20546/ijcrbp.2016.301.004.

40 Sayar, N.A., Pinar, O., Kazan, D., and Sayar, A.A. (2019). Bioethanol production from Turkish hazelnut husk process design and economic evaluation. *Waste Biomass Valorization* 10: 909–923. https://doi.org/10.1007/s12649-017-0103-y.

41 Madu, J.O. and Agboola, B.O. (2017). Bioethanol production from rice husk using diferent pretreatments and fermentation conditions. *Biotech* 8: https://doi.org/10.1007/s13205-017-1033-x.

42 Trakulvichean, S., Chaiprasert, P., Otmakhova, J., and Songkasiri, W. (2019). Integrated economic and environmental assessment of biogas and bioethanol production from cassava cellulosic waste. *Waste Biomass Valorization* 10: 691–700. https://doi.org/10.1007/s12649-017-0076-x.

43 Karimi Alavijeh, M. and Yaghmaei, S. (2016). Biochemical production of bioenergy from agricultural crops and residue in Iran. *Waste Management* 52: 375–394. https://doi.org/10.1016/J.WASMAN.2016.03.025.

44 Meegoda, J.N., Li, B., Patel, K., and Wang, L.B. (2018). A review of the processes, parameters, and optimization of anaerobic digestion. *International Journal of Environmental Research and Public Health* 15: 2224. https://doi.org/10.3390/IJERPH15102224.

45 Anekwe, I.M.S., Armah, E.K., Tetteh, E.K. et al. (2022). Bioenergy production: emerging technologies. In: *Biomass, Biorefineries and Bioeconomy* (ed. M. Samer). IntechOpen https://doi.org/10.5772/INTECHOPEN.102692.

46 Sharma, N., Kaur, L., and Sharma, S.G. (2024). Biochemical approach to biomass conversion: biofuel production. In: *Biomass Energy for Sustainable Development* (ed. M.P. Shah and P. Kaur), 62–96. CRC Press https://doi.org/10.1201/9781003406501-4.

47 Kaur, G., Yadav, N., and Kumar, S. (2024). Anaerobic digestion: a sustainable biochemical approach to convert biomass to bioenergy. In: *Biomass Energy for Sustainable Development* (ed. M.P. Shah and P. Kaur), 111–126. Boca Raton: CRC Press https://doi.org/10.1201/9781003406501-6.

48 Singh, P. and Singh, N. (2019). Political economy of bioenergy transitions in developing countries: a case study of Punjab, India. *World Development* 124: 104630. https://doi.org/10.1016/J.WORLDDEV.2019.104630.

49 Arndt, C., Henley, G., and Hartley, F. (2019). Bioenergy in Southern Africa: an opportunity for regional integration? *Development Southern Africa* 36: 145–154. https://doi.org/10.1080/0376835X.2018.1447363.

50 Oyegoke, T., Folake Oyegoke, A., Fasanya, O.O., and Gambo Ibrahim, A.-A. (2023). State of Agro-wastes management in Nigeria: status, implications, and way forward. In: *Waste Recovery and Management*, vol. 1 (ed. P. Ajay,

A. Kumar, R.K. Mittal, and R. Goel), 241–272. CRC Press https://doi.org/10.1201/9781003359784-14.

51 Qyyum, M.A., Ali Shah, S.F., Qadeer, K. et al. (2022). Biowaste to bioenergy options for sustainable economic growth opportunities in developing countries: product space model analysis and policy map development. *Renewable and Sustainable Energy Reviews* 169: 112832. https://doi.org/10.1016/J.RSER.2022.112832.

52 Tolessa, A. (2023). Bioenergy production potential of available biomass residue resources in Ethiopia. *Journal of Renewable Energy* 2023: 1–12. https://doi.org/10.1155/2023/2407300.

53 Dirisu, J.O., Salawu, E.Y., Ekpe, I.C. et al. (2023). Promoting the use of bioenergy in developing nations: a CDM route to sustainable development. *Frontiers in Energy Research* 11: 1184348. https://doi.org/10.3389/FENRG.2023.1184348/BIBTEX.

54 UN (2024). Goal 7: Affordable and clean energy. The Global Goals. https://www.globalgoals.org/goals/7-affordable-and-clean-energy/ (accessed 19, April 2024).

55 UN (2024). Goal 7 | Department of economic and social affairs. The SDGs. https://sdgs.un.org/goals/goal7 (accessed 19, April 2024).

56 UN (2024). Goal 11: Sustainable cities and communities. The Global Goals. https://www.globalgoals.org/goals/11-sustainable-cities-and-communities/ (accessed 19, April 2024).

57 Alsaleh, M., Abdul-Rahim, A.S., Mohd-Shahwahid, H.O. et al. (2016). An empirical analysis for technical efficiency of bioenergy industry in EU28 region based on data envelopment analysis method. *International Journal of Energy Economics and Policy* 6: 290–304.

58 Falcone, P.M. (2023). Sustainable energy policies in developing countries: a review of challenges and opportunities. *Energies (Basel)* 16: 6682. https://doi.org/10.3390/EN16186682.

59 Zhen, H, Man, L (2018). A SWOT analysis of development of biomass energy in Heilongjiang Province in China. *Proceedings of the 4th International Conference on Economics, Management, Law and Education (EMLE 2018)*, Moscow, Russia (25–26 December 2018). Atlantis Press, p. 329–32. https://doi.org/10.2991/EMLE-18.2018.60.

60 Okello, C., Pindozzi, S., Faugno, S., and Boccia, L. (2014). Appraising bioenergy alternatives in Uganda using strengths, weaknesses, opportunities and threats (SWOT)-analytical hierarchy process (AHP) and a desirability functions approach. *Energies (Basel)* 7: 1171–1192. https://doi.org/10.3390/EN7031171.

61 Soimu, O. (2014). A SWOT analysis of bio-energy production in Republic of Moldova. *Procedia—Social and Behavioral Sciences* 149: 896–900. https://doi.org/10.1016/J.SBSPRO.2014.08.277.

62 Niyibizi, A. (2015). SWOT analysis for renewable energy in Africa: challenges and prospects. *Renewable Energy Law and Policy Review* 6: 276–293. https://www.jstor.org/stable/26256470 (accessed 19, April 2024).

63 Shaikh, A., Hameed Shaikh, P., Kumar, L. et al. (2022). A SWOT analysis for a roadmap towards sustainable electric power generation. *International Transactions on Electrical Energy Systems* 2022: https://doi.org/10.1155/2022/1743570.

64 Betancur, S., Ortega-Avila, N., and López-Vidaña, E.C. (2023). Strengths, weaknesses, opportunities, and threats analysis for the strengthening of solar thermal energy in Colombia. *Resources* 13: 3. https://doi.org/10.3390/RESOURCES13010003.

65 Benzaghta, M.A., Elwalda, A., Mousa, M.M. et al. (2021). SWOT analysis applications: an integrative literature review. *Journal of Global Business Insights* 6: 55–73. https://doi.org/10.5038/2640-6489.6.1.1148.

66 Shahani, S., Gao, Z., Korai, M.S. et al. (2023). Assessment of bioenergy potential of waste biomass through stoichiometry analysis in Pakistan. *Journal of Biobased Materials and Bioenergy* 17: 257–269. https://doi.org/10.1166/JBMB.2023.2281.

67 Mahmood, A., Wang, X., Shahzad, A.N. et al. (2021). Perspectives on bioenergy feedstock development in Pakistan: challenges and opportunities. *Sustainability* 13: 8438. https://doi.org/10.3390/SU13158438.

68 Khan, S., Nisar, A., Wu, B. et al. (2022). Bioenergy production in Pakistan: potential, progress, and prospect. *Science of the Total Environment* 814: 152872. https://doi.org/10.1016/J.SCITOTENV.2021.152872.

69 Guzmán-Bello, H., López-Díaz, I., Aybar-Mejía, M. et al. (2023). Biomass energy potential of agricultural residues in the Dominican Republic. *Sustainability* 15: 15847. https://doi.org/10.3390/SU152215847.

70 Parra, C.R., Ramirez, A.D., Navas-Gracia, L.M. et al. (2023). Prospects for bioenergy development potential from dedicated energy crops in Ecuador: an agroecological zoning study. *Agriculture (Switzerland)* 13: 186. https://doi.org/10.3390/AGRICULTURE13010186/S1.

71 Boechat, C.L., de S, A.A.M., Duda, R.M. et al. (2017). Solid waste in agricultural soils: an approach based on environmental principles, human health, and food security. In: *Solid Waste Management in Rural Areas* (ed. F.-C. Mihai). IntechOpen https://doi.org/10.5772/intechopen.69701.

72 Kumar, A., Bhansali, S., Gupta, N., and Sharma, M. (2019). *Bioenergy and Climate Change: Greenhouse Gas Mitigation*, 269–289. Cham: Springer https://doi.org/10.1007/978-3-030-14463-0_10.

73 Hanaki, K. and Portugal-Pereira, J. (2018). *The Effect of Biofuel Production on Greenhouse Gas Emission Reductions*, 53–71. Tokyo: Springer https://doi.org/10.1007/978-4-431-54895-9-6.

74 Jetter, J.J. and Kariher, P. (2009). Solid-fuel household cook stoves: characterization of performance and emissions. *Biomass Bioenergy* 33: 294–305. https://doi.org/10.1016/J.BIOMBIOE.2008.05.014.

75 Phillip, E., Langevin, J., Davis, M. et al. (2023). Improved cookstoves to reduce household air pollution exposure in sub-Saharan Africa: a scoping review of intervention studies. *PLoS One* 18: https://doi.org/10.1371/JOURNAL.PONE.0284908.

76 Moskalenko, N., Löffler, K., Hainsch, K. et al. (2024). Europe's independence from Russian natural gas—effects of import restrictions on energy system development. *Energy Reports* 11: 2853–2866. https://doi.org/10.1016/J.EGYR.2024.02.035.

77 Richard, J. A. (2008). Europe's dependence on Russian natural gas: perspectives and recommendations for a long-term strategy | George C. *Marshall European Center for Security Studies*. Marshall Center. https://www.marshallcenter.org/en/publications/occasional-papers/europes-dependence-russian-natural-gas-perspectives-and-recommendations-long-term-strategy-0 (accessed 22, April 2024).

78 Rokicki, T., Bórawski, P., and Szeberényi, A. (2023). The impact of the 2020–2022 crises on EU countries' independence from energy imports, particularly from Russia. *Energies (Basel)* 16: 6629. https://doi.org/10.3390/EN16186629.

79 Maltsoglou, I., Koizumi, T., and Felix, E. (2013). The status of bioenergy development in developing countries. *Global Food Security* 2, Elsevier: 104–109. https://doi.org/10.1016/J.GFS.2013.04.002.

80 Rutz, D. and Janssen, R. (2014). Summary: The contribution of bioenergy to energy access and energy security. In: *Socio-Economic Impacts of Bioenergy Production*, vol. 9783319038292, 285–290. Springer International Publishing http://dx.doi.org/10.1007/978-3-319-03829-2_17.

81 Bunmi, A. (2023). Power outage in Niger as Nigeria cuts electricity supply amid coup. The Cable.

82 Africanews (2008). *Nigeria cuts Electricity to Niger after Coup*. Africanews.

83 World Bank (2024). Development projects: program to support development of renewable bioenergy in India - P179178. World Bank. https://projects.worldbank.org/en/projects-operations/project-detail/P179178 (accessed 22, April 2024).

84 World Bank (2024). Africa - Biomass energy initiative for Africa project. World Bank. https://documents.worldbank.org/en/publication/documents-reports/documentdetail/673641468005409923/africa-biomass-energy-initiative-for-africa-project (accessed 22, April 2024).

85 Fatma, E.E. (2017) Bangladesh: Developing governance and infrastructure in urban slums - OPEC Fund for International Development. OPEC Fund for International Development. https://opecfund.org/news/bangladesh-developing-governance-and-infrastructure-in-urban-slums (accessed 22, April 2024).

86 Reuters (2008). Bangladesh infrastructure seen hurdle to growth | Reuters. Reuters Reports. https://www.reuters.com/article/markets/oil/bangladesh-infrastructure-seen-hurdle-to-growth-idUSDHA90552/ (accessed 22, April 2024).

87 ICCI (2010). Poor infrastructure hampering economic growth. ICCI Report. Business Recorder Publisher. https://www.brecorder.com/news/3726035 (accessed 22, April 2024).

88 Abdul Ghafoor, A. and Vashma, A. (2014). Impact of infrastructure development on economic growth- a case study of Pakistan. *International Journal of Development and Economic Sustainability* 2: 1–15.

89 Adeyemi A. (2022). Nigeria can't continue to depend on imported technology, says Ibrahim | The Guardian Nigeria News - Nigeria and World News—Technology. The Guardian Nigeria News – Nigeria and World News. https://guardian.ng/technology/nigeria-cant-continue-to-depend-on-imported-technology-says-ibrahim/ (accessed 22, April 2024).

90 Saima S. (2023). Pakistan says 'heavily reliant' on expensive imports amid public outcry over record energy prices. Arab News PK. https://www.arabnews.pk/node/2375106/pakistan (accessed 22, April 2024).

91 Saeed A-B. (2024). Yemeni riyal hits fresh low of 1,630 per dollar. Arab News.

92 IMF (2023). African Currencies Are Under Pressure Amid Higher-for-Longer US Interest Rates. IMF. https://www.imf.org/en/Blogs/Articles/2023/05/15/african-currencies-are-under-pressure-amid-higher-for-longer-us-interest-rates (accessed 22, April 2024).

93 EPC (2023). Emirates Policy Center | Dollar Crisis in Iraq: An Economic Dilemma with Geopolitical Implications. EPC. https://epc.ae/en/details/brief/dollar-crisis-in-iraq-an-economic-dilemma-with-geopolitical-implications (accessed 22, April 2024).

94 Ramoni-Perazzi, J. and Romero, H. (2022). Exchange rate volatility, corruption, and economic growth. *Heliyon* 8: e12328. https://doi.org/10.1016/J.HELIYON.2022.E12328.

95 The East African (2023). DR Congo currency drops amid war spending and arrears payments. The East African.

96 Mbuyi, A.K., Kakasi, C.K.-K., Ntumba, C.M. et al. (2022). Exchange rate volatility and economic growth in the democratic republic of congo (DRC). *Modern Economy* 13: 729–746. https://doi.org/10.4236/ME.2022.135039.

97 Popp, J., Lakner, Z., Harangi-Rákos, M., and Fári, M. (2014). The effect of bioenergy expansion: food, energy, and environment. *Renewable and Sustainable Energy Reviews* 32: 559–578. https://doi.org/10.1016/J.RSER.2014.01.056.

98 Graham-Rowe, D. (2011). Agriculture: beyond food versus fuel. *Nature* 474: S6–S8. https://doi.org/10.1038/474s06a.

99 Kline, K.L., Msangi, S., Dale, V.H. et al. (2017). Reconciling food security and bioenergy: priorities for action. *GCB Bioenergy* 9: 557–576. https://doi.org/10.1111/GCBB.12366.

100 Philip, D.D. and Moses, A.O. (2013). Corruption as a bane for underdevelopment in Nigeria: issues and challenges. *International Affairs and Global Strategy* 15: 76–83.

101 Kirkpatrick, C. and Parker, D. (2004). Regulatory impact assessment and regulatory governance in developing countries. *Public Administration and Development* 24: 333–344. https://doi.org/10.1002/PAD.310.

102 Chiodelli, F. and Moroni, S. (2015). Corruption in land-use issues: a crucial challenge for planning theory and practice. *Town Planning Review* 86: 437–455. https://doi.org/10.3828/TPR.2015.27.

103 Nithya, D.J., Raju, S., Pradhan, A. et al. (2018). Effect of agricultural seasons on energy intake: evidence from two agro-ecologically different rural districts of India. *Food and Nutrition Sciences* 9: 542–555. https://doi.org/10.4236/FNS.2018.95041.

104 Blackmore, I., Rivera, C., Waters, W.F. et al. (2021). The impact of seasonality and climate variability on livelihood security in the Ecuadorian Andes. *Climate Risk Management* 32: 100279. https://doi.org/10.1016/J.CRM.2021.100279.

105 Chinedu A. (2024). Nigeria's currency has fallen to a record low as inflation surges. How did things get so bad? | AP News. AP News.

106 Thornley, P. and Cooper, D. (2008). The effectiveness of policy instruments in promoting bioenergy. *Biomass Bioenergy* 32: 903–913. https://doi.org/10.1016/J.BIOMBIOE.2008.01.011.

107 FasterCapital (2024). Government Policies and Incentives for Biomass Energy Production. FasterCapital Report. https://fastercapital.com/topics/government-policies-and-incentives-for-biomass-energy-production.html (accessed 22, April 2024).

108 Qadir, S.A., Al-Motairi, H., Tahir, F., and Al-Fagih, L. (2021). Incentives and strategies for financing the renewable energy transition: a review. *Energy Reports* 7: 3590–3606. https://doi.org/10.1016/J.EGYR.2021.06.041.

109 Chen, H., He, L., Chen, J. et al. (2019). Impacts of clean energy substitution for polluting fossil-fuels in terminal energy consumption on the economy and environment in China. *Sustainability* 11: 6419. https://doi.org/10.3390/SU11226419.

110 Twumasi, Y.A., Ning, Z.H., Namwamba, J.B. et al. (2022). Bioenergy crops as a promising alternative to fossil fuels in Louisiana: a geographic information system (GIS) perspective. *Journal of Sustainable Bioenergy Systems* 12: 57–81. https://doi.org/10.4236/JSBS.2022.124005.

111 Muhammad, U.L., Shamsuddin, I.M., Danjuma, A. et al. (2018). Biofuels as the starring substitute to fossil fuels. *Petroleum Science and Engineering* 2: 44. https://doi.org/10.11648/J.PSE.20180201.17.

112 Bell, J.E., Brown, C.L., Conlon, K. et al. (2018). Changes in extreme events and the potential impacts on human health. *Journal of the Air and Waste Management Association* 68: 265. https://doi.org/10.1080/10962247.2017.1401017.

113 Rocha, J., Oliveira, S., Viana, C.M., and Ribeiro, A.I. (2022). Climate change and its impacts on health, environment and economy. In: *One Health: Integrated Approach to 21st Century Challenges to Health* (ed. J.C. Prata, A.I. Ribeiro, and T. Rocha-Santos), 253–279. Elsevier https://doi.org/10.1016/B978-0-12-822794-7.00009-5.

114 Bolan, S., Padhye, L.P., Jasemizad, T. et al. (2024). Impacts of climate change on the fate of contaminants through extreme weather events. *Science of the Total Environment* 909: 168388. https://doi.org/10.1016/J.SCITOTENV.2023.168388.

115 Kipkoech, R., Takase, M., and Amankwa Afrifa, E.K. (2022). Renewable energies in Ghana in relation to market condition, the environment, and food security. *Journal of Renewable Energy* 2022: 1–8. https://doi.org/10.1155/2022/8243904.

116 Malode, S.J., Prabhu, K.K., Mascarenhas, R.J. et al. (2021). Recent advances and viability in biofuel production. *Energy Conversion and Management: X* 10: 100070. https://doi.org/10.1016/J.ECMX.2020.100070.

117 Valenga, M.G.P., Boschen, N.L., Rodrigues, P.R.P., and Maia, G.A.R. (2019). Agro-industrial waste and Moringa oleifera leaves as antioxidants for biodiesel. *Industrial Crops and Products* 128: https://doi.org/10.1016/j.indcrop.2018.11.031.

118 Garg, A., Niazi, K.R. (2023). A robust energy management strategy for multi carrier energy system. *2023 7th International Conference on Computer Applications in Electrical Engineering-Recent Advances: Sustainable Transportation Systems*, CERA 2023, http://dx.doi.org/10.1109/CERA59325.2023.10455511.

119 Tiwari, S., Singh, J.G., and Garg, A. (2021). A static robust energy management approach for modelling low emission multi-vectored energy hub including emission markets and power-to-gas units. *Energy* 294: 130827. https://doi.org/10.1016/J.ENERGY.2024.130827.

120 Tiwari, S. and Singh, J.G. (2024). A cooperation based transactive energy management for networked energy hubs considering improved payoff allocation mechanism. *Sustainable Energy Technologies and Assessments* 65: 103777. https://doi.org/10.1016/J.SETA.2024.103777.

121 Eladl, A.A., El-Afifi, M.I., El-Saadawi, M.M., and Sedhom, B.E. (2023). A review on energy hubs: models, methods, classification, applications, and future trends. *Alexandria Engineering Journal* 68: https://doi.org/10.1016/j.aej.2023.01.021.

122 Barz, M. (2008). Biomass technology for electricity generation in community. *International Journal of Renewable Energy* 3: 1–10.

123 Vongchan, P., Chompunth, C., and Phoochinda, W. (2020). Green business model of biomass very small power producers in Thailand. *International Journal of GEOMATE* 19: https://doi.org/10.21660/2020.72.9495S.

124 Oliveira, M.O., Somariva, R., Ando Junior, O.H. et al. (2012). Biomass electricity generation using industry poultry waste. *Renewable Energy and Power Quality Journal* 1: https://doi.org/10.24084/repqj10.791.

125 Alidrisi, H. and Demirbas, A. (2016). Enhanced electricity generation using biomass materials. *Energy Sources, Part A: Recovery, Utilization and Environmental Effects* 38: https://doi.org/10.1080/15567036.2014.948647.

126 Masud, M.H., Ananno, A.A., Arefin, A.M.E. et al. (2019). Perspective of biomass energy conversion in Bangladesh. *Clean Technologies and Environmental Policy* 21: 719–731. https://doi.org/10.1007/S10098-019-01668-2/METRICS.

127 IRENA (2017). *Renewable Energy in District Heating and Cooling: A Sector Roadmap for REmap*. International Renewable Energy Agency.

128 Bagherian, M.A., Mehranzamir, K., Rezania, S. et al. (2021). Analyzing utilization of biomass in combined heat and power and combined cooling, heating, and power systems. *Processes* 9: https://doi.org/10.3390/pr9061002.

129 Abbas, T., Issa, M., Ilinca, A., and El-Ali, A. (2020). Biomass combined heat and power generation for Anticosti Island: a case study. *Journal of Power and Energy Engineering* 8: https://doi.org/10.4236/jpee.2020.83005.

130 Asri, N.P., Podjojono, B., and Fujiani, R. (2017). Utilization of eggshell waste as low-cost solid base catalyst for biodiesel production from used cooking oil. In: *IOP Conference Series: Earth and Environmental Science*, vol. 67, No. 1, 012021. IOP Publishing. http://dx.doi.org/10.1088/1755-1315/67/1/012021.

131 Oyegoke, T. and Ibraheem, K.A. (2021). Kinetic modeling of the biodiesel production process using neem seed oil: an alternative to petroleum-diesel. *European Journal of Chemistry* 12: https://doi.org/10.5155/eurjchem.12.3.242-247.2085.

132 Umaru, M. and Aberuagba, F. (2012). Characteristics of a typical Nigerian Jatropha *curcas oil* seeds for biodiesel production. *Research Journal of Chemical Sciences* 2: 7–12.

133 Oyegoke, T. and Dabai, F. (2018). Techno-economic feasibility study of bioethanol production from a combined cellulose and sugar feedstock in Nigeria: 2-economic analysis. *Nigerian Journal of Technology* 37: 921–926. https://doi.org/10.4314/njt.v37i4.9.

134 Oyegoke, T. and Dabai, F. (2018). Techno-economic feasibility study of bioethanol production from a combined cellulose and sugar feedstock in Nigeria: 1-modeling, simulation and cost evaluation. *Nigerian Journal of Technology* 37: 913–920. https://doi.org/10.4314/njt.v37i4.8.

135 Oyegoke, T., Sardauna, M.Y., Abubakar, H.A., and Obadiah, E. (2021). Exploration of biomass for the production of bioethanol: "a process modelling and simulation study." *Renewable Energy Research and Application* 2: 51–69.

136 Gebreegziabher, Z., Mekonnen, A., Ferede, T., and Kohlin, G. (2017). Profitability of bioethanol production: the case of Ethiopia. *Ethiopian Journal of Economics* 26: 101–122.

137 Oyegoke, T., Obadiah, E., Mohammed, Y.S. et al. (2022). Exploration of biomass for the production of bioethanol: "a techno-economic feasibility study of using rice (*Oryza Sativa*) husk." *Renewable Energy Research and Application* 3 (1): 1–19. https://doi.org/10.22044/rera.2020.10288.1043.

138 Gunatilake, S.K. and Gunatilake, S.K. (2016). Removal of Cr (III) ions from wastewater using sawdust and rice husk biochar pyrolyzed at low temperature. *International Journal for Innovation Education and Research* 4: 44–54. https://doi.org/10.31686/ijier.vol4.iss4.531.

139 Song, C., Zhang, C., Zhang, S. et al. (2020). Thermochemical liquefaction of agricultural and forestry wastes into biofuels and chemicals from circular economy perspectives. *Science of the Total Environment* 749: 141972. https://doi.org/10.1016/J.SCITOTENV.2020.141972.

10

Integration of Carbon Reduction Techniques (CDR) and Emission Trading Mechanisms Among MV-EHs

Energy System Modelling with Carbon Reduction Techniques (CDR), Carbon Taxes, Carbon Market, and Trading Strategies to Ensure a Low-Emission Energy System Modeling

Mohammad Parhamfar[1] *and Saeed Khorrami*[2]

[1] *Independence Power and Energy Consultant, Isfahan, Iran*
[2] *Department of Electrical Engineering, Sapienza University of Rome, Rome, Italy*

10.1 Introduction

Carbon trading is a mechanism where the government establishes an upper bound on how much carbon can be released and then divides it into various units. These units are awarded to different groups, industries, or careers and can be traded like any other product. Initially, this chapter will look at energy system models that adopt carbon reduction, followed by an examination and an exploration of blockchain, carbon trading, and applications of artificial intelligence (AI).

Achieving zero greenhouse gas (GHG) emissions by 2025 is a favorite objective of the European Union. As it appears, there is no mechanism around this target yet. Among the best alternatives in this case would be to allow carbon trading within Europe. This scheme represents 17% of global GHG emissions, according to World Bank data for 2020. It is increasingly acknowledged that CO_2 elimination is a critical and indispensable part of achieving the long-term goal of the Paris Agreement, which is to reduce global temperature levels. Table 10.1 depicts a study on the carbon removal pathway [1].

Another study [2] examined energy models for the 21st century. In this work, the relevant models are classified into four categories: energy system optimization models, energy system simulation models, power system and electricity market models, and mixed qualitative models and scenarios.

Hamming [3] asserted that the main goal of his calculations is to provide understanding, rather than just generating numerical values. The formulation of energy policy was initiated in response to the oil crisis in the 1970s when policymakers and business leaders recognized the significance of developing a comprehensive and sustainable strategy [4]. The development of energy system models can be

Energy Management Strategies for Multi-Vectored Energy Hubs to Achieve Low Carbon Societies, First Edition.
Edited by Shubham Tiwari, Jai Govind Singh, Sivaraman Palanisamy, Sharmeela Chenniappan, Rupendra Kumar Pachauri, and Sanjeevikumar Padmanaban.

Table 10.1 Overview of potential, breakeven cost, and duration of storage.

Type	Techniques	Removal ($GtCO_2$/year)	Cost (USD/tCO_2)	Storage	Storage permanence
Biological-based	Forestry techniques	0.5–3.6	−40 to 10	Standing forests and long-lived wood products (decades to centuries)	Low
	Land management	2.3–5.3	−90 to −20	Soil organic carbon (years to decades)	Low
	Biochar	0.3–2.0	−70 to −60	Black carbon (decades to centuries)	Medium
	BECCS	0.5–5.0	60–160	Geological sequestration (millennia)	High
Chemical-based	Enhanced weathering	2.0–4.0	<200	Aqueous carbonate (centuries)	High
	DACCS	0.5–5.0	200–600 100	Geological sequestration (millennia)	High

attributed to the increasing importance of scenario planning throughout the 20th century [5]. None of the emerging technologies existed during the 20th century.

a) The increase in flexible demand through emerging technologies such as smart meters and distributed generation.
b) The importance of power supply and intermittent availability, considering additional details.
c) The evolving model of distribution networks and variable renewable energies requires a higher level of spatial precision.

Table 10.2 presents an analysis of multiple studies conducted in the field of diverse energy models [2]. Most models analyzed in recent research on the modeling of electricity markets, small-scale and decentralized energy planning, integrating renewable energy into existing systems, agent-based models, random models, urban energy systems, and the accessibility and transparency of codes and data [6–14].

To achieve a full picture of low-emission energy systems, the integration of energy hubs (EHs) is very important. EHs are key points in the energy infrastructure that aid in linking together different types of energy, such as renewable energy generation, storage systems, and demand-side management technologies.

Table 10.2 Recent studies on energy system models (authors).

Publication	Focus	Coverage
Jebaraj and Iniyan [6]	Overview of energy models	252 publications
Bazmi and Zahedi [7]	Overview of power sector optimization models	277 publications
Parhamfar [48]	Electricity market modeling	8 models
Foley et al. [39]	Electricity system models	7 models
Hiremath [9]	Decentralized energy planning	74 models
SensfuB et al. [11]	Agent-based electricity market models	14 groups of models
Möst and Keles [12]	Stochastic electricity market models	20 models
Connolly et al. [10]	Renewables integration	37 models
Keirstead et al. [13]	Urban energy system models	219 publications

The use of EHs could be utilized by communities to optimize energy consumption, mitigate network constraints, and promote local energy resilience. In addition, EHs play a major role in enabling linkages between carbon reduction techniques (CDR) and emission trading schemes (ETS) for microgrid energy hubs (MV-EHs). Through efficient organization within EHs, carbon trading can be harnessed by stakeholders for economic gains, which also improves the effectiveness of CDR strategies on both environmental and cost fronts at once. In this regard, it is intended to examine how far EHs may contribute to moving towards a sustainable, low-carbon future through their inclusion in power system models.

10.2 Energy System Models

It focuses on four paradigms of models and discusses representative examples to illustrate their salient features. The four types of models used in energy research are (1) optimization models for energy systems; (2) simulation models for energy systems; (3) models for power systems and electricity markets; and (4) scenarios based on qualitative and hybrid methods approaches. In the review of the models, in general, four cases are studied, as mentioned:

10.2.1 Energy System Optimization Models

Long-standing utilization of extensive bottom-up optimization models has been used for a long time. These models can be developed by providing a detailed and intricate description of the complex technological specifications they require.

For example, simplifying for maintainability is a result of the abundance of intricate elements. For example, implementing restrictions on the annual or seasonal production of domestic mass technologies while considering supply and demand factors, MARKAL/TIMES are two widely applied models. MARKEL is primarily used in energy optimization systems.

The study utilized the TIMES (Integrated MARKAL-EFOM System) model, a bottom-up energy system model, to analyze the energy consumption and CO_2 emissions in the transportation and service sectors of China and the United States. The findings of this model indicate that the transportation sectors of China and the United States will follow distinct trajectories over the next few decades. The adoption of advanced technologies, efficient fuel consumption policies, and a transition to low-energy modes can significantly mitigate energy consumption. The TIMES model provides a technology-rich economic model generator for long-term estimation [15].

MARKAL is an acronym for MARKAL Allocation, which represents a mathematical energy model. Energy estimates covering end-user and service demands, including automobiles, trucks, home lighting, and heat required in the paper industry, can be analyzed. This model calculates the energy balance at each level of an energy system. The objective of this concept is to provide energy services while minimizing global costs [16]. The MARKAL/TIMES and MESSAGE models share significant similarities and serve the purpose of illustrating the potential evolution of the energy system at national, regional, or global levels over several decades.

10.3 Energy System Simulation Models

In contrast to the bottom-up energy optimization model, this approach prioritizes forecasting the potential of the system rather than generating a hypothetical future. In contrast to the optimization model with complex mathematical calculations, this approach is modular and incorporates several methods. Significant examples of this family include NEMS (the U.S. Energy Information Administration's National Energy Modeling System) and PRIMES (a comparable model that covers the EU). These two models have been in use since 1990, making them more recent than the energy optimization models. LEAP, a practical tool, was developed by the Stockholm Environment Institute and is used in both the private and public sectors.

10.4 Power Systems and Electricity Market Models

These models are derived from complex energy networks and primarily focus on a set of models that target a specific issue. Power system models commonly used in electrical sector facilities and organizations serve as decision-making tools for various purposes, including investment planning and operational plans, such as the utilization of generators. This model can be used to modify, optimize, or predict the simulation of other models. Power system models typically have a higher level of complexity and include temporal variations. A critical element of an efficient power system is a consistent factor that maintains a balance between energy supply and demand. Energy market models are different from power system models because they focus on analyzing the energy market rather than the physical aspects of power and load balance within the network. Consequently, they integrate multiple models to improve the efficiency of various strategies in the power market. The California 2050 model projects GHG emissions in California from 2010 to 2050. It incorporates multi-sector economic and physical infrastructure models. This model includes an analysis of various strategies designed to reduce GHG emissions in six sectors: residential, commercial, industrial, agriculture, transportation, and the oil industry. Several statistical analyses of this model have been conducted using the SAS program [17]. Another model that has been studied is the automated Wien system planning, a computer program designed to strategize the expansion of the electrical production system [18]. Additional models worth mentioning include PLEXOS and ELMOD [19, 20].

10.5 Qualitative and Mixed-Methods Scenarios

According to the bottom-up concept, large-scale energy systems have complex intricacies that are paradoxically seen as both advantageous and disadvantageous. The complexity of these models arises from the need to generate numerous interactions. Due to the demand for high resolution and the complexity of technological specifications, the computations required for these models can extend over several hours and even days. Employing a quantitative methodology is one of these solutions. However, from an alternative perspective, the use of both quantitative and qualitative approaches is more flexible and responsive. An illustrative recent example is the British Department of Energy's Project Quantitative Scenario for 2050 and its proposed strategy for addressing climate change [21]. In addition to the four methods, others are not included in this classification. However, most of the included methods are those that were described. To explore alternative approaches, refer to Ref. [23].

10.6 A Review of Commitments and Protocols

The Climate Change Conference was convened in 1992 to address the issue of GHGs and global warming. Subsequently, the Kyoto Protocol was ratified in 1997. The Intergovernmental Panel on Climate Change (IPCC) was established in 1988 and subsequently released its first report in 1990, presenting scientific data to support the reality of climate change. The establishment of the United Nations Framework Convention on Climate Change (UNFCCC) in 1992 was prompted by the actions and warnings of the IPCC regarding climate change and its consequences. During the inaugural session of the Conference of the Parties (COP1), it was determined that the industrialized nations' commitment to reduce GHG emissions to the 1990 level by the year 2000 falls short of the convention's long-term objectives. According to the Kyoto Protocol in 1997, developed nations were required to reduce their combined GHG emissions by at least 5% compared to the 1990 emission levels from 2008 to 2012. The Kyoto Protocol has addressed the reduction of GHG emissions for a total of 39 developed nations. In this protocol, the term "greenhouse gases" includes carbon dioxide (CO_2), methane (CH_4), nitrous oxide (N_2O), hydrofluorocarbons (HFCs), perfluorocarbons (PFCs), and sulfur hexafluoride (SF_6).

Forestry and reforestation initiatives established after 1990 are eligible carbon balance projects under the Clean Development Mechanism (CDM) in the Kyoto Treaty for industrialized countries. Trees play a significant role in reducing the level of CO_2 by absorbing the primary GHG and storing it as carbon within their leaves, branches, trunk, and roots. Every ton of carbon stored in trees represents the removal of approximately 3.67 tons of carbon dioxide from the atmosphere.

The 2015 United Nations Climate Change Conference, also known as COP21, took place in Paris from November 30 to December 12, 2015. This conference marks the 21st conference of the annual gathering of the parties to the United Nations Climate Change Convention since 1992 and serves as the 11th summit specifically dedicated to the Kyoto Protocol. The main goal of this conference is to achieve a comprehensive international agreement and treaty on current climate change, with the active participation of all governments worldwide. UN Secretary-General Ban Ki-moon has emphasized the significance of restricting the global temperature increase to less than 2 °C. He emphasized the importance of taking swift and proactive measures to prevent the release of environmental pollutants and GHGs.

10.7 Glasgow COP26 Carbon Trading Agreement

The Glasgow Climate Change Conference in November 2021 led to the establishment of regulations for a global carbon market. This marked the first time that a comprehensive global strategy was implemented under the 2015 Paris Climate Agreement. Article 6 of the agreement comprises two distinct systems: a centralized system and a bilateral system. The centralized system serves both the public and private sectors, while the bilateral system is specifically designed for countries to participate in the trade of carbon offset credits, helping them achieve their emission reduction goals. According to the new agreement, individuals or entities that generate carbon credits will contribute 5% of their earnings to a fund designed to help developing nations address climate change. In addition, 2% of the credit will be retired from the system to ensure a broader reduction in pollutants. The updated regulations allow participants to use credits earned from 2013 to 2020. This has raised concerns that they could flood the market and put pressure on prices.

Proponents of this paradigm argue that it establishes financial incentives for nations and corporations to develop and improve technology and activities aimed at reducing emissions. These activities include mechanical systems for carbon absorption and reforestation, all of which contribute to the reduction of carbon in the atmosphere.

10.7.1 Carbon ETS Design

The ETS is an emissions cap-and-trade system designed to reduce GHG emissions by establishing a limit, or cap, on GHG emissions for specific sectors of the economy. Each ETS grants companies the authority to release GHG emissions equivalent to the global warming potential of one ton of CO_2.

The GHG market system, also known as the carbon ETS and including other GHG, is a trading framework created to regulate and facilitate the trading of GHG emissions. The goal of this approach is to reduce the market dominance of fossil fuels and accelerate the allocation of resources toward low-carbon energy alternatives such as solar and wind power.

This technique refers to a market-based process that allows entities (such as countries, companies, enterprises, and factories) emitting GHGs to participate in the trading of these emissions. Governments allow the use of carbon credits and carbon trading to gradually reduce overall carbon emissions and lessen their impact on climate change. Carbon trading agreements facilitate the exchange of carbon dioxide emission credits among nations as part of a global agreement aimed at gradually reducing GHG levels.

Countries that use fossil fuels and emit carbon dioxide do not immediately bear financial responsibility for the resulting environmental impacts. They incur expenses such as the direct cost of fuel, but there are additional expenses that are not included in the fuel price. These elements are commonly known as external factors. When it comes to the use of fossil fuels, these external effects usually result in negative consequences, indicating that the use of such products has harmful impacts on third parties.

The Glasgow Climate Change Conference in November 2021 established regulations for the international carbon market and, for the first time, endorsed an agreement to achieve the goals of the 2015 Paris Climate Agreement. The goal is to encourage nations and corporations to reduce their carbon emissions by adopting technology to shift towards environmentally friendly manufacturing, thereby increasing their surplus permits available for trading. Wealthier nations are effectively providing financial support to less affluent countries with higher levels of pollution by purchasing their carbon credits. However, over time, wealthier nations will gradually reduce their GHG emissions, thereby decreasing their reliance on purchasing them in the market.

Currently, there are three major carbon ETSs [23]:

1) European Union ETS (EU ETS)
2) The California Global Warming Solution Act (the United States)
3) The Chinese National Emission Trading System

10.7.2 European Union ETS

The EU ETS is currently operational in all 28 EU member states, as well as in Iceland, Liechtenstein, and Norway (three European Economic Area countries). The EU ETS operates on a cap-and-trade system. In Fig. 10.1, the mechanism of the EU ETS has been seen. EU ETS depending on the percentage of emissions on voyages as shown in Figure 10. 2

The EU ETS, a cap-and-trade system for GHGs from energy and industry, was implemented to ensure that the EU meets its GHG emission targets at a reasonable cost. The system offers flexibility to businesses participating in the scheme

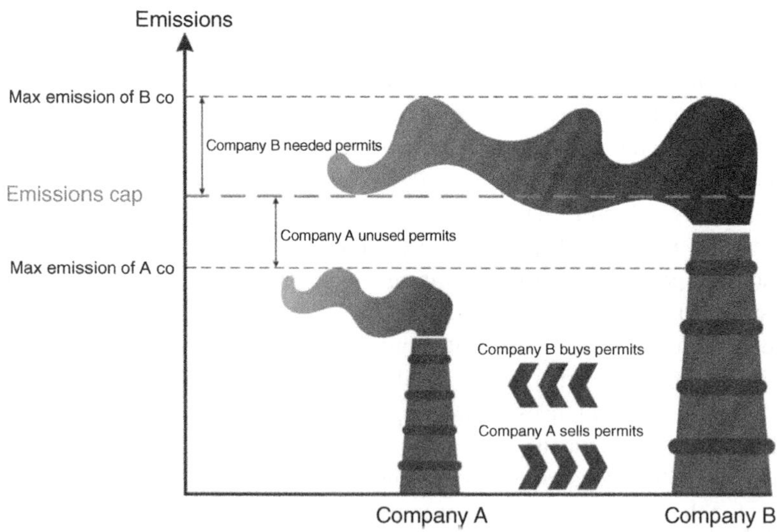

Figure 10.1 Mechanism of the EU ETS.

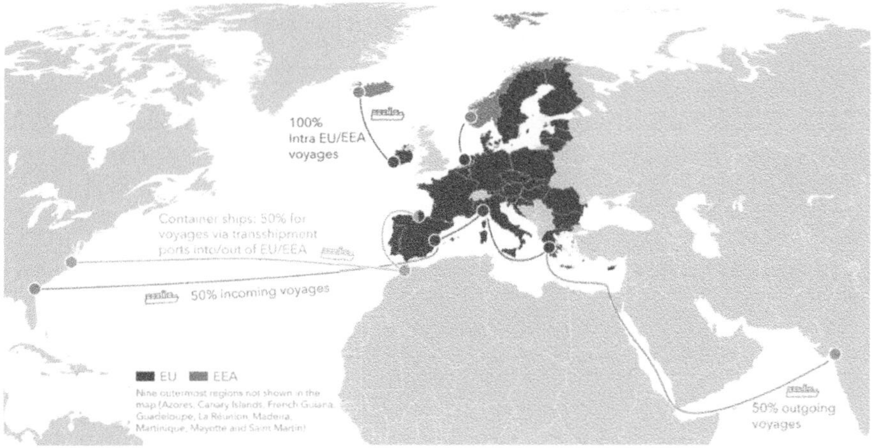

Figure 10.2 EU ETS based on percentage of emissions on voyages. *Source:* [24] / Det Norske Veritas.

by allowing them to choose between reducing emissions or purchasing emissions from other companies, based on carbon prices. This approach encourages the implementation of low-cost measures to reduce GHG emissions while postponing the adoption of more expensive reduction measures. There are approximately 330 registered covered or opt-in entities. These entities represent approximately 400 registered emitting sources or facilities.

This was the first carbon ETS, which began in 2005, and it is currently the most liquid exchange system in the world. The ETS includes 27 EU member countries, as well as Iceland, Liechtenstein, and Norway. During the initial phase of its adaptation, this system regulated the following departments:

- Power plants and other combustion facilities with more than 20 MW of thermal input.
- Industries include oil refineries, coke ovens, iron and steel factories, cement, glass, lime, brick, ceramic, pulp, paper, and cardboard manufacturing operations.

The aviation industry was the only additional sector regulated during the second stage of ETS compliance. Finally, in the third stage of compliance, additional sectors were incorporated into the EU ETS, including carbon absorption and storage facilities, as well as the manufacturing of petrochemicals, ammonia, nonferrous and ferrous metals, gypsum, aluminum, nitric, glyoxylic, and adipic.

The average sales price of each metric ton of CO_2 in the European Market [25] is shown in Figure 10.3. The average sales price of each metric ton of CO_2 in the California Market is shown in Figure 10.4.

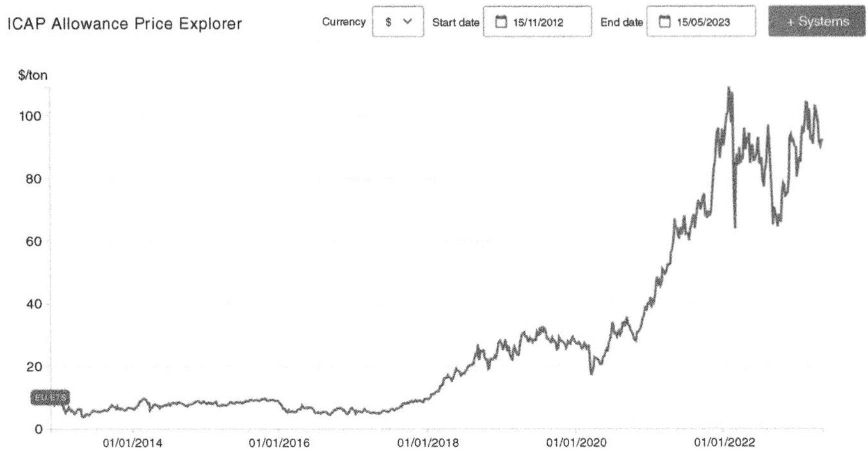

Figure 10.3 The average sales price of each metric ton of CO_2 in the European Market. *Source:* [25] / International Carbon Action Partnership.

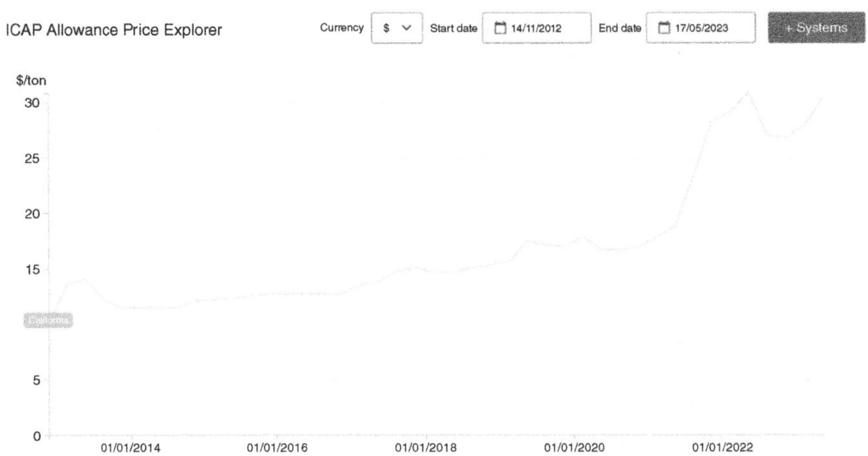

Figure 10.4 The average sales price of each metric ton of CO_2 in the California Market. *Source:* [25] / International Carbon Action Partnership.

10.8 The California Global Warming Solution Act (the United States)

This system has been operational since 2012 and falls under the jurisdiction. The annual GHG emissions in this area total 425 million tons. The California ETS involves the participation of 330 registered entities.

In 2012, the California Cap-and-Trade Program launched its tracking system for allocating compliance instruments, distributing auction proceeds, and facilitating trading. Compliance obligations began in January 2013. The program addresses about 75% of the state's GHG emissions.

The program covers approximately 400 facilities and emissions from the power, industry, transportation, and building sectors. Allowances are distributed through a combination of auctions, free allocation, and free allocation with consignment. Auction proceeds are reinvested in projects that reduce emissions, thereby strengthening the economy, public health, and the environment, especially in disadvantaged communities.

The California Air Resources Board (CARB) oversees the implementation of the California Cap-and-Trade program. California has been a member of the Western Climate Initiative (WCI) since 2007, and its program was officially linked with Quebec's in January 2014. The average auction price for 2022 was USD 28.08.

First Compliance Period (2013–2014): The system started in 2013 with a cap of 162.8 $MtCO_2e$, declining to 159.7 $MtCO_2e$ in 2014, at a rate of ~2% annually.

Second Compliance Period (2015–2017): With the program expanding to include fuel distribution, the cap rose to 394.5 $MtCO_2e$ in 2015. The cap decline factor averaged 3.1% per year in the second compliance period (2015–2017), reaching 370.4 $MtCO_2e$.

Third Compliance Period (2018–2020): The cap in the third compliance period started at 358.3 $MtCO_2e$ and declined at an average annual rate of 3.3% to 334.2 $MtCO_2e$ in 2020.

Fourth Compliance Period (2021–2023) and Beyond: During the 2021–2030 period, the cap declines by about 13.4 $MtCO_2e$ each year, averaging ~4% per year, to reach 200.5 $MtCO_2e$ in 2030.

10.9 The Chinese Market

The Chinese demonstrate a deep commitment to this issue and have ambitious goals, as shown by the launch of China's national ETS in 2021. The jurisdiction of this entity covers the entire territory of China. The annual GHG emissions in this region amount to 12,301 megatons. This will surpass the European Union's nearly two-decade-old ETS.

China, as the world's second-largest economy in terms of GDP and a significant regional power, is actively involved in negotiations aimed at achieving a practical and widely accepted solution for managing climate change. However, when it came to implementing specific measures, it adopted a cautious and sometimes unassertive approach. China, as a signatory to the Kyoto Protocol, receives international funds and offers low-cost reduction certificates, currently known as the CDM or certified emission reductions (CERs).

10.10 Certified Emissions Reductions

The introduction aptly portrays CERs in China. However, this could be improved by explaining further how CERs fit into strategies for reducing carbon and their place in the international trading of emissions. Although China's discussion on CERs is informative, a more detailed assessment of the efficiency and challenges of the CDM would have been more enlightening. Conversely, it would be more practical to give cases of successful CER projects and what they have achieved. Moreover, a holistic approach to climate change demands an understanding of how CERs are embedded in carbon reduction strategies. As such, CERs are imperative since they allow countries to invest in emission reduction projects in developing nations that form a major part of international emissions trading. This not only motivates cutting down on carbon but also promotes sustainable development within developing countries, as well as creating room for innovations aimed at reducing GHG emissions. Nevertheless, it is critical to assess the effectiveness of the mechanism that governs the issuance of CERs relating to how efficiently this has been applied within the CDM. Besides making it more practical and relatable for readership, use additional examples from actual events showing some successfully implemented projects involving the use of CERs [26, 27].

10.10.1 Clean Development Mechanism

Just like the CERs section, this part of the CDM section gives an overview of China's adoption of this mechanism. However, it is not clear how CDM is integrated into the national emissions trading scheme and what its gains for reducing carbon are. A critical analysis of the strengths and limitations of CDM in promoting sustainable development and reducing carbon emissions would improve the section. Finally, a discussion on challenges experienced during the implementation of CDM projects, as well as possible solutions, will make this part more meaningful. Moreover, discussing the potential synergies between CDM and other international climate change mechanisms, such as the Paris Agreement, can contribute to global efforts toward reducing carbon emission rates [28].

In addition, an inclusive examination of how the Chinese government promotes CDME is essential in understanding how effective it has been overall. China's GHG scheme is unique due to (1) its combination of a bottom-up structure, with pilot projects at the provincial and city levels, and elements of strong command and control that operate from top to bottom. (2) If a national emissions trading system were to be implemented, it would be on an unprecedented scale, given that the country is the largest carbon emitter in the world. (3) In contrast to many emissions trading systems worldwide, China's system was not developed through democratic processes or political conflicts but rather established under a single-party government [29].

China's national ETS was implemented in 2021 to effectively control and gradually reduce carbon emissions. China's national ETS is the world's largest in terms of covered emissions, estimated to encompass more than 4 billion tons of CO_2 and accounting for over 40% of the country's carbon emissions [29].

The China National Emissions Trading Scheme regulates over 2000 power companies with annual emissions of more than 26,000 metric tons of CO_2, including combined heat and power plants and captive power plants in other industries. It is an intensity-based system in which allowances are freely allocated based on benchmarks and actual production levels. Compliance obligations are currently limited and vary depending on the type of power generation. The system's coverage will gradually expand to include other sectors.

The national ETS builds on the successful experience of piloting carbon markets implemented in eight regions. The pilots will continue to operate alongside the national ETS, encompassing sectors and entities not included in the national system. As the system's coverage expands, it is expected that entities currently covered by regional systems will integrate into the national ETS. The power entities included in the Chinese regional ETS pilots have transitioned into the national market.

The cap is set bottom-up, meaning that the sum of the total allowance allocation to all covered entities forms the cap. It is an intensity-based cap that adjusts according to the actual production levels. The national ETS is estimated to have a cap of 4500 $MtCO_2$ in 2021. Figure 10.5 depicts the average sales price per metricton of CO_2 in the Chinese market.

Figure 10.5 The average sales price of each metric ton of CO_2 in the Chinese Market. *Source:* [25] / International Carbon Action Partnership.

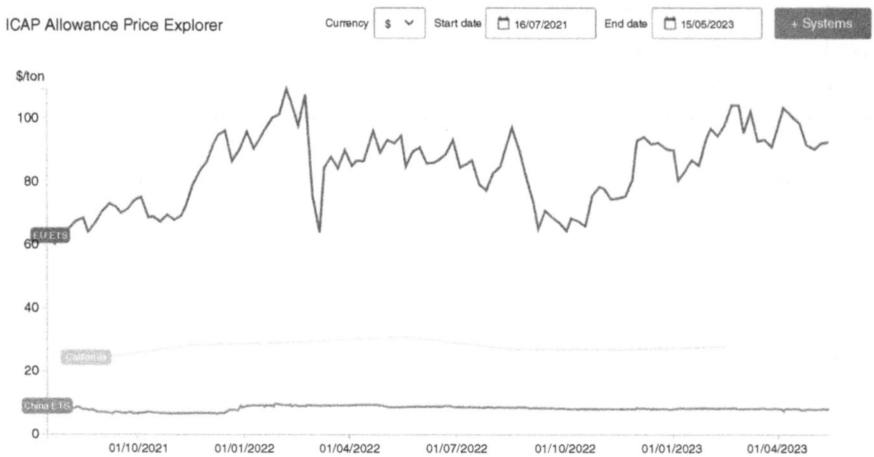

Figure 10.6 Comparison of Three ETSs: Europe, the United States, and China. *Source:* [25] / International Carbon Action Partnership.

The Draft Interim Regulations published by the Ministry of Ecology and Environment (MEE) in 2021 outline the possibility of centralized development of a cap and allocation plan, suggesting the potential for a top-down process of setting caps in the future [25].

Figure 10.6 illustrates a graphical representation of the comparative auction prices of carbon dioxide in the market mentioned earlier.

10.11 Analysis of Risks/Challenges of the Main ETS

Investing in carbon credit markets carries inherent risks, with the primary concern being the potential volatility and uncertainty associated with policies aimed at ensuring price stability. Inadequate pricing discourages investors and hampers the financing of credit-generating projects, while excessive pricing presents challenges for supervised entities and impairs their efficiency.

The second risk is associated with geopolitical tensions and their significant impact on various categories.

10.11.1 Carbon Pricing Scheme, Carbon Tax, and Emission Trading

Typically, carbon sources can be classified into two categories: direct and indirect. The primary source of carbon emissions is the direct release of carbon through on-site chemical reactions, including fuel combustion (such as crude oil, paraffin, coal, and natural gas), decomposition, fermentation, decay of organic matter, and the production of chemical products (such as oil and coal). The latter primarily consists of carbon located outside a specified area [33]. Carbon emissions originate from either the energy source or its utilization. The key factor in reducing carbon emissions is the use of environmentally friendly energy sources and the optimization of energy efficiency [34].

Carbon emission trading specifically refers to the exchange of pollutants, where rights to emit GHGs are acquired through a market mechanism known as emission trading [35].

The World Bank's 2020 trend report reveals that there are currently 61 carbon pricing schemes being implemented or scheduled for implementation worldwide. These include both systems for trading GHG emissions and taxes on carbon. The institutions that oversee the carbon market, such as the CDM, are currently operational. The parties to the Kyoto Protocol committed to obtaining carbon credits from developing nations through the CDM.

The consumption of products contributes to pollution. The production of steel, electricity, and cigarettes generates pollution that adversely affects individuals who are not directly involved in the manufacturing or use of these products. This type of pollution is known as a negative externality. To regulate the costs associated with externalities, political intervention is necessary. Postponing the implementation of emission reduction policies will result in significant costs due to market failure.

There are three methods to control these negative impacts:

1) **Traditional Method (Command and Control):** Concentration standards should be set at a specific threshold to prevent a significant increase in the overall costs of emission reduction and to avoid diminishing motivation for this endeavor.
2) The subsidy policy approach involves discontinuing subsidies for measures or technologies that have carbon emissions and are typically not economically efficient.
3) The market-based approach to addressing pollution involves implementing a pricing mechanism, such as a carbon tax or emission permit, to regulate and control the emission of pollutants. In this scenario, financial incentives are established, and a standard threshold is considered. Publishers are required to either minimize their consumption or bear the burden of taxes or the expenses associated with emission reduction. The main advantage is its cost-effectiveness.

The carbon pricing methods encompass the following types:

1) Carbon Tax

   A carbon tax is levied on each metric ton of carbon dioxide equivalent, with the key feature being the definite cap on the cost per ton of emission and the indeterminate final emission amount.

2) Emission trading
   Instead of imposing taxes on emitters, an alternative approach called cap and trade is implemented. The lawmaker establishes a maximum limit for allowable emissions, usually with a decreasing trend. These permits are then allocated among emitters, either through reduction or self-purchase. By the end of the year, emitters must have the necessary permits. We are aware of the precise price of the final emission. In an emissions trading program, the ability to buy and sell emission rights gives polluting entities the autonomy to determine the most cost-effective approach to meeting the emission limit.
3) The combined mode (Price Cap)
   The combined mode (Price Cap) operates similarly to emission trading, but with the added flexibility that emitters can purchase additional permits at the same maximum price if the permit prices exceed a certain limit.

The carbon tax does not establish a definitive threshold for emissions, but it does set a known cost for GHG emissions. In emission trading, the maximum emission threshold is specified, while the ultimate cost remains uncertain. When operating in the combined mode, it considers the benefits and drawbacks of both systems. The combined mode considers the advantages and disadvantages of both systems:

- Cap-and-trade program
- Offset or credit program
- Rate-based program
- Performance of the cap-and-trade program

1) The legislature sets a maximum limit for the total amount of a pollutant that can be emitted by a group of emitters within a specific period, such as one year.
2) The legislatures convert the emission ceiling into a quota, with each quota representing a specific emission entitlement for the targeted pollutant.
3) The legislature allocates these quotas to the emitters.
4) During the designated time frame, each emitter quantifies and documents the quantity of their emissions.
5) At the end of each designated time interval, each emitter is required to offer a permit that is commensurate with their emission.
6) If an emitter fails to meet the requirement of emitting a certain amount of emissions, it will be subject to a fine imposed by the legislator.

10.11.2 Cap-and-Trade Program

The primary benefit of cap-and-trade programs is the reduction of overall costs associated with emissions.

An explanation of emission cap-and-trade programs: several advantages of cap-and-trade programs are:

- Minimizing cap-and-trade programs.
- Cost reduction minimization.
- Greater environmental certainty is achieved compared to other methods, as the determined emission ceiling is met.
- Companies are urged to expedite their efforts in reducing emissions, which will consequently lead to earlier attainment of environmental advantages.
- Providing financial incentives to promote the implementation and advancement of emission reduction technologies, thereby fostering innovation in this domain.
- Enhancing the potential for attaining greater environmental advantages.

10.11.3 Offset or Credit Programs

Example: CDM Under the Kyoto Protocol

Typically, they are not employed as a standalone application (complementary to other emission trading programs). The potential to acquire emission rights beyond the legal boundaries of the cap-and-trade program is determined by calculating the credit amount for each project, which is based on the comparison of emissions before and after the project with a baseline state. The challenges are as follows:

- Determining the baseline state
- Redundancy
- Leakage (emission beyond the project's boundaries)
- high official and contract costs

It is not a standalone system, but it is a complementary system.

10.11.4 Rate-Based Program

This program establishes the performance standard instead of determining the emission ceiling. For instance, the quantity of carbon dioxide emitted by power plants per unit of electricity generated. The legislator establishes the benchmark for the performance level of the resources in every department. For instance, the level of pollution emitted per unit of electricity produced is measured in megawatt hours. Credits are given to sources that emit less than the prescribed standard, while sources that exceed the standard are required to provide credits.

Hint: A resource can receive credit if the rate of activity increase exceeds the standard rate of decrease.

10.11.5 Performance of Cap-and-Trade Program

- Minimizing cap-and-trade programs.
- Cost reduction minimization.
- Greater environmental certainty is achieved compared to other methods, as the determined emission ceiling is met.
- Companies are urged to expedite their efforts in reducing emissions, which will consequently lead to earlier attainment of environmental advantages.
- Providing financial incentives to promote the implementation and advancement of emission reduction technologies, thereby fostering innovation in this domain.
- Enhancing the potential for attaining greater environmental advantages.

10.12 Carbon Tax

Carbon pricing is predicated on this goal, which does not encompass the external expenses associated with GHG emissions and the climate ramifications for those who emit them. Nations should collaborate to achieve the emission targets. Indeed, it is the general populace that bears the responsibility for mitigating emissions.

- Carbon pricing is a market-based instrument.
- Market-based Instruments
- Market-based instruments are policy instruments that leverage market prices and economic variables to incentivize emitters to decrease emissions.

Summary: A range of instruments has been developed thus far:

Establish a connection between external costs and emitters. The matter at hand pertains solely to financial considerations and cost reduction of emissions. Alternatively, the user can cover the expenses associated with emissions. Expenses related to flooding and additional expenses.

The seasons have undergone alterations, resulting in the emergence of early spring around mid-March, followed by a subsequent onset of cold weather and the disappearance of blossoms after a few weeks. The costs incurred by heavy rainfall, which is a consequence of climate change, lead to the destruction of crops, and it is necessary to allocate funds to address or mitigate these occurrences.

In European countries with a high level of sensitivity, the root cause is not cultural but rather the need for people to exert control over expenses. At elevated levels, individuals show decreased consumption when exposed to heated silver. The cost of electricity and gas is not burdensome for a segment of the population, and it is possible to develop methods to reduce its impact. Second example: The cost of electricity for industries such as steel and petrochemicals has increased, and these sectors are known for their high energy consumption and significant carbon emissions impact. Therefore, these industries should consider strategies to minimize energy consumption and reduce their carbon footprint. At the very least, referring to the FOB price at the Persian Gulf.

- Pricing tools include:
- Carbon Tax
- ETS
- Offset or crediting mechanisms
- Results-based climate finance (RBCF)

If we want to offer compensation for the cost in some way, we are engaging in offsetting. For example, the steel company in Kerman has created a man-made forest to offset 9% of its GHG emissions.

10.13 Crediting

Carbon prices are determined by the difference between emission permits (emission quotas) and actual emissions, whether the investment is made elsewhere or the certificate is bought. The principle revolves around the valuation of carbon as a tradable asset. For example, if the permit allows for 100 tons of emissions and the actual emissions are either 150 or 105 tons, it is considered acceptable, although with different implications in the market. Greater distances will lead to increased demand, subsequently driving up the price of carbon.

A carbon tax is a direct levy on the emission of GHGs that result from the carbon content of fossil fuels. A single metric ton of carbon dioxide carries a monetary value that is equivalent to that of other GHGs, as determined by a coefficient (tCO_2).

10.13.1 GWP: Global Warming Potential

This potential factor mitigates global warming. One metric ton of any GHG can be considered equivalent to carbon dioxide (CO_2) gas.

Comparison of carbon tax and emission trading:

Both approaches aim to mitigate the societal costs linked to carbon emissions by implementing a carbon pricing mechanism. The main difference in this approach lies in the fact that, with the carbon tax method, the government sets a fixed price for carbon and allows the market to determine the level of emissions. Conversely, in the emission trading method, the government sets a maximum allowable emission limit and lets the market determine the corresponding price.

10.13.2 Advantages and Disadvantages of a Carbon Tax

- The effectiveness of emission reduction through a carbon tax is not assured.
- The implementation of a small carbon tax will yield no impact.
- Carbon tax revenues can exceed those generated by emissions trading.
- The price of carbon under a tax system is well-defined and transparent.
- The implementation of a carbon tax is generally less costly and complex.
- In economically vulnerable countries where carbon pricing could have a substantial adverse effect on the economy, a carbon tax may not be the most suitable option.
- There is a potential for a leak. We may reduce emissions in one location while simultaneously increasing emissions in another location, or have a specific location for release. For example, a product is manufactured in a factory situated in an area where the implementation of a carbon tax is not a concern.

10.13.3 Advantages and Disadvantages of Emission Trading

- Verification of Measurement and Reporting.
- The emission industry requires access to a wide-ranging market.
- The emission reduction market occurs in the location with the most economical cost.
- It offers flexibility, allowing for the sale of a certificate in one year or the purchase of a certificate in another year.

In the cap-and-trade system, the reduction of emissions may not be directly linked to the implementation of improvement measures (economic crisis). In emission ceiling-free systems such as CDM, there is no correlation between the expected and actual emission reductions, and no restrictions are placed on emissions. Additionally, these systems involve greater technical complexity. There is a potential for leakage.

10.14 How Is the Offset Market in Developed Countries?

Organizations and individuals in many countries are actively pursuing carbon offsets, either as a voluntary choice or because of regulatory requirements. An individual or organization has the option to pay a broker to remove a specific amount of carbon dioxide from the atmosphere, usually in a different geographical location. The customer determines the extent of their pollutants, and the broker receives compensation proportional to that extent. Subsequently, the broker allocates a portion of these funds to the project, thereby reducing carbon emissions.

For example, an individual may operate an aircraft, leading to the release of a specific amount of GHGs into the atmosphere. An individual uses a tool to calculate the amount of GHG emissions produced during a specific flight. Subsequently, he buys carbon credits from a broker to offset his emissions. The broker deducts a portion of his commission and allocates the remaining funds to invest in a project aimed at mitigating GHG emissions, such as reforestation.

10.15 Peer-to-Peer Energy Trading System

Due to the emergence of modern energy technologies, the smart energy market is shifting from a centralized to a decentralized model. The decentralized model is commonly referred to as the peer-to-peer (P2P) model in technical terminology. The main challenges in a P2P energy market are scalability, robustness, and concerns about security and privacy [33].

A blockchain-based solution is proposed in Ref. [34] to establish a unified energy trading market. This study illustrates the full decentralization of a blockchain-based system, allowing market participants to directly interact and conduct energy trading without the need for intermediaries. Furthermore, a scholarly article [35] introduces a P2P electricity trading framework that utilizes blockchain technology specifically designed for plug-in hybrid electric vehicles. Instead of depending on the traditional method of importing electricity from a distant source, this model functions by dynamically responding to demand and encouraging consumers to actively participate through a variety of incentives. Every participant in this system can synchronize their electricity demand with the supply to maximize their incentive.

Blockchains are usually managed by a P2P computer network to function as a publicly accessible distributed ledger. All nodes in this network can use the consensus algorithm protocol to add and validate new transaction blocks [36]. P2P energy trading can effectively increase internal consumption by taking advantage of the widespread adoption of renewable energies. This, in turn, reduces the strain on the electricity grid and enables the transition to a decarbonized system. Figure 10.7 illustrates the process of P2P energy trading within a specific geographical area.

10.16 Blockchain Role in Carbon Trading

Carbon trading is a market-based instrument employed to mitigate climate change resulting from the release of GHGs. As stated in the article's introduction, this tool was initially implemented as a component of the Kyoto Protocol in 1997.

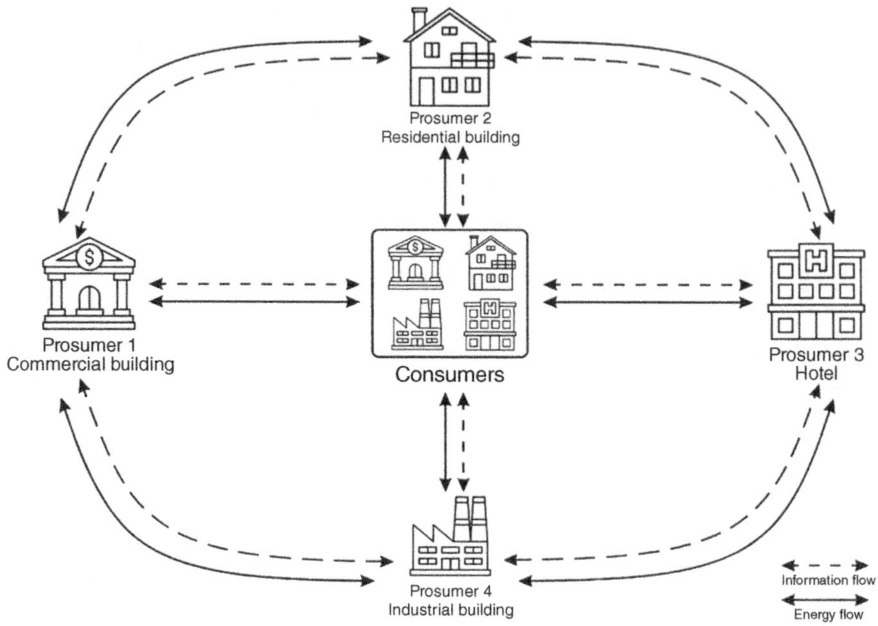

Figure 10.7 P2P Energy Trading Within a Confined Geographical Region.

Presently, there exist 64 national carbon taxes and ETSs that encompass approximately 20% of worldwide GHG emissions [37]. Blockchain's application in the carbon industry is unavoidable. The carbon trading market shares numerous similarities with the blockchain mechanism. Blockchain is characterized by a decentralized database. Financial experts and senior economists worldwide assert that blockchain technology, which facilitates the decentralized sharing of data and information through a public ledger, enables companies and commercial institutions to conduct business without encountering structural obstacles, is free from the influence of authorities, and Possesses the ability to obtain primary data.

Blockchain technology offers significant benefits to businesses by improving operational efficiency, reducing energy consumption, and encouraging the use of renewable energy sources. The use of blockchain technology to protect environmental indicators and tackle ecological issues has seen significant growth in recent years. During the 2019 United Nations General Assembly, blockchain technology was recognized as a strategic tool to tackle global crises, especially environmental challenges.

Greta Thunberg, a 16-year-old environmental activist, strongly criticized the political leaders of the member countries of the organization for their failure to

Figure 10.8 Blockchain-based Carbon Trading.

take decisive action in addressing environmental crises during the previous year's United Nations General Assembly. Greta Thunberg argues that blockchain technology has the potential to be a viable solution for addressing climate change and environmental crises.

Carbon trading involves the assessment, management, exchange, and regulation of carbon emissions. Blockchain is a form of digital information, while carbon trading involves the use of data [38]. Figure 10.8 illustrates the simplified process of conducting carbon trading on a blockchain platform. In this process, project owners create tokens for carbon-saving projects, which are then traded on a decentralized marketplace.

As a result, the field of blockchain research and its practical implementation are experiencing significant growth. The use of blockchain technology has entered a new era, and its adoption in China is steadily increasing. The Ministry of Industry and Information Technology of the People's Republic of China is currently developing national standards for blockchain technology. Undoubtedly, the integration of blockchain technology in the carbon industry is inevitable. The carbon trading market shares numerous similarities with the blockchain mechanism. Carbon trading involves the assessment, storage, exchange, and management of carbon. Given that blockchain is a data framework and carbon trading relies on data, it has the potential to significantly facilitate the growth of this industry.

The benefits of utilizing blockchain technology in carbon trading include:

1) Inherent security and reliability.
2) Achieving optimal effectiveness and providing a high level of comfort.
3) Being open and inclusive.

Soheil Saraji and his colleagues from the University of Wyoming presented a model for this business and the utilization of blockchain in a white paper.

This project necessitates a minimum of four smart contracts:

Figure 10.9 illustrates the design specifications for each smart contract.

Smart contract 1: A blockchain-based registry system designed to record essential information for individuals participating in a specific project or organization.

A) Approvers are responsible for approving the credit holders' carbon credits. They also confirm that burned carbon tokens are equivalent to an appropriate amount of burned carbon.

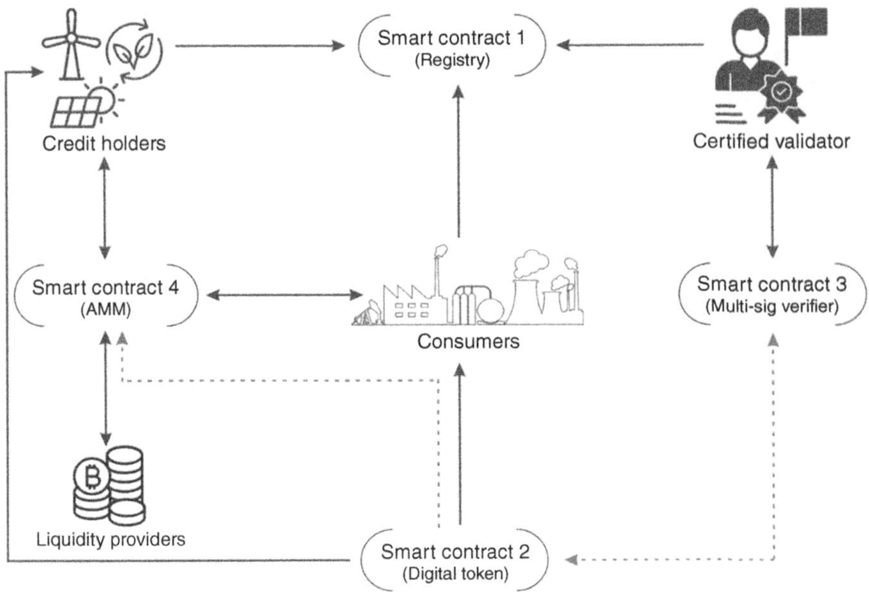

Figure 10.9 Carbon credit ecosystem on a blockchain.

B) Credit holders are organizations that possess carbon credits, which are units traded within the GHG ecosystem.
C) Customers: Individuals and organizations who are interested in acquiring their products or services. Carbon footprint offsetting involves the purchase of carbon credits and the subsequent retirement of carbon tokens.

Smart contract 2: The process of studying and approving this smart contract, which generates carbon tokens using carbon credits, involves a sequence of functions.

A) Approved credits verified by credit holders.
B) Engaging in the act of counterfeiting or producing counterfeit carbon currency.
C) Incinerating carbon credits.
D) The irreplaceable creation of tokens serves as a symbol of burned tokens that represent the offsetting of carbon emissions.

Smart contract 3: A smart contract requires the endorsement of multiple signatories. Create or incinerate carbon tokens. A minimum of 70% approval is necessary for this contract. Validators are executed automatically.

An Automatic Market Maker Smart Contract

A) The automated exchange of carbon tokens using digital currencies, such as stablecoins or potential central bank digital currencies in the future.

B) Encouraging liquidity providers by imposing a transaction fee of 0.3%, for instance.
C) Uncovering fluctuating prices for carbon tokens in the open market.

Smart Contract 4: An automated market maker (AMM) smart contract that allows (a) automated trading of Carbon tokens with digital money (e.g., stablecoins or future central bank digital currencies).

In summary, the market has reached a specific threshold that will lay the groundwork for the implementation of blockchain technology. However, blockchain technology is still in a phase of continuous development and evolution, and it has not yet achieved a mature level. Additionally, there are still challenges related to computing power and response speed. Conversely, the energy system is complex enough to require extensive research and support. Significant progress still needs to be made before energy blockchain technology becomes a reality.

10.17 Application of AI in Blockchain-Based Emission Trading

The government's recent goal is to increase the proportion of renewable energy sources and influence consumer behavior to promote a financially efficient balance between energy supply and demand. The accomplishment of this important goal is achievable using advanced information and control infrastructure in smart power grids, which enables cooperation among various stakeholders. Blockchain and AI are advanced technologies that tackle these two factors. Blockchain offers decentralized trading platforms for energy markets, while AI facilitates optimal operational control of power systems. AI has the potential to help operate carbon credit markets. AI can improve efficiency and productivity by automating tasks such as matching buyers and sellers, as well as managing invoicing and payment processing. AI in digital carbon credit markets offers the advantage of advanced decision-making. Through the analysis of historical transactions, market trends, customer preferences, and project details, AI can provide valuable insights and recommendations. These insights and recommendations can help make well-informed decisions about buying or selling credit. It helps financial investors more accurately assess the risks and potential returns of various projects. Moreover, incorporating or avoiding the use of AI in carbon sequestration initiatives can improve the effectiveness and efficiency of project development and management, ultimately resulting in decreased carbon emissions and helping to mitigate the impacts of climate change. AI can now be used to detect and prevent fraudulent activities in the market, thereby ensuring the trustworthiness and reliability of transactions.

Furthermore, the sophisticated measurement infrastructures employed by smart grids generate a substantial quantity of valuable data. AI can utilize this data to enhance the understanding of current conditions and optimize the functioning of power systems. This is particularly advantageous when small- or medium-sized consumers engage in the functioning of power systems and autonomously make decisions, considering the financial constraints of their control systems.

10.17.1 Game Theory

Game theory is a collection of analytical methods used to model strategic decision-making and interactions among stakeholders in power systems. Optimization is a method used to determine the most efficient decisions to reach specific objectives. Moreover, the imprecisions resulting from the sporadic nature of renewable energy sources (RES) and adaptable demand have an impact on the precision of power system modeling.

The market has seen a positive response to carbon trading discussions, witnessing a significant surge in carbon goods' value from $270 billion to $851 million between 2020 and 2021. Projections indicate further growth to a remarkable $22 trillion by 2050. Notably, the rising demand for carbon credits due to inflation is being met by increasing their supply, potentially aligning yearly availability with demand by 2030.

The emergence of AI tools and technologies has notably bolstered the popularity of carbon credits by simplifying their management and monitoring processes. These tools play pivotal roles in tracking and compensating for carbon footprints, predicting emissions, overseeing credit control, and assessing potential investments. Through AI integration, companies can confidently monitor their investments' progress and ensure compliance, consequently enhancing the value of carbon credits and stimulating investments in GHG reduction [39].

The Cournot and Stackelberg models are both well-established frameworks for examining the decision-making processes of stakeholders. The Cournot model characterizes a market where participants offer identical products and engage in independent and simultaneous decision-making regarding the quantity of products supplied [40]. The Stackelberg model involves a decision-making process that is hierarchical and sequential, consisting of two or more levels [41]. In the context of two-level decision-making, participants are categorized into a leader level, responsible for initiating the first decision, and a follower level, responsible for subsequent decisions based on the leader's strategies. In multi-level decision-making, once the initial level of followers has made their responsive decisions, they ascend to a leadership level to prioritize the decisions of the subsequent level of followers. This process persists until the final level of followers reaches their decision-making stage.

AI and machine learning are indeed potent instruments for enhancing the efficiency and efficacy of carbon credit projects. These projects can assist developers in attaining their objectives by automating the process of monitoring and evaluation, optimizing project design and management, and enhancing risk assessment. AI can improve carbon credit projects by automating the monitoring and evaluation of project performance. Consider, for instance, a project that entails the cultivation of trees on an extensive piece of land. AI can examine live data collected by sensors in trees to detect any deviations from the anticipated level of performance. If the AI system detects soil moisture levels below the anticipated threshold, it can promptly notify the developer about potential drought stress, enabling the developer to promptly respond and resolve the problem. This can also be accomplished through a straightforward automation scheme.

However, it is crucial to consider factors such as overall weather conditions, geographical location, humidity level, product characteristics, and time of year. AI can utilize machine learning models to accurately identify projects with a high-risk factor. This identification is based on various factors, such as location, project type, carbon sequestration technology, carbon reduction estimates, and development experience. This thorough examination, through ongoing monitoring of climate variations, market fluctuations, and policy shifts, assists developers in comprehending the influence of external factors. It aids in identifying potential risks and uncertainties associated with the project and empowers developers to undertake the requisite measures to mitigate potential hazards and enhance the project's sustained prosperity. The utilization of AI and machine learning holds immense potential to bring about a significant transformation in carbon credit initiatives. This technological advancement can revolutionize efforts to combat climate change and contribute to the establishment of a green planet.Many Major applications of AI in carbon trading are as follows [42].

10.17.2 AI for Monitoring Carbon Footprint

Put simply, the carbon footprint refers to the cumulative amount of GHGs, such as carbon and methane, generated by any given activity. In the present day, due to the ongoing rise in carbon emissions, it has become imperative for companies to diminish their carbon footprint. Nevertheless, manually calculating and monitoring carbon footprints is not a prudent approach due to its inherent inaccuracies, inefficiencies, and potential for errors. Nevertheless, the advent of IoT-enabled devices can assist companies in tracking and monitoring emissions throughout their entire carbon footprint. IoT devices offer companies the advantage of gathering and organizing data from their activities and operations. And from every aspect of your value chain, encompassing materials as well.

1) AI for Tracking Materials Containing Carbon Emissions
Embodied carbon encompasses the carbon dioxide emissions that arise from the materials and processes involved in the entire lifespan of a building or any infrastructure. This encompasses all carbon dioxide emissions generated throughout the entire lifecycle of the material, starting from its extraction, transportation to the manufacturer, manufacturing process, and transportation to the job site. Measuring embodied carbon is a highly intricate process, as it necessitates the meticulous tracing of materials throughout intricate manufacturing supply chains. In 2018, materials used for construction accounted for 11% of global energy and process-related GHG emissions. This is the domain where AI can be utilized. AI can facilitate the computation of the comprehensive embodied carbon emissions of materials, a task that would otherwise pose challenges in monitoring large-scale workplaces.
2) AI for Tracking Fuel Carbon Emission
AI may be utilized to monitor and measure carbon emissions from fuel in the future. Fuel carbon emissions are a direct measure of the carbon emissions produced when various types of fuel are burned. AI, in conjunction with other technologies like the Internet of Things (IoT), can be programmed to monitor and measure carbon emissions from different sources within work environments. It aids companies in identifying fuels with high and low emissions, enabling them to establish targets, make decisions regarding their usage, and decrease emissions. As an illustration, cars powered by compressed natural gas (CNG) generally release a lower amount of carbon emissions compared to cars powered by diesel fuel. IoT devices can be utilized to detect GHG emissions in such circumstances. An AI-integrated record-keeping dashboard can offer a comprehensive overview of carbon emissions. This enables the planning of fuel consumption to effectively manage carbon credits.
3) AI for Optimizing the Utilization of Machinery
Machines are significant contributors to the release of GHGs and other detrimental substances, including carbon monoxides, nitrogen oxides, and suspended particles. Hence, to control and oversee this, it is crucial to prioritize effective fleet management. Due to the extensive use of machines in workshops, manual management proves to be highly inefficient, as it is impractical for humans to constantly monitor each machine. AI and computer vision technologies can have a significant impact in this context. These technologies enable continuous monitoring of working hours, fuel consumption, and non-essential usage of machinery, ensuring no loss occurs. This facilitates the optimization of machinery utilization.

4) AI for Monitoring Carbon Offset
Carbon offsets are actions taken to counterbalance the release of carbon dioxide into the atmosphere resulting from industrial or other human activities. Effective monitoring of carbon offsets requires careful and precise tracking of the various activities undertaken by a company to counterbalance carbon emissions. This task involves intricate computations, meticulous analysis, thorough comparisons, and careful distinctions between carbon emissions and carbon offsets, which cannot realistically be done manually. Nevertheless, industries can leverage technologies such as AI, computer vision, cloud computing, and similar advancements to automatically gather and analyze diverse data with limited human intervention.
5) AI for Monitoring Air Quality
AI can also be utilized to audit carbon credits to monitor air quality. The utilization of AI can enhance air quality through its ability to identify and analyze minuscule airborne particles. The primary contributors to air pollution, such as the construction, mining, oil, and natural gas industries, can mitigate their impact by employing AI. This technology enables industries to accurately measure and forecast air quality and pollution levels, as well as monitor and anticipate fluctuations in air pollution levels. Workplace air pollution AI can also assist in simulating chemical reactions involving pollutants. For instance, algorithms like the atmospheric transport modeling system (ATMoS) can aid in comprehending PM 2.5 levels and consequently facilitate air quality management.
6) AI for Construction and Demolition Waste Management
The construction industry alone accounts for 35% of the world's industrial waste, according to statistics. Most of the carbon is emitted into the atmosphere as a result of the construction and demolition (C&D) waste that accumulates at disposal sites. Hence, it is crucial to implement effective administration and on-site categorization of these C&D wastes. Technologies such as AI, computer vision, and machine learning can be utilized to effectively address on-site waste segregation and prevent illegal dumping. This, in turn, contributes to the reduction of carbon emissions and environmental pollution.
7) AI for Carbon Credits Management
A carbon credit is a transferable authorization or document that grants the holder the privilege to release a designated quantity of CO_2 or any other GHG. It is crucial to closely monitor carbon credits, as the credit-holding company is obligated to adhere to these restrictions, which could lead to receiving compensation. The manual administration of carbon credits is laborious and ineffective. However, when used collectively, technologies such as AI, the IoT, and cloud computing can fulfill a crucial function by autonomously monitoring the carbon credits of companies.

8) AI for Carbon Emission Prediction
Predictive AI can assist companies in forecasting future emissions throughout the company's carbon footprint. This includes considering current mitigation efforts, the adoption of new CDR, and anticipated future demand. This facilitates their ability to establish, modify, and attain reduction objectives with greater precision.
9) AI for Boosting Overall Carbon Credit Trading
Carbon credits are exchangeable in both private and public markets. Companies with surplus or unused carbon credits have the option to sell them to other companies in need. The process of give-and-take in carbon credit trading is intricate, and the utilization of technologies such as AI and predictive analytics can facilitate a smooth and efficient trading experience, ultimately empowering the entire carbon credit trade.

10.18 Optimization Approaches

Optimization approaches can be categorized into programming techniques and heuristic algorithms. Programming techniques encompass linear programming, integer linear programming, mixed-integer linear programming, and non-linear programming. Linear programming is an optimization problem where the objective functions and constraints are all linear functions of the decision variables [43]. Integer linear programming restricts decision variables to binary values and integers exclusively [44]. Integer and non-integer numbers can both be utilized as decision variables in mixed integer linear programming [45]. Nonlinear programming is a type of optimization problem where at least one objective or constraint function is a nonlinear function of the decision variables [46, 47].

References

1 Rickels, W., Proelß, A., Geden, O. et al. (2021). Integrating carbon dioxide removal into European emissions trading. *Frontiers in Climate* 3: 690023.
2 Pfenninger, S., Hawkes, A., and Keirstead, J. (2014). Energy systems modeling for twenty-first-century energy challenges. *Renewable and Sustainable Energy Reviews* 33: 74–86.
3 Hamming, R. (2012). *Numerical Methods for Scientists and Engineers*. Courier Corporation.
4 Helm, D. (2002). Energy policy: security of supply, sustainability and competition. *Energy Policy* 30 (3): 173–184.

5 Chermack, T.J., Lynham, S.A., and Ruona, W.E. (2001). A review of scenario planning literature. *Futures Research Quarterly* 17 (2): 7–32.

6 Jebaraj, S. and Iniyan, S. (2006). A review of energy models. *Renewable and Sustainable Energy Reviews* 10 (4): 281–311.

7 Bazmi, A.A. and Zahedi, G. (2011). Sustainable energy systems: role of optimization modeling techniques in power generation and supply—a review. *Renewable and Sustainable Energy Reviews* 15 (8): 3480–3500.

8 Fodstad, M., del Granado, P.C., Hellemo, L. et al. (2022). Next frontiers in energy system modeling: A review on challenges and the state of the art. *Renewable and Sustainable Energy Reviews* 160: 112246.

9 Hiremath, R.B., Shikha, S., and Ravindranath, N.H. (2007). Decentralized energy planning; modeling and application—a review. *Renewable and Sustainable Energy Reviews* 11 (5): 729–752.

10 Connolly, D., Lund, H., Mathiesen, B.V., and Leahy, M. (2010). A review of computer tools for analyzing the integration of renewable energy into various energy systems. *Applied Energy* 87 (4): 1059–1082.

11 Sensfuß, F., Ragwitz, M., Genoese, M., and Möst, D. (2007, 2007). Agent-based simulation of electricity markets. *A Literature Review* 15 (2).

12 Möst, D. and Keles, D. (2010). A survey of stochastic modeling approaches for liberalized electricity markets. *European Journal of Operational Research* 207 (2): 543–556.

13 Keirstead, J., Jennings, M., and Sivakumar, A. (2012). A review of urban energy system models: Approaches, challenges, and opportunities. *Renewable and Sustainable Energy Reviews* 16 (6): 3847–3866.

14 DeCarolis, J.F., Hunter, K., and Sreepathi, S. (2012). The case for repeatable analysis with energy economy optimization models. *Energy Economics* 34 (6): 1845–1853.

15 Zhang, H., Yin, X., and Chen, W. (2014). A bottom-up model analysis of transport sector: a study of China and USA. *Energy Procedia* 61: 895–898.

16 Loulou, R., Goldstein, G., and Noble, K. (2004). Energy technology systems analysis program. *Documentation for the MARKAL Family of Models*. PART I July 2016 (Last update: February 2021).

17 Williams, J.H., DeBenedictis, A., Ghanadan, R. et al. (2012). The technology path to deep greenhouse gas emissions cuts by 2050: the pivotal role of electricity. *Science* 335 (6064): 53–59.

18 Version IV, W. A. S. P. (2001). Wien Automatic System Planning (WASP) Package. Version WASP-IV User's Manual International Atomic Energy Agency, Vienna, The originating Section of this document in the IAEA was: Planning and Economic Studies Section International Atomic Energy Agency Wagramer Strasse 5 P.O. Box 100 A-1400 Vienna, Austria.

19 Leuthold, F.U., Weigt, H., and von Hirschhausen, C. (2012). A large-scale spatial optimization model of the European electricity market. *Networks and Spatial Economics* 12: 75–107.

20 Energy Exemplar, Power market modeling software; (2013). https://www.energyexemplar.com/.

21 DECC (2050). Pathways Analysis. Technical Report, HM Government, London; 2010. url: https://www.gov.uk/government/publications/2050-pathways-analysis.

22 Kahouli-Brahmi, S. (2008). Technological learning in energy–environment–economy modelling: A survey. *Energy Policy* 36 (1): 138–162.

23 “Carbon emission trading systems”. https://www.carboncredits.com.

24 “Pathway To NetZero.” https://www.dnv.com.

25 “ICAP allowance price explorer.” https://icapcarbonaction.com/en/ets-prices, 2023.

26 Francois, M. and Hamaide, B. (2011). Certified emission reductions weights for improved CDM projects. *Environmental Policy and Governance* 21 (1): 31–41.

27 Mohammad, Parhamfar, and Iman Sadeghkhani .”Towards the net zero carbon future: a review of blockchain-enabled peer-to-peer carbon trading”14.2.2024

28 “Impleneting Effrctive Emissions Trading Systems.” https://www.iea.org/reports/implementing-effective-emissions-trading-systems.

29 Goron, C. and Cassisa, C. (2017). Regulatory institutions and market-based climate policy in China. *Global Environmental Politics* 17 (1): 99–120.

30 Al Sadawi, A., Madani, B., Saboor, S. et al. (2021). A comprehensive hierarchical blockchain system for carbon emission trading utilizing blockchain of things and smart contract. *Technological Forecasting and Social Change* 173: 121124.

31 Levin, R.B., Waltz, P., and LaCount, H. (2018). Betting blockchain will change everything–SEC and CFTC regulation of blockchain technology. In: *Handbook of Blockchain, Digital Finance, and Inclusion*, vol. 2, 187–212. Academic Press.

32 Zhang, S. and Chen, W. (2022). Assessing the energy transition in China towards carbon neutrality with a probabilistic framework. *Nature communications* 13 (1): 87.

33 Zanghi, E., Filho, D.C., and M. B., & Stacchini de Souza, J. C. (2019). Conceptual framework for blockchain-based metering systems. *Multiagent and Grid Systems* 15 (1): 77–97.

34 Khalid, R., Javaid, N., Javaid, S. et al. (2020, June). A blockchain-based decentralized energy management in a P2P trading system. In: *ICC 2020–2020 IEEE International Conference on Communications (ICC)*, 1–6. IEEE.

35 Kang, J., Yu, R., Huang, X. et al. (2017). Enabling localized peer-to-peer electricity trading among plug-in hybrid electric vehicles using consortium blockchains. *IEEE Transactions on Industrial Informatics* 13 (6): 3154–3164.

36 Iansiti, M. and Lakhani, K.R. (2017). The truth about blockchain. *Harvard Business Review* 95 (1): 118–127.

37 Sipthorpe, A., Brink, S., Van Leeuwen, T., and Staffell, I. (2022). Blockchain solutions for carbon markets are nearing maturity. *One Earth* 5 (7): 779–791.

38 Pan, Y., Zhang, X., Wang, Y. et al. (2019). Application of blockchain in carbon trading. *Energy Procedia* 158: 4286–4291.

39 https://www.foley.com.

40 Varian, H.R. (2014). *Intermediate Microeconomics with Calculus: A Modern Approach*. WW norton & company.

41 Von Stackelberg, H. (2010). *Market Structure and Equilibrium*. Springer Science & Business Media.

42 "AI in carbon Trading Systems." https://www.viact.ai/.

43 Dantzig, G. (1963). *Linear Programming and Extensions*. Princeton university press.

44 Schrijver, A. (1998). *Theory of Linear and Integer Programming*. John Wiley & Sons.

45 Williams, H.P. (2009). Integer programming. In: *Logic and Integer Programming*, 25–70. Boston, MA: Springer US.

46 Bazaraa, M.S., Sherali, H.D., and Shetty, C.M. (2013). *Nonlinear programming: theory and algorithms*. Wiley.

47 Parhamfar, M. (2025). *Renewable Energy Transition: A Pathway to Energy Security and Global Peace*.

48 Parhamfar, M. (2025). *Innovations in Energy Trading: AI-Powered Models and Digital Platforms*.

Index

Energy Management Strategies for Multi-Vectored Energy Hubs to Achieve Low Carbon Societies, First Edition.
Edited by Shubham Tiwari, Jai Govind Singh, Sivaraman Palanisamy, Sharmeela Chenniappan, Rupendra Kumar Pachauri, and Sanjeevikumar Padmanaban.